Fundamental Trends in Fluid-Structure Interaction

Series on Contemporary Challenges in Mathematical Fluid Dynamics and Its Applications

Founder and Editor: Giovanni P. Galdi

Vol. 1 Fundamental Trends in Fluid-Structure Interaction
Editors: G. P. Galdi and R. Rannacher

Contemporary Challenges in Mathematical Fluid Dynamics and Its Applications

Volume 1

Fundamental Trends in Fluid-Structure Interaction

Editors

Giovanni P Galdi
University of Pittsburgh, USA

Rolf Rannacher
Heidelberg University, Germany

World Scientific

NEW JERSEY • LONDON • SINGAPORE • BEIJING • SHANGHAI • HONG KONG • TAIPEI • CHENNAI

Published by

World Scientific Publishing Co. Pte. Ltd.

5 Toh Tuck Link, Singapore 596224

USA office: 27 Warren Street, Suite 401-402, Hackensack, NJ 07601

UK office: 57 Shelton Street, Covent Garden, London WC2H 9HE

British Library Cataloguing-in-Publication Data
A catalogue record for this book is available from the British Library.

FUNDAMENTAL TRENDS IN FLUID-STRUCTURE INTERACTION
Contemporary Challenges in Mathematical Fluid Dynamics and Its Applications — Vol. 1

ISBN-13 978-981-4299-32-9
ISBN-10 981-4299-32-4

Typeset by Stallion Press
Email: enquiries@stallionpress.com

Printed in Singapore.

PREFACE

Interaction of a fluid with a solid body is a widespread phenomenon in Nature. It occurs at different scales and in different applied disciplines: swimming of fish, flight of an airplane, transport of material through water pipelines and blood flow in human arteries are just few significant examples.

Interestingly enough, even though the mathematical theory of the motion of bodies in a liquid is one of the oldest and most classical problems in fluid mechanics, owed to the seminal contributions of Stokes, Kirchhoff, and Thomson (Lord Kelvin), only very recently have mathematicians become interested in a systematic study of the basic problems related to fluid-structure interaction, from both analytical and numerical viewpoints.

In fact, contributions to the subject are nowadays growing at such a fast pace that it is highly desirable to have an updated information on the state of the art.

This book is a unique collection of fundamental papers written by world renowned experts aimed at furnishing the highest level of development in several significant areas of fluid-structure interaction.

Specifically, the contribution of Th. Dunne *et al.* is devoted to a numerical analysis of the problem of a viscous fluid interacting with a deformable elastic body. In particular, it reviews the pros and cons of whether it is more appropriate to use a Lagrangean or Eulerian formulation.

The article by V. Heuveline and P. Wittwer provides a detailed survey on the progress over the recent years made on the problem of the interaction of an exterior Navier–Stokes flow with a rigid structure at low Reynolds number.

The paper of M. Razzaq *et al.* centers around the use of the Arbitrary Lagrangean Eulerian formulation in the numerical resolution of the problem of fluid-solid interaction. As an application, the influence of endovascular stent implantation onto cerebral aneurysm hydrodynamics is investigated.

J. San Martin and M. Tucsnak consider the coupled problem of the interaction of a fluid with a number of rigid bodies. Their contribution surveys the fundamental mathematical analysis that is at the basis of the problem.

Finally, the article of A. Quarteroni presents some of the basic models that are used to describe blood flow dynamics in local arterial environments and to predict the vessel wall deformation in compliant arteries.

We hope that the diversity of the topics along with the different approaches will allow the reader to have a global and updated view on the latest results on the subject and on the relevant open questions.

Editors
Giovanni P. Galdi and Rolf Rannacher

CONTENTS

CHAPTER 1

NUMERICAL SIMULATION OF FLUID-STRUCTURE INTERACTION BASED ON MONOLITHIC VARIATIONAL FORMULATIONS

Th. Dunne, R. Rannacher* and Th. Richter

Institute of Applied Mathematics
University of Heidelberg, INF 293/294
D-69120 Heidelberg, Germany
**rannacher@iwr.uni-heidelberg.de*

The dilemma in modeling the coupled dynamics of fluid-structure interaction (FSI) is that the fluid model is normally based on an Eulerian perspective in contrast to the usual Lagrangian formulation of the solid model. This makes the setup of a common variational description difficult. However, such a variational formulation of FSI is needed as the basis of a consistent Galerkin discretization with *a posteriori* error control and mesh adaptation, as well as the solution of optimal control problems based on the Euler–Lagrange approach. This article surveys recent developments in the numerical approximation of FSI problems based on "monolithic" variational formulations. The modeling is based either on an arbitrary Lagrangian–Eulerian (ALE) or a fully Eulerian–Eulerian (Eulerian) description of the (incompressible) fluid and the (elastic) structure dynamics. These global one-field formulations constitute a strongly implicit coupling of the dynamics of fluid and structure which, in contrast to the commonly used weakly coupled two-field formulations, provides the basis for a robust and efficient solution process. In this context a fully consistent treatment of mesh adaptation (DWR method) and optimal control ("all-at-once" approach) becomes possible within a Galerkin finite element discretization.

1. Introduction

Computational fluid dynamics and computational structure mechanics are two major areas of numerical simulation of physical systems. With the introduction of high performance computing it has become possible to tackle systems with a coupling of fluid and structure dynamics. General examples of such fluid-structure interaction (FSI) problems are flow transporting elastic particles (particulate flow), flow around elastic structures (airplanes, submarines) and flow in elastic structures (haemodynamics, transport of fluids in closed containers). In all these settings the dilemma in modeling the coupled dynamics is that the fluid model is normally based on an Eulerian perspective in contrast to the usual Lagrangian approach for the solid model. This makes the setup of a common variational description difficult. However, such a variational formulation of FSI is needed as the basis of a consistent approach to residual-based *a posteriori* error estimation and mesh adaptation as well as to the solution of optimal control problems by the Euler–Lagrange method. This is the subject of the present paper, which is largely based on the doctoral dissertation of the first author Dunne[22] and the survey article Dunne and Rannacher.[24]

Combining the Eulerian and the Lagrangian setting for describing FSI involves conceptional difficulties. On one hand the fluid domain itself is time-dependent and depends on the deformation of the structure domain. On the other hand, for the structure the fluid boundary values (velocity and the normal stress) are needed. In both cases values from one problem are used for the other, which is costly and can lead to a drastic loss of accuracy. A common approach to dealing with this problem is to separate the two models, solve each separately, and so converge iteratively to a solution which satisfies both together with the interface conditions (Fig. 1). Solving the separated problems serially multiple times is referred to as a "partitioned (or segregated) approach".

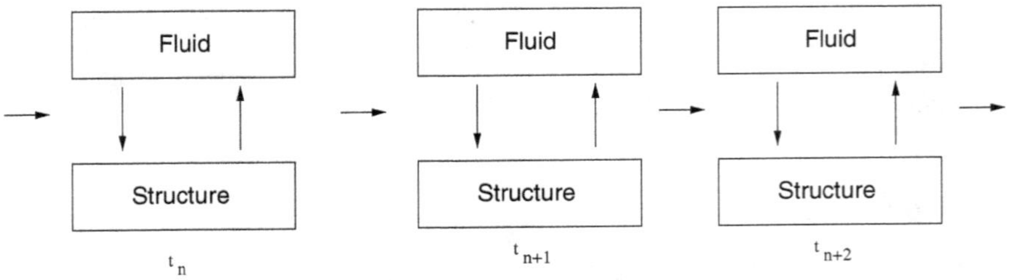

Fig. 1. Partitioned approach, Lagrangian and Eulerian frameworks coupled.

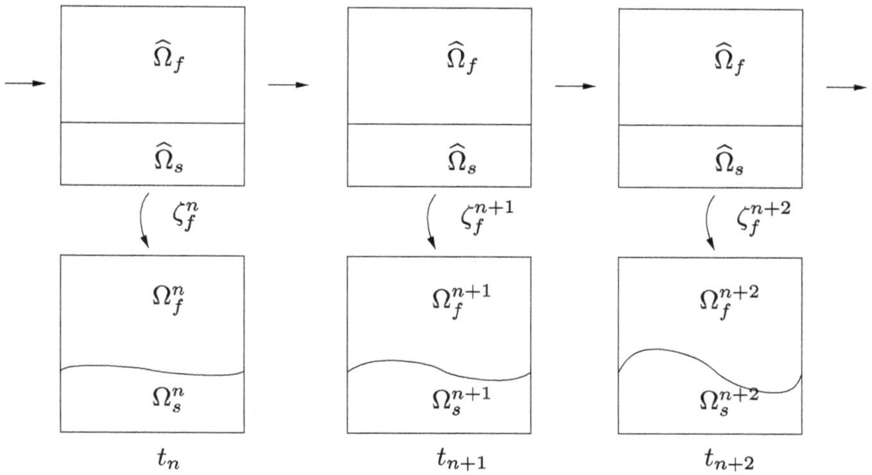

Fig. 2. Transformation approach, both frameworks Lagrangian.

A partitioned approach does not contain a variational equation for the fluid-structure interface. To achieve this, usually an auxiliary unknown coordinate transformation function T_f is introduced for the fluid domain. With its help the fluid problem is rewritten as one on the transformed domain which is fixed in time. Then, all computations are done on the fixed reference domain and as part of the computation the auxiliary transformation function T_f has to be determined at each time step. Figure 2 illustrates this approach for the driven cavity problem considered in Sec. 8. Such so-called "arbitrary Lagrangian–Eulerian" (ALE) methods are used in Huerta and Liu,[35] Wall,[58] Hron and Turek,[33] and corresponding transformed space-time finite element formulations in Tezduyar, Behr and Liou[51,52] and Tezduyar, Sathe, Stein and Aureli.[53] For other ways of dealing with implicit coupling in FSI models, we refer to Vierendeels[56] and Wall, Gerstenberger, Gamnitzer, Forster and Ramm.[59] Computational comparisons of partitioned and monolithic approaches have recently been made in Heil, Hazel and Boyle.[30]

Both, the partitioned and the transformation approach to overcome the Euler–Lagrange discrepancy explicitly track the fluid-structure interface by mesh adjustment and are generally referred to as "interface tracking" methods. Both methods leave the structure problem in its natural Lagrangian setting.

However, one may follow the alternative way of posing the fluid as well as the structure problem in a fully Eulerian–Eulerian (Eulerian) framework. A similar approach has been used by Lui and Walkington[43] in the context of the transport of visco-elastic bodies in a fluid. In the Eulerian setting a phase variable is employed on the fixed mesh to distinguish between the different phases, liquid and solid. This approach to identifying the fluid-structure interface is generally referred to as "interface capturing", a method commonly used in the simulation of multiphase flows, Joseph and Renardy.[40] Examples for the use of such a phase variable are the Volume of Fluid (VoF) method[32] and the Level Set (LS) method — Chang, Hou, Merriman and Osher,[16] Osher and Sethian,[44] Sethian.[49] In the classical LS approach the distance function has to continually be reinitialized, due to the smearing effect by the convection velocity in the fluid domain. This makes the use of the LS method delicate for modeling FSI problems particularly in the presence of cornered structures. To cope with this difficulty, in Dunne[21,22] a variant of the LS method, the Initial Position (IP) method, has been proposed that makes reinitialization unnecessary and which easily copes with cornered structures. This approach does not depend on the specific structure model.

The key variable in structure dynamics is the deformation, and since this depends on the deflection, it is understandable why structure dynamics is preferably described in the Lagrangian frame. The set of "initial positions" (IP set) of all structure points enables us to describe the deformations in the Eulerian frame. This set is then transported with the structure velocity in each time step. Based on this concept the displacement is now available in an Eulerian sense. Also its gradient has to be rewritten appropriately, which will be explained in Sec. 4.2. Since the fluid-structure interface will be crossing through cells, we will have to also transport the IP set in the fluid domain.

If we were to use the fluid velocity for the advection of the IP set, this would lead to entanglement of the respective displacements, which would "wreak havoc" on the interface cells. This is a known problem with LS approaches. A common way for fixing this problem has been to occasionally fix the LS field between the time steps. The problem with this approach is that the variational formulation is no longer consistent. As an alternative, we harmonically continue the structure velocity into the fluid domain. In the fluid domain, we then use this velocity for advecting the IP set. Since an IP set is available in both domains, we can always at each point determine if it belongs to the fluid or solid part of the model.

Again this approach is similar to the LS approach. Actually, it is possible to also develop a model for FSI using the level set approach, Legay, Chessa and Belytschko.[41] But when developing a complete variational formulation the two key characteristics of the LS approach also become the main cause of concern: reinitialization and the signed distance function. Although the problem of reinitialization here can also be avoided by using an harmonically extended velocity, the trouble concerning corner approximation persists. In contrast to this, by using an initial position set, we are deforming a virtual mesh of the structure which is extended into the whole domain.

The equations we use are based on the momentum and mass conservation equations for the flow of an incompressible Newtonian fluid and the deformation of a compressible St. Venant–Kirchhoff or likewise incompressible neo-Hookean solid. The spatial discretization is by a second-order finite element method with conforming equal-order (bilinear) trial functions using "local projection stabilization", Becker and Braack.[4,5] The time discretization uses the second-order "Fractional-Step-θ" scheme originally proposed by Bristeau, Glowinski and Periaux.[13] This method has the same complexity as the Crank–Nicolson scheme but better stability properties, see Rannacher.[46]

Based on the Eulerian variational formulation of the FSI system, we use the "dual weighted residual" (DWR) method, described in Becker and Rannacher,[7–9] Becker, Heuveline and Rannacher,[6] Bangerth and Rannacher,[2] Braack and Richter,[10] to derive "goal-oriented" *a posteriori* error estimates. The evaluation of these error estimates requires the approximate solution of a linear dual variational problem. The resulting *a posteriori* error indicators are then used for automatic local mesh adaptation. The full application of the DWR method to FSI problems requires a Galerkin discretization in space as well as in time. Due to the use of a difference scheme in time, in this paper we are limited to "goal-oriented" mesh adaptation in computing steady states or (somewhat heuristically) to quasi-steady states within the time stepping process. The incorporation of automatic time-step control will be the subject of forthcoming work.

The method for computing FSI described in this paper is validated at a stationary model problem that is a lid-driven cavity involving the interaction of an incompressible Stokes fluid with a linearized incompressible neo-Hookean solid. Then, as a more challenging test the self-induced oscillation of a thin elastic bar immersed in an incompressible fluid is treated (FLUSTRUK-A benchmark, see Hron and Turek.[34]) For this test problem, our method is also compared against a standard "arbitrary Lagrange Eulerian" (ALE) approach. The possible potential of the fully Eulerian formulation of the FSI problems is indicated by its good behavior for large structure deformations. All computations and visualizations are done using the flow-solver package GASCOIGNE[26] and the graphics package VISUSIMPLE.[57] The details on the software implementation can be found in Dunne.[21–23]

The outline of this paper is as follows. Section 2 ("Notation") introduces the basic notation for the ALE as well as the Eulerian formulation of the FSI problem and Sec. 3 ("Reference frameworks") discusses the reference frameworks, *Lagrangian* and *Eulerian*, which will be used throughout this paper. The corresponding variational formulations are developed in detail, first separately for the structure and fluid parts in Sec. 4 ("Variational formulations of fluid and structure problems") and then for the coupled FSI problem in Sec. 5 ("Variational formulations of the FSI problem"). Section 6 ("Discretization") describes the discretization in space and time as well as the techniques for solving the algebraic systems and for evaluating directional derivatives. The derivation of *a posteriori* error estimates and strategies for mesh adaptation is explained in Sec. 7 ("Mesh adaptation"). In Sec. 8 ("Numerical test 1: elastic flow cavity") the newly proposed Eulerian method is validated at a stationary test problem "elastic flow cavity". Then, Sec. 9 ("Numerical test 2: FSI benchmark FLUSTRUK-A") contains the results obtained by the two approaches, ALE and Eulerian, for the solution of the benchmark problem FLUSTRUK-A (oscillations of a thin elastic bar) for various combinations of material models and flow conditions. The paper is closed by Sec. 10 ("Summary and future development") which gives a summary and points at some directions of ongoing and future research on the basis of the approaches described in this paper.

2. Notation

We begin with introducing some notation which will be used throughout this paper. By $\Omega \subset \mathbb{R}^d$ ($d = 2$ or $d = 3$), we denote the domain of definition of the FSI problem. The domain Ω is supposed to be *time independent* but to consist of two possibly time-dependent subdomains, the fluid domain $\Omega_f(t)$ and the structure domain $\Omega_s(t)$. Unless needed, the explicit time dependency will be skipped in this notation. The boundaries of Ω, Ω_f, and Ω_s are denote by $\partial\Omega$, $\partial\Omega_f$, and $\partial\Omega_s$, respectively. The common interface between Ω_f and Ω_s is $\Gamma_i(t)$, or simply Γ_i.

The initial structure domain is denoted by $\widehat{\Omega}_s$. Spaces, domains, coordinates, values (such as pressure, displacement, velocity) and operators associated to $\widehat{\Omega}_s$ (or $\widehat{\Omega}_f$) will likewise be indicated by a "hat".

Partial derivatives of a function f with respect to the i-th coordinate are denoted by $\partial_i f$, and the total time-derivative by $d_t f$. The divergence of a vector and tensor is written as $\operatorname{div} f = \sum_i \partial_i f_i$ and $(\operatorname{div} F)_i = \sum_j \partial_j F_{ij}$. The gradient of a vector valued function v is the tensor with components $(\nabla v)_{ij} = \partial_j v_i$.

By $[f]$, we denote the jump of a (possibly discontinuous) function f across an interior boundary, where n is always the unit vector n at points on that boundary.

For a set X, we denote by $L^2(X)$ the Lebesque space of square-integrable functions on X equipped with the usual inner product and norm

$$(f, g)_X := \int_X fg\,dx, \quad \|f\|_X^2 = (f, f)_X,$$

respectively, and correspondingly for vector- and matrix-valued functions. Mostly the domain X will be Ω, in which case we will skip the domain index in products and norms. For Ω_f and Ω_s, we similarly indicate the associated spaces, products, and norms by a corresponding index "f" or "s".

We will generally use roman letters, V, for denoting spaces of functions depending only on spatial variables and calligraphic letters, $\mathcal{V}$, for spaces of functions depending additionally on time. Let $L_X := L^2(X)$ and $L_X^0 := L^2(X)/\mathbb{R}$. The functions in L_X (with $X = \Omega$, $X = \Omega_f(t)$, or $X = \Omega_s(t)$) with first-order distributional derivatives in L_X make up the Sobolev space $H^1(X)$. Further, $H_0^1(X) = \{v \in H^1(X) : v_{|\partial X_D} = 0\}$, where ∂X_D is that part of the boundary ∂X at which Dirichlet boundary conditions are imposed. Further, we will use the function spaces $V_X := H^1(X)^d$, $V_X^0 := H_0^1(X)^d$, and for time-dependent functions

$$\mathcal{L}_X := \mathcal{L}^2[0, T; L_X], \quad \mathcal{V}_X := \mathcal{L}^2[0, T; V_X] \cap \mathcal{H}^1[0, T; V_X^*],$$
$$\mathcal{L}_X^0 := \mathcal{L}^2[0, T; L_X^0], \quad \mathcal{V}_X^0 := \mathcal{L}^2[0, T; V_X^0] \cap \mathcal{H}^1[0, T; V_X^*],$$

where V_X^* is the dual of V_X^0, and $\mathcal{L}^2$ and $\mathcal{H}^1$ indicate the corresponding properties in time. Again, the X-index will be skipped in the case of $X = \Omega$, and for $X = \Omega_f$ and $X = \Omega_s$ a corresponding index "f" or "s" will be used.

3. Reference Frameworks

In modeling the variation of a spatial continuum in time two approaches are commonly used. The *Lagrangian* or *material* framework and the *Eulerian* or *spatial* framework. Both approaches have the simple goal of describing how a certain scalar quantity of interest $f : \mathbb{R}^d \times I \to \mathbb{R}$ changes in space and with time. The choice of the "reference point" of the value f is what distinguishes the two frameworks. We denote by $x \in \mathbb{R}^d$ and $t \in I$ the spatial and temporal coordinates, respectively. The function f is assumed to be sufficiently smooth with respect to space and time.

3.1. *Lagrangian framework*

In the Lagrangian framework one observes the value at a preselected point that is moving (and possibly accelerating) steadily through space. The initial position of the point at the initial time t_0, we define as $\hat{x}$. Thus, the position of the point is a function of the initial position $\hat{x}$ and time t,

$$x = x(\hat{x}, t).$$

We define the velocity v of this point as the total time derivative of its position x,

$$v(x, t) := d_t x(\hat{x}, t) = \partial_t x + \hat{\nabla} x \, d_t \hat{x}. \tag{3.1}$$

Since $\hat{x}$ is the position of the point at an initial time it follows that it does not change in time, i.e., $d_t \hat{x} = 0$ and $v = \partial_t x$.

To be more precise, in the Lagrangian framework we should refer to $f(x, t)$ as $\hat{f}(\hat{x}, t) := f(x(\hat{x}, t), t)$. Visually one can imagine that we are observing the value at a *material point* that was initially at the position $\hat{x}$ and is moving through space with velocity v. The total time derivative of $\hat{f}$ in the Lagrangian framework can thus be written:

$$d_t \hat{f}(\hat{x}, t) = \partial_t \hat{f}(\hat{x}, t) + \hat{\nabla} \hat{f}(\hat{x}, t)\, d_t \hat{x} = \partial_t \hat{f}(\hat{x}, t). \tag{3.2}$$

Since the Lagrangian approach describes the movement and deformation of individual particles and volumes it follows that this framework is the natural approach for modeling structure dynamics.

3.2. *Eulerian framework*

In the Eulerian framework one observes the value at a fixed point x in space. Hence this framework is also referred to as a *spatial framework*. Looking back at the Lagrangian framework one can imagine that at the point x at different times there will continuously be different material points moving through. Each such material points will have a respective initial position $\hat{x}$. Thus, the velocity v at this space-time position (x, t) is still to be understood as the velocity of the material point with the initial position $\hat{x}$, i.e., $v(x, t) := d_t x(\hat{x}, t) = \partial_t x$.

In an Eulerian framework the quantity of interest is written as $f(x, t)$ with x and t being anywhere within the permitted space-time continuum. Taking the total time derivative of f leads to

$$d_t f(x, t) = \partial_t f(x, t) + \nabla f(x, t)\, d_t x = \partial_t f(x, t) + v \cdot \nabla f(x, t). \tag{3.3}$$

The second term is referred to as the "transport" or "convection term". This term is a characteristic difference between the Eulerian and Lagrangian frameworks. In the Lagrangian framework, when the total time derivative is expanded into all its partial derivatives, there is no convective term due to the spatial parameter being constant in time. In contrast, in Eulerian frameworks convection can generally be expected in the expanded total time derivative.

The Eulerian framework presents itself as the natural approach for modeling fluid flow. This follows as a consequence that one is less interested in the individual behavior of particles and more interested in flow properties at certain spatial points in the flow domain. In viscous fluids with behavior similar to soft materials, a Lagrangian approach would be plausible.

Generally though particle movement in fluids is considerable and their initial positions in relation to each other have effectively nothing in common with their later relative positions. Hence the Eulerian framework presents itself as the natural approach to modeling fluid flow.

The key formula for deriving the basic conservation equations in continuum mechanics is provided by Reynold's transport theorem. Let $\hat{T}: \hat{\Omega} \times I \to \Omega$ be a C^2 diffeomorphism and the scalar function $f(x, t)$: $\Omega \times I \to \mathbb{R}$ be differentiable. Then, for any subset $\hat{V} \subset \hat{\Omega}$, with $V(t) = \hat{T}(\hat{V}, t)$ and the velocity field v as defined above, there holds

$$d_t \int_{V(t)} f\, dx = \int_{V(t)} \{\partial_t f + \operatorname{div}(fv)\}\, dx. \tag{3.4}$$

3.3. *Arbitrary Lagrangian framework*

The Lagrangian and Eulerian frameworks introduced above are *natural* frameworks of references. It is more common though that one will need a framework of reference that is *arbitrary* and independent of the initial particle positions or the spatial domain. A common example (one we will later also encounter in the numerical tests) is a fluid flow in a domain that changes with time, $\Omega(t)$. Instead of modeling and simulating the flow in $\Omega(t)$ one assumes the existence of a (for a fixed time t) C^2 diffeomorphic mapping $\hat{T}(\hat{x}, t) : \hat{\Omega} \times I \to \Omega(t)$ continuous in space and time, with $\hat{\Omega}$ as the reference (and usually the initial) domain $\Omega(t)$. The requirement of C^2 diffeomorphism means that the mapping is (in addition to being diffeomorphic) also two times continuously differentiable. An approach that uses such an arbitrary frame of reference is called *"arbitrary Lagrangian–Eulerian (ALE)"*, see Hughes, Liu and Zimmermann.[38] We will also refer to an arbitrary framework as an *"ALE framework"*. For an overview of the various methods of applications using an ALU framework see Bungartz and Schäfer[14] and Heil, Hazel and Boyle.[30] With the help of the mapping $\hat{T}$ functions and operators on $\Omega(t)$ can be rewritten as such on the domain $\hat{\Omega}$. For this reason, as a preparatory measure, we introduce the most commonly needed transformation identities developed below.

By $\hat{F}$ and $\hat{J}$, we denote the Jacobian matrix and determinant of $\hat{T}$, respectively,

$$\hat{F} := \hat{\nabla}\hat{T}, \quad \hat{J} := \det \hat{F}. \tag{3.5}$$

In the context of material deformations later the mapping $\hat{T}$ will also be referred to as the "*deformation*" and accordingly $\hat{f}$ as the "*deformation gradient*". Since $\hat{T}$ is a deformation it must for each fixed time preserve orientation and not annihilate volume. This follows from the requirement that it is continuous and diffeomorphic. Thus, $\hat{J} > 0$. Let $f(x, t)$ and $v(x, t)$ denote scalar- and vector-valued functions that are differentiable in space and time. With $\hat{T}$ we define

$$\hat{f}(\hat{x}, t) := f(\hat{T}(\hat{x}, t), t), \quad \hat{v}(\hat{x}, t) := v(\hat{T}(\hat{x}, t), t). \tag{3.6}$$

The respective reference-based spatial derivatives of $\hat{f}$ can be obtained by the chain rule,

$$\hat{\partial}_i \hat{f}(\hat{x}, t) = \sum_{j=1}^{n} \partial_j f(\hat{T}(\hat{x}), t) \frac{\hat{\partial}\hat{T}_j(\hat{x}, t)}{\hat{\partial}\hat{x}_i}. \tag{3.7}$$

Thus, we can write the gradient of $\hat{f}$ as

$$\hat{\nabla}\hat{f} = \nabla f \hat{F}. \tag{3.8}$$

For the partial and total time derivatives of a scalar function $f(x, t)$ there holds

$$\begin{aligned} \partial_t f &= \partial_t \hat{f} - (\hat{F}^{-1}\partial_t \hat{T} \cdot \nabla)\hat{f}, \\ d_t f &= \partial_t \hat{f} + (\hat{F}^{-1}(\hat{v} - \partial_t \hat{T}) \cdot \hat{\nabla})\hat{f}, \end{aligned} \tag{3.9}$$

which can be seen by considering the derivative as limit of difference quotients. Analogous relations hold for vector-valued functions.

Next, we derive transformation formulas for volume and surface integrals. Let $V \subset \Omega$ be an arbitrary volume in Ω and $\hat{V} = \hat{T}^{-1}(V)$ the corresponding subset in $\hat{\Omega}$. With $\hat{f}(\hat{x}) = f(\hat{T}(\hat{x})) = f(x)$ and $dx = \det \hat{T} d\hat{x} = \hat{J} d\hat{x}$, we transform the volume integral over V to an integral over $\hat{V}$,

$$\int_V f(x)\, dx = \int_{\hat{V}} \hat{f}(\hat{x}) \hat{J}\, d\hat{x}. \tag{3.10}$$

The "Piola transform" of v is $\hat{J}\hat{F}^{-1}\hat{v}$. For its divergence there holds

$$\hat{J}\,\mathrm{div} v = \widehat{\mathrm{div}}(\hat{J}\hat{F}^{-1}\hat{v}), \quad x \in \hat{T}(\hat{x}) \in \Omega. \tag{3.11}$$

Further there holds

$$d_t \hat{J} = \widehat{\mathrm{div}}(\hat{J}\hat{F}^{-1}\hat{v}). \tag{3.12}$$

There hold the transformation formulas

$$\int_V \mathrm{div} v \, dx = \int_{\hat{V}} \widehat{\mathrm{div}}(\hat{J}\hat{F}^{-1}\hat{v}) \, d\hat{x}. \tag{3.13}$$

and

$$\int_{\partial V} v \cdot n \, do_x = \int_{\partial \hat{V}} (\hat{J}\hat{F}^{-1}\hat{v}) \cdot \hat{n} \, d\hat{x}. \tag{3.14}$$

3.4. *Conservation equations and boundary conditions*

Fluid flow is modeled based on the assumption that mass, momentum, angular momentum and energy are preserved. These are naturally expressed in an Eulerian framework. Here, we will only consider "incompressible" Newtonian fluids with constant density. The constitutive relation for the (symmetric) Cauchy stress tensor in the case of an incompressible Newtonian fluid is

$$\sigma := -pI + \rho\nu(\nabla v + \nabla v^T), \tag{3.15}$$

with the kinematic viscosity ν, the hydrodynamic pressure p, the density ρ, and the flow velocity v. Conservation of angular momentum is automatically fulfilled for incompressible Newtonian fluid flow. Based on this the conservation equations for momentum and mass decouple from the energy conservation equation. For modeling our FSI problems, we will not need the temperature or the specific internal energy-density state variables and therefore omit the energy conservation equation. Thus, we only consider the conservation equations for momentum and mass, i.e., the classical "Navier–Stokes equations":

$$\begin{aligned} \rho\partial_t v + \rho(v \cdot \nabla)v + \mathrm{div}\sigma &= \rho f \quad \text{in } \Omega_f \times I, \\ \mathrm{div} v &= 0 \quad \text{in } \Omega_f \times I. \end{aligned} \tag{3.16}$$

These equations are derived by using Reynold's transport theorem (3.4) for $f = \rho$ ("continuity equation") and $f = \rho v_i$ ("momentum equation").

The Navier–Stokes equations are usually not written with the full stress tensor σ in Eq. (3.15) but instead with a reduced version of the tensor $\tilde{\sigma} := -pI + \rho\nu\nabla v$, which is justified by the relation $\mathrm{div}\nabla v^T = 0$. We refrain from using the reduced tensor $\tilde{\sigma}$ since this would lead to an incorrect representation of the boundary forces. The proper calculation of these forces is most important since fluid-structure interaction is essentially driven by just these forces at the interface.

We will also need the momentum equation in an ALE framework with the respective reference frame $\hat{V}$. We use the mapping $\hat{T}$ and the notation and transformation formulas described above. There holds

$$\int_{\hat{V}} \{\hat{\rho}\hat{J}\partial_t\hat{v} + \hat{\rho}\hat{J}((\hat{v} - \partial_t\hat{T}) \cdot \hat{\nabla})\hat{v} - \widehat{\mathrm{div}}(\hat{J}\sigma\hat{F}^{-T})\}\, d\hat{x} = \int_{\hat{V}} \hat{J}\hat{\rho}\hat{f}\, d\hat{x}.$$

Since this relation holds for arbitrary volumes V and respective $\hat{V}$, we conclude

$$\hat{\rho}\hat{J}\partial_t\hat{v} + \hat{\rho}\hat{J}((\hat{v} - \partial_t\hat{T}) \cdot \hat{\nabla})\hat{v} - \widehat{\mathrm{div}}(\hat{J}\sigma\hat{F}^{-T}) = \hat{J}\hat{\rho}\hat{f}. \tag{3.17}$$

Generally when modeling flows using an Eulerian framework the boundaries are fixed and not moving. As a boundary condition in time an initial value v_0 for v at the initial time t_0 is prescribed. Spatially the boundary $\partial\Omega_f$ can be split into three non-overlapping parts, $\partial\Omega_f = \Gamma_{fD} \cup \Gamma_{fN} \cup \Gamma_i$ with each part relating to a different boundary condition. The first two parts are the well-known conditions

$$v = v_{fD} \quad \text{on } \Gamma_{fD} \quad \text{(Dirichlet)},$$

$$\sigma n_f = g_f \quad \text{on } \Gamma_{fN} \quad \text{(Neumann)}.$$

In the FSI problem there is a moving interface Γ_i which is the common boundary to the structure. We assume that on this boundary momentum is conserved and that the velocity of the fluid and material particles just at the boundary are equal. This leads to the FSI boundary conditions on Γ_i that must be fulfilled simultaneously:

$$v_f = v_s, \quad \sigma_f n_f = \sigma_s n_f \quad \text{on } \Gamma_i. \tag{3.18}$$

Here and below, to distinguish the fluid and structure values, we have added a respective "f" or "s" suffix.

4. Variational Formulations of Fluid and Structure Problems

In this section, we prepare for the "monolithic" variational formulations of the FSI problem by developing formulations of the structure and fluid parts in the different frameworks considered.

4.1. *Fluid model*

For the liquid part, we assume Newtonian incompressible flow governed by the usual Navier–Stokes equations, i.e., the equations describing conservation of mass and momentum. The (constant) density and kinematic viscosity of the fluid are ρ_f and ν_f, respectively.

4.1.1. *Fluid model in Eulerian formulation*

The equations are written in an Eulerian framework in the time-dependent domain $\Omega_f(t)$. The physical unknowns are the scalar pressure field $p_f \in \mathcal{L}_f$ and the vector velocity field $v_f \in v_f^D + \mathcal{V}_f^0$. Here, v_f^D is a suitable extension of the prescribed Dirichlet data on the boundaries (both moving or stationary) of Ω_f, and g_1 is a suitable extension to all of $\partial\Omega_f$ of the Neumann data for $\sigma_f \cdot n$ on the boundaries. We have "hidden" the fluid-structure interface conditions of continuity of velocity and normal stress in parts of the boundary data v_f^D and g_1.

The variational form of the Navier–Stokes equations in an Eulerian framework is obtained by multiplying them with suitable test functions from the test space V_f^0 for the momentum equations and L_f for the mass conservation equation.

Problem 1 (Fluid model in Eulerian formulation). *Find* $\{v_f, p_f\} \in \{v_f^D + \mathcal{V}_f^0\} \times \mathcal{L}_f$, *such that* $v_f(0) = v_f^0$, *and*

$$\begin{aligned} (\rho_f(\partial_t + v_f \cdot \nabla)v_f, \psi^v)_f + (\sigma_f, \varepsilon(\psi^v))_f &= (g_1, \psi^v)_{\partial\Omega_f} + (f, \psi^v)_f, \\ (\mathrm{div} v_f, \psi^p)_f &= 0, \end{aligned} \tag{4.1}$$

for all $\{\psi^v, \psi^p\} \in V_f^0 \times L_f$, *where*

$$\sigma_f := -p_f I + 2\rho_f \nu_f \varepsilon(v_f), \quad \varepsilon(v) := \frac{1}{2}(\nabla v + \nabla v^T).$$

4.1.2. *Fluid model in ALE formulation*

In fluid-structure interaction problems that we will later be observing the FSI domain Ω is time independent, but it is composed of the fluid domain Ω_f and the structure domain Ω_s, which will be changing with time. An approach to modeling a fluid flow in a dynamic domain is assuming that a reference domain $\hat{\Omega}_f$ and piecewise continuously differentiable invertible mapping $\hat{T}$ exist so that $\hat{T}(\hat{x}, t) : \hat{\Omega}_f \times I \to \Omega(t)$. Based on this assumption, we rewrite the Navier–Stokes equations in an ALE framework with the reference frame $\hat{\Omega}_f$ where we use the mapping $\hat{T}$ and the notation described above,

$$\begin{aligned} \hat{J}\rho\partial_t\hat{v} + \hat{J}\rho(\hat{F}^{-1}(\hat{v} - \partial_t\hat{T}) \cdot \hat{\nabla})\hat{v} - \widehat{\mathrm{div}}(\hat{J}\hat{\sigma}\hat{F}^{-T}) = \hat{J}\rho\hat{f} \quad \text{in } \hat{\Omega}_f, \\ \widehat{\mathrm{div}}(\hat{J}\hat{F}^{-1}\hat{v}) = 0 \quad \text{in } \hat{\Omega}_f, \end{aligned} \tag{4.2}$$

with $\hat{\sigma} := -\hat{p}I + \rho\nu(\hat{\nabla}\hat{v}\hat{F}^{-1} + \hat{F}^{-T}\hat{\nabla}\hat{v}^T)$, $\hat{F} := \hat{\nabla}\hat{T}$, and $\hat{J} := \det\hat{F}$.

Similarly the boundary conditions must be set in the ALE framework. As a boundary condition in time the same initial value is prescribed $\hat{v}_0(\hat{x}, 0) = v_0(\hat{T}(\hat{x}, 0)) = v_0$ for $\hat{v}$, now set in the ALE framework, at the initial time t_0. The fluid boundary $\partial\hat{\Omega}_f$ can be split into three non-overlapping components, $\partial\hat{\Omega}_f = \hat{\Gamma}_{fD} \cup \hat{\Gamma}_{fN} \cup \hat{\Gamma}_i$, with each part relating to a different boundary condition. The first two parts are the well-known conditions

$$\begin{aligned} \hat{v} = \hat{v}_D \quad &\text{on } \hat{\Gamma}_{fD} \quad \text{(Dirichlet)}, \\ \hat{J}\hat{\sigma}\hat{F}^{-T}\hat{n} = \hat{g} \quad &\text{on } \hat{\Gamma}_{fN} \quad \text{(Neumann)}. \end{aligned}$$

The moving boundary Γ_i corresponds to a fixed boundary $\hat{\Gamma}_i$ on the reference domain. In an Eulerian framework the boundary conditions on the moving boundary Γ_i are the same as in the fluid flow case: continuity of velocity and normal-flux of the stress. In the reference configuration the velocity is not transformed. The stress though is transformed, since not the Cauchy stress tensor is used in the momentum conservation equations, but instead the first Piola–Kirchhoff stress tensor. This leads to the boundary conditions

$$\hat{v}_f = \hat{v}_s, \quad \hat{J}_f\hat{\sigma}_f\hat{F}_f^{-T}\hat{n}_f = \hat{J}_s\hat{\sigma}_s\hat{F}_s^{-T}\hat{n}_f \quad \text{on } \hat{\Gamma}_i. \tag{4.3}$$

The variational form of the Navier–Stokes equations in an ALE framework is again obtained by multiplying them with suitable test functions from the test space $\hat{V}_f^0$ for the momentum equations and $\hat{L}_f$ for the mass conservation equation. For later purposes, we write all fluid specific variables with a respective "f" suffix. This includes the domain mapping now referred to as $\hat{T}_f$. The physical unknowns are the scalar pressure $\hat{p}_f \in \hat{\mathcal{L}}_f$ and the velocity $\hat{v}_f \in \hat{v}_f^D + \hat{\mathcal{V}}_f$.

Problem 2 (Fluid model in ALE formulation). *Find* $\{\hat{v}_f, \hat{p}_f\} \in \{\hat{v}_f^D + \hat{\mathcal{V}}_f^0\} \times \hat{\mathcal{L}}_f$ *such that* $\hat{v}_f(0) = \hat{v}_f^0$ *and*

$$
\begin{aligned}
(\hat{J}_f \rho_f \partial_t \hat{v}_f + \hat{J}_f \rho_f (\hat{F}_f^{-1}(\hat{v}_f - \partial_t \hat{T}_f) \cdot \hat{\nabla}) \hat{v}_f, \hat{\psi}^v)_{\hat{f}} + (\hat{J}_f \hat{\sigma}_f \hat{F}_f^{-T}, \hat{\nabla}\hat{\psi}^v)_{\hat{f}} \\
= (\hat{g}_f, \hat{\psi}^v)_{\hat{\Gamma}_{fN}} + (\hat{J}_f \hat{\sigma}_f \hat{F}_f^{-T} \hat{n}_f, \hat{\psi}^v)_{\hat{\Gamma}_i} + (\rho_f \hat{J}_f \hat{f}_f, \hat{\psi}^v)_{\hat{f}}, \\
(\widehat{\mathrm{div}}(\hat{J}_f \hat{F}_f^{-1} \hat{v}_f, \hat{\psi}^p)_{\hat{f}} = 0, \qquad (4.4)
\end{aligned}
$$

for all $\{\hat{\psi}^v, \hat{\psi}^p\} \in \hat{V}_f^0 \times \hat{L}_f$ *where*

$$
\hat{\sigma}_f := -\hat{p}_f I + \rho_f \nu_f (\hat{\nabla}\hat{v}_f \hat{F}_f^{-1} + \hat{F}_f^{-T} \hat{\nabla}\hat{v}_f^T), \quad \hat{F}_f := \hat{\nabla}\hat{T}_f, \quad \hat{J}_f := \det \hat{F}_f.
$$

Here, $\hat{v}_f^D$ is a suitable extension of the prescribed Dirichlet data on the boundaries of $\hat{\Omega}_f$ and $\hat{g}_f$ is the Neumann boundary condition on $\hat{\Gamma}_{fN}$. We have hidden the fluid-structure interface condition of continuity of velocity in part of the boundary data $\hat{v}_f^D$. The fluid-structure interface condition of continuity of the force $\hat{J}_f \hat{\sigma} \hat{F}^{-T} \hat{n}_f$, we have let stand for later purposes.

4.2. *Structure model*

Material deformations are modeled based on the assumption of conservation of momentum and optionally volume. The main quantity of interest is the vector field describing the displacement of the body from its initial state. Consequently the Lagrangian formulation is the natural frame of reference. Here, we will consider elastic materials, that is to say the observed material returns to its initial state once all applied forces are removed. We refer to the domain of the initial state as $\hat{\Omega}_s$ ("reference configuration") and use the mapping $\hat{T}$ and the notation introduced above. The displacement $\hat{u}$ and mapping $\hat{T}$ ("deformation") satisfy the relation

$$\hat{T}(\hat{x}, t) = \hat{x} + \hat{u}(\hat{x}, t). \tag{4.5}$$

The gradient of $\hat{T}$ is the deformation gradient $\hat{F} = \hat{\nabla}\hat{T} = I + \hat{\nabla}\hat{u}$. The state variables are the density $\hat{\rho}$ in the initial state, the velocity $\hat{v}$, the displacement $\hat{u}$, and the Cauchy stress tensor $\hat{\sigma}$, which is a function of $\hat{u}$ and optionally a pressure $\hat{p}$. The external force is denoted by $\hat{f}$.

In the examples presented below, we consider two different types of materials, a standard compressible elastic material described by the "St. Venant–Kirchhoff (STVK)" model and an "incompressible neo-Hookean (INH)" material. These two models will be described in the next two subsections.

The density of the structure is ρ_s. The material elasticity is usually described by a set of two parameters, the Poisson ration ν_s and the Young modulus E_s, or alternatively, the Lamé coefficients λ_s and μ_s. These parameters satisfy the following relations:

$$\nu_s = \frac{\lambda_s}{2(\lambda_s + \mu_s)}, \qquad E_s = \mu_s \frac{3\lambda_s + 2\mu_s}{\lambda_s + \mu_s},$$

$$\mu_s = \frac{E_s}{2(1 + \nu_s)}, \qquad \lambda_s = \frac{\nu_s E_s}{(1 + \nu_s)(1 - 2\nu_s)},$$

where $\nu_s = \frac{1}{2}$ for incompressible and $\nu_s < \frac{1}{2}$ for compressible material. Common notation for the stress tensor and various constituents are $P := \hat{J}\hat{\sigma}\hat{F}^{-T}$ and $S := \hat{F}^{-1}P$ for the "1st and 2nd Piola–Kirchhoff stress tensor", $\hat{E} := \frac{1}{2}(\hat{F}^T\hat{F} - I)$ for the "Green–Lagrange strain tensor", and $\hat{F}\hat{F}^T$ and $\hat{F}^T\hat{F}$ for the "left and right Cauchy–Green deformation tensor". We encounter the 1st Piola–Kirchhoff stress tensor as the "transformed" stress tensor on the reference domain $\hat{\Omega}$.

Principally the momentum conservation equations here are the same as with fluid flows, the only differences are that they are commonly set in a Lagrangian framework and the constitutive equation for the Cauchy stress tensor is based on the displacement field and not the velocity field. The equations for the elastic materials below differ slightly due to the different constitutive laws for the stress tensor.

Generally when modeling material deformation the boundaries will be moving in time. In the Lagrangian framework boundary conditions can be posed in the reference configuration. As a boundary condition in time initial values $\hat{u}^0$ and $\hat{v}^0$ for $\hat{u}$ and $\hat{v}$ are prescribed at the initial time t_0. Similar to the fluid boundary conditions the material boundary $\partial\hat{\Omega}_s$ can be split into three non-overlapping components, $\partial\hat{\Omega}_s = \hat{\Gamma}_{sD} \cup \hat{\Gamma}_{sN} \cup \hat{\Gamma}_i$, with each part relating to a different boundary condition. The first two parts are the well-known conditions

$$\hat{u} = \hat{u}^D, \quad \hat{v} = \hat{v}^D \quad \text{on } \hat{\Gamma}_{sD} \quad \text{(Dirichlet)},$$
$$\hat{J}\hat{\sigma}\hat{F}^{-T}\hat{n} = \hat{g} \quad \text{on } \hat{\Gamma}_{sN} \quad \text{(Neumann)}.$$

Here, $\hat{u}^D$ and $\hat{v}_s^D$ are assumed to be given as suitable extensions of the prescribed Dirichlet data on the boundaries of $\hat{\Omega}_s$, and $\hat{g}_2$ is a suitable extension to all of $\partial\hat{\Omega}_s$ of the Neumann data for $\hat{\sigma}_s \cdot n$ on the boundaries. For the sake of simplicity, we assume that the only boundary displacement that take place are on $\hat{\Gamma}_i$, i.e.,

$$\hat{u}^D = \hat{v}_s^D = 0 \quad \text{on } \partial\hat{\Omega}_s \backslash \hat{\Gamma}_i. \tag{4.6}$$

The moving boundary $\hat{\Gamma}_i$ is of course fixed on the reference domain. We assume that an appropriate mapping of the initial fluid domain $\hat{\Omega}_f$ on the present domain Ω_f is provided. With this in mind, we can rewrite the fluid quantities v and σ in an ALE framework. In an Eulerian framework the boundary conditions on the moving boundary Γ_i are the same as in the fluid flow case: continuity of velocity v and the normal-flux of the stress σn. In the reference configuration the velocity is not transformed. The stress though is transformed since not the Cauchy stress tensor is used in the momentum equations but instead the 1st Piola–Kirchhoff stress tensor. This leads to the boundary conditions

$$\hat{v}_s = \hat{v}_f, \quad \hat{J}_s\hat{\sigma}_s\hat{F}_s^{-T}\hat{n}_s = \hat{J}_f\hat{\sigma}_f\hat{F}_f^{-T}\hat{n}_s \quad \text{on } \hat{\Gamma}_i. \tag{4.7}$$

Similar to the structure variables the fluid variables are also denoted with a "hat", which indicates that they are set in an ALE framework. Again, similarly as for the fluid problem, we will hide the fluid-structure interface conditions of continuity of velocity and normal stress in parts of the boundary data $\hat{v}_s^D$ and $\hat{g}_2$.

4.2.1. *St. Venant–Kirchhoff (STVK) material*

The St. Venant–Kirchhoff (STVK) model is a classical (geometrically) nonlinear model for compressible elastic materials. It can be used for large displacements but requires small strains $\hat{E}$. This model is governed by the equations for conservation of mass and momentum. Again these equations are written in Lagrangian form using the same notation as above. The sought unknowns are the displacement $\hat{u}$ and the velocity $\hat{v}$, which are determined by the following system:

$$\begin{aligned} \hat{\rho} d_t \hat{v} - \widehat{\mathrm{div}}(\hat{J}\hat{\sigma}\hat{F}^{-T}) &= \hat{\rho}\hat{f} \quad \text{in } \hat{\Omega}_s, \\ d_t \hat{u} - \hat{v} &= 0 \quad \text{in } \hat{\Omega}_s, \end{aligned} \tag{4.8}$$

where $\hat{F} := I + \hat{\nabla}\hat{u}_s$, $\hat{J} := \det\hat{F}$, $\hat{E} := \frac{1}{2}(\hat{F}^T\hat{F} - I)$, and

$$\hat{\sigma} = \hat{J}^{-1}\hat{F}(\lambda_s(\mathrm{tr}\hat{E})I + 2\mu_s\hat{E})\hat{F}^T.$$

The variational formulation is as follows.

Problem 3 (STVK structure model in Lagrangian formulation). *Find* $\{\hat{u}_s, \hat{v}_s\} \in \{\hat{u}^D + \hat{V}_s^0\} \times \{\hat{v}_s^D + \hat{V}_s^0\}$, *such that* $\hat{u}_s(0) = \hat{u}_s^0$, $\hat{v}_s(0) = \hat{v}_s^0$, *and*

$$\begin{aligned} (\rho_s d_t \hat{v}_s, \hat{\psi}^u)_{\hat{S}} + (\hat{J}\,\hat{\sigma}_s\,\hat{F}^{-T}, \hat{\varepsilon}(\hat{\psi}^u))_{\hat{S}} &= (\hat{g}_2, \hat{\psi}^u)_{\partial\hat{\Omega}_s} + (\hat{f}_2, \hat{\psi}^u)_{\hat{S}}, \\ (d_t \hat{u}_s - \hat{v}_s, \hat{\psi}^v)_{\hat{S}} &= 0, \end{aligned} \tag{4.9}$$

for all $\{\hat{\psi}^u, \hat{\psi}^v\} \in \hat{V}_s^0 \times \hat{V}_s^0$, *where* $\hat{\varepsilon}(\hat{\psi}^u) := \frac{1}{2}(\hat{\nabla}\hat{\psi}^u + \hat{\nabla}\hat{\psi}^{uT})$.

4.2.2. *Incompressible Neo–Hookean (INH) material*

Numerous materials can be subjected to strains without a notable change of volume. Typical examples of such materials are plastics and rubber-like substances. A common idealization in continuum mechanics is to regard such materials as generally "incompressible" that only permit so-called "isochoric" deformations. The incompressibility of the material is ensured by demanding that the deformation conserve volume, hence the additional constraint $\hat{J} = 1$. The sought unknowns are the displacement $\hat{u}$, the velocity $\hat{v}$, and the (hydrostatic) pressure $\hat{p}$, which are determined by the system

$$\begin{aligned} \hat{\rho} d_t \hat{v} - \widehat{\mathrm{div}}(\hat{\sigma} \hat{F}^{-T}) &= \hat{\rho} \hat{f} \quad \text{in } \hat{\Omega}_s, \\ d_t \hat{u} - \hat{v} &= 0 \quad \text{in } \hat{\Omega}_s, \\ \hat{J} - 1 &= 0 \quad \text{in } \hat{\Omega}_s, \end{aligned} \tag{4.10}$$

with $\hat{\sigma} = -\hat{p}I + \mu_s(\hat{F}\hat{F}^T - I)$ and $\hat{F} = I + \hat{\nabla}\hat{u}$.

With suitable function spaces $\hat{\mathcal{V}}_s^0$ for the vector displacement and velocity fields and $\hat{\mathcal{L}}_s$ for the scalar pressure field $\hat{p}_s$, we write these equations in the following variational form.

Problem 4 (INH structure model in Lagrangian formulation). *Find* $\{\hat{u}_s, \hat{v}_s, \hat{p}_s\} \in \{\hat{u}^D + \hat{\mathcal{V}}_s^0\} \times \{\hat{v}_s^D + \hat{\mathcal{V}}_s^0\} \times \hat{\mathcal{L}}_s$, *such that* $\hat{u}_s(0) = \hat{u}_s^0$, $\hat{v}_s(0) = \hat{v}_s^0$, *and*

$$\begin{aligned} (\rho_s d_t \hat{v}_s, \hat{\psi}^u)_{\hat{S}} + (\hat{\sigma}_s \hat{F}^{-T}, \hat{\varepsilon}(\hat{\psi}^u))_{\hat{S}} &= (\hat{g}_2, \hat{\psi}^u)_{\partial\hat{\Omega}_s} + (\hat{f}_2, \hat{\psi}^u)_{\hat{S}}, \\ (d_t \hat{u}_s - \hat{v}_s, \hat{\psi}^v)_{\hat{S}} &= 0, \\ (\det \hat{F}, \hat{\psi}^p)_{\hat{S}} &= (1, \hat{\psi}^p)_{\hat{S}}, \end{aligned} \tag{4.11}$$

for all $\{\hat{\psi}^u, \hat{\psi}^v, \hat{\psi}^p\} \in \hat{V}_s^0 \times \hat{V}_s^0 \times \hat{L}_s$, *where* $\hat{F} := I + \hat{\nabla}\hat{u}_s$ *and*

$$\hat{\sigma}_s := -\hat{p}_s I + \mu_s(\hat{F}\hat{F}^T - I), \quad \hat{\varepsilon}(\hat{\psi}^u) := \frac{1}{2}(\hat{\nabla}\hat{\psi}^u + \hat{\nabla}\hat{\psi}^{uT}).$$

4.2.3. *Structure model in Eulerian framework*

In fluid-structure interaction problems the FSI domain Ω is usually time independent but it is composed of the fluid domain Ω_f and the structure domain Ω_s, which are changing with time. One approach, already mentioned, to treating this problem is to introduce a mapping $\hat{T}(\hat{x}, t) : \hat{\Omega} \times I \to \Omega(t)$. With this mapping the fluid problem is rewritten in an ALE framework. As an alternative, one may change the reference frame of the structure from Lagrangian to Eulerian which will eventually lead to a fully Eulerian formulation of the whole FSI problem.

All material stress quantities occurring above are based on the Lagrangian deformation gradient $\hat{F} := I + \hat{\nabla}\hat{u}$. In an Eulerian framework, we will still have the deformation since this is simply a quantity being specified in another reference frame, i.e., $u(x) = \hat{u}(\hat{x})$. What is not immediately available though is the deformation gradient $\hat{\nabla}\hat{u}$ since $\hat{\nabla}\hat{u} \neq \nabla u$. Therefore, we introduce the "inverse deformation"

$$T(x,t) : \Omega_s(t) \times I \to \hat{\Omega}_s, \quad T(x,t) := \hat{x} = x - u(x,t).$$

Together with the deformation $\hat{T}(\hat{x}, t)$ this leads to the identity

$$T(\hat{T}(\hat{x},t),t) = \hat{x}.$$

Differentiating this with respect to $\hat{x}$ gives us $(I - \nabla u)(I + \hat{\nabla}\hat{u}) = I$ or $(I + \hat{\nabla}\hat{u}) = (I - \nabla u)^{-1}$, and, thus,

$$\hat{\nabla}\hat{u} = (I - \nabla u)^{-1} - I. \tag{4.12}$$

Hence the gradients and Jacobi determinants of the deformation and inverse deformation relate to each other in the following manner:

$$F := I - \nabla u = \hat{F}^{-1}, \quad J := \det F = \det \hat{F}^{-1} = \hat{J}^{-1}. \tag{4.13}$$

The total time derivatives of the velocity and displacement are in the usual way written as

$$d_t v = \partial_t v + v \cdot \nabla v, \quad d_t u = \partial_t u + v \cdot \nabla u.$$

Using the above relations, we can rewrite the structure equations for St. Venant–Kirchhoff (STVK) materials and incompressible neo-Hookean (INH) materials in Eulerian framework. The unknowns are the displacement u, the velocity v and in the INH case the (hydrostatic) pressure p, which are determined by the following system:

$$\begin{aligned} \hat{\rho} J \partial_t v + \hat{\rho} J v \cdot \nabla v - \operatorname{div}\sigma &= \hat{\rho} J f \quad \text{in } \Omega_s, \\ \partial_t u + v \cdot \nabla u - v &= 0 \quad \text{in } \Omega_s, \\ 1 - J &= 0 \quad \text{in } \Omega_s \quad \text{(INH material)}, \end{aligned} \tag{4.14}$$

where $E := \frac{1}{2}(F^{-T}F^{-1} - I)$, $F := I - \nabla u$, $J := \det F$, and

$$\sigma := \begin{cases} JF^{-1}(\lambda_s \operatorname{tr} E\, I + 2\mu_s E)F^{-T} & \text{(STVK material)}, \\ -pI + \mu_s(F^{-1}F^{-T} - I) & \text{(INH material)}. \end{cases}$$

Similarly the boundary conditions must be set in the Eulerian framework. As a boundary condition in time the same initial value v^0, now in

Eulerian framework, is prescribed for v at the initial time t_0. The fluid boundary $\partial\Omega_s$ is again split into three non-overlapping components $\partial\Omega_s = \Gamma_{sD} \cup \Gamma_{sN} \cup \Gamma_i$, with each part relating to a different boundary condition,

$$v = v^D \quad \text{on } \Gamma_{sD} \quad \text{(Dirichlet)},$$

$$\sigma n = g \quad \text{on } \Gamma_{sN} \quad \text{(Neumann)}.$$

The fixed boundary $\hat{\Gamma}_i$ on the reference domain now becomes the moving boundary Γ_i just as in the fluid flow case. The boundary conditions on Γ_i are similar to the fluid flow case: continuity of velocity v and the normal flux of the stress σn,

$$v_s = v_f, \quad \sigma_s n_s = \sigma_f n_s \quad \text{on } \Gamma_i. \tag{4.15}$$

Just as in the Lagrangian framework u_s^D and v_s^D are suitable extensions of the prescribed Dirichlet data on the boundaries of Ω_s, and g_s is the Neumann boundary condition on Γ_{sN}. Similarly as for the fluid problems, we have hidden the fluid-structure coupling condition of continuity of velocity in part of the boundary data v_s^D. The condition of continuity of normal stresses $\sigma_s n_s$, we have let stand.

The variational form of the structure equations in an Eulerian framework is again obtained by multiplying them with suitable test functions from the test space V_s^0 for the momentum equations and L_s for velocity/displacement and the optional incompressibility equations. The equations are written in an Eulerian framework in the domain Ω_s. The physical unknowns are the displacement $u_s \in u_s^D + \mathcal{V}_s^0$, the velocity $v_s \in v_s^D + \mathcal{V}_s^0$, and the optional pressure $p_s \in \mathcal{L}_s$, which are determined by the following problem.

Problem 5 (Structure model in Eulerian formulation). *Find* $\{u_s, v_s\} \in \{u_s^D + \mathcal{V}_s^0\} \times \{v_s^D + \mathcal{V}_s^0\}$ *and, in the INH case,* $p_s \in \mathcal{L}_s$, *such that* $u_s(0) = u_s^0$, $v_s(0) = v_s^0$, *and*

$$\begin{aligned}
&(\hat{\rho}_s J_s \partial_t v_s, \psi^v)_s + (\hat{\rho}_s J_s v_s \cdot \nabla v_s, \psi^v)_s + (\sigma_s, \nabla \psi^v)_s \\
&\qquad = (g_s, \psi^v)_{\Gamma_{sN}} + (\sigma_s n_s, \psi^v)_{\Gamma_i} + (\hat{\rho}_s J_s f_s, \psi^v)_s, \\
&(\partial_t u_s + v_s \cdot \nabla u_s - v_s, \psi^u)_s = 0, \\
&(1 - \det F_s, \psi^p) = 0 \quad (\textit{in the INH case}),
\end{aligned} \tag{4.16}$$

for all $\{\psi^u, \psi^v\} \in V_s^0 \times V_s^0$ *and, in the INH case,* $\psi^p \in L_s$ *where* $F_s := I - \nabla u_s$, $J_s := \det F_s$, $E := \frac{1}{2}(F_s^{-T}F_s^{-1} - I)$, *and*

$$\sigma_s := \begin{cases} J_s F_s^{-1}(\lambda_s \mathrm{tr} E\, I + 2\mu_s E)F_s^{-T} & \textit{(in the STVK case)}, \\ -p_s I + \mu_s(F_s^{-1}F_s^{-T} - I) & \textit{(in the INH case)}. \end{cases}$$

5. Variational Formulations of the FSI Problem

Based on the foregoing preparations, we can now state the variational formulations of the FSI problem in the ALE and the fully Eulerian frameworks.

5.1. *The ALE formulation of the FSI problem*

The ALE formulation of the FSI problem uses the natural Lagrangian framework for the structure and the fluid model is transformed from its Eulerian description into an arbitrary Lagrangian framework. Accordingly, the variational ALE formulation of the fluid part is stated on the (arbitrary) reference domain $\hat{\Omega}_f$, while the structure part uses the domain $\hat{\Omega}_s$, where $\hat{\Omega} = \hat{\Omega}_f \cup \hat{\Gamma}_i \cup \hat{\Omega}_s$ forms the common reference domain. In this setting the continuity of velocity across the fluid-structure interface $\hat{\Gamma}_i$ is strongly enforced by requiring one common continuous field for the velocity on $\hat{\Omega}$. The stress interface condition

$$\hat{J}_f \hat{\sigma}_f \hat{F}_f^{-T} \hat{n}_f = \hat{J}_s \hat{\sigma}_s \hat{F}_s^{-T} \hat{n}_f \quad \text{on } \hat{\Gamma}_i,$$

is still present in the form of a jump of the 1st Piola–Kirchhoff normal stresses of both subsystems,

$$(\hat{J}_f \hat{\sigma}_f \hat{F}_f^{-T} \hat{n}_f, \hat{\psi}^v)_{\hat{\Gamma}_i} + (\hat{J}_s \hat{\sigma}_s \hat{F}_s^{-T} \hat{n}_f, \hat{\psi}^v)_{\hat{\Gamma}_i},$$

on the right hand side. By omitting these integral terms the continuity of the normal stress becomes imposed "weakly", i.e., a condition implicitly contained in the combined variational formulation.

The combined formulation though implies that a domain mapping function $\hat{T}_f$ for the fluid domain is known. Such a mapping is obtained by adding an auxiliary equation to the fluid and structure equations. The boundary conditions for this mapping are clear: There is no deformation on all "outer" boundaries $\partial\hat{\Omega}_f \backslash \hat{\Gamma}_i$, and the deformation on $\hat{\Gamma}_i$ should be equal to $\hat{u}_s$. Thus, the global deformation $\hat{u}$ with $\hat{u}|_{\hat{\Omega}_s} = \hat{u}_s$ must have a trace on $\hat{\Gamma}_i$ such that $\hat{u} \in \hat{u}^D + \hat{\mathcal{V}}^0$.

The deformation itself can be sought as the solution to various "deformation problems" (i.e., "extension problems" of the boundary data $\hat{u}|_{\hat{\Omega}_s} = \hat{u}_s$), the simplest being the harmonic extension. If it is necessary that the deformation preserves volume an incompressibility condition can be added in the form

$$\hat{J}_f = \det(I + \hat{\nabla}\hat{u}_s) = 1, \tag{5.1}$$

or in a simplified form $\widehat{\operatorname{div}}\hat{u}_s = 0$, leading to the "Stokes extension". If the deformation should be more "smooth", then as an alternative the biharmonic extension can be used. In what follows, we use the harmonic extension of the structure displacement $\hat{u}_s$ into the fluid domain.

The remaining parts of the Neumann data $\hat{g}_f$ and $\hat{g}_s$ form the Neumann boundary data on $\hat{\Gamma}_N := \hat{\Gamma}_{fN} \cup \hat{\Gamma}_{sN}$ and are combined to $\hat{g}$. The right hand side functions $\hat{f}_f$ and $\hat{f}_s$ are combined to $\hat{f}$. We define the density $\hat{\rho}$ and the Cauchy stress tensor $\hat{\sigma}$ for the whole domain by

$$\hat{\rho}(\hat{x}) := \begin{cases} \hat{\rho}_f(\hat{x}), & \hat{x} \in \hat{\Omega}_f, \\ \hat{\rho}_s(\hat{x}), & \hat{x} \in \hat{\Omega}_s \cup \Gamma_i, \end{cases} \qquad \hat{\sigma}(\hat{x}) := \begin{cases} \hat{\sigma}_f(\hat{x}), & \hat{x} \in \hat{\Omega}_f, \\ \hat{\sigma}_s(\hat{x}), & \hat{x} \in \hat{\Omega}_s \cup \Gamma_i. \end{cases} \tag{5.2}$$

Further, we introduce the characteristic functions $\hat{\chi}_f$ and $\hat{\chi}_s$ defined by

$$\hat{\chi}_f(\hat{x}) := \begin{cases} 1, & \hat{x} \in \hat{\Omega}_f, \\ 0, & \hat{x} \in \hat{\Omega}_s \cup \hat{\Gamma}_i, \end{cases} \qquad \hat{\chi}_s := 1 - \hat{\chi}_f. \tag{5.3}$$

With this notation the variational ALE formulations of the nonstationary as well as the stationary FSI problem read as follows.

Problem 6 (ALE formulation of FSI problem). *Find* $\{\hat{v}, \hat{p}, \hat{u}\} \in \{\hat{v}^D + \hat{\mathcal{V}}^0\} \times \hat{\mathcal{L}} \times \{\hat{u}^D + \hat{\mathcal{V}}^0\}$, *such that* $\hat{v}(0) = \hat{v}^0$, $\hat{u}(0) = \hat{u}^0$, *and*

$$\begin{aligned}
&(\hat{\chi}_s\hat{\rho}d_t\hat{v}, \hat{\psi}^v) + (\hat{\chi}_f\hat{J}\hat{\rho}(\partial_t\hat{v} + ((\hat{F}^{-1}(\hat{v} - \partial_t\hat{T}) \cdot \hat{\nabla})\hat{v}, \hat{\psi}^v) \\
&\qquad + (\hat{J}\hat{\sigma}\hat{F}^{-T}, \hat{\nabla}\hat{\psi}^v) = (\hat{g}, \hat{\psi}^v)_{\hat{\Gamma}_N} + (\hat{\rho}\hat{J}\hat{f}, \hat{\psi}^v), \\
&(\hat{\chi}_f\widehat{\operatorname{div}}(\hat{J}\hat{F}^{-1}\hat{v}), \hat{\psi}^p) + \hat{\alpha}_p\{(\hat{\chi}_s\hat{\nabla}\hat{p}, \hat{\nabla}\hat{\psi}^p) - (\partial_n\hat{p}, \hat{\psi}^p)_{\hat{\Gamma}_i}\} = 0, \qquad (STVK)
\end{aligned}$$

$$(\hat{\chi}_f \widehat{\mathrm{div}}(\hat{J}\hat{F}^{-1}\hat{v}), \hat{\psi}^p) + (\hat{\chi}_s(\hat{J}-1), \hat{\psi}^p) = 0, \quad (INH)$$
$$(\hat{\chi}_s(d_t\hat{u} - \hat{v}), \hat{\psi}^u) + \hat{\alpha}_u\{(\hat{\chi}_f\hat{\nabla}\hat{u}, \hat{\nabla}\hat{\psi}^u) - (\hat{\partial}_n\hat{u}, \hat{\psi}^u)_{\hat{\Gamma}_i}\} = 0, \tag{5.4}$$

for all $\{\hat{\psi}^v, \hat{\psi}^p, \hat{\psi}^u\} \in \hat{V}^0 \times \hat{L}^* \times \hat{V}^0$, *where* $\hat{\alpha}_p, \hat{\alpha}_u$ *are small positive constants,* $\hat{T} := \mathrm{id} + \hat{u}$, $\hat{F} := \hat{\nabla}\hat{T}$, $\hat{J} := \det\hat{F}$, *and*

$$\hat{\sigma}|_{\hat{\Omega}_f} := -\hat{p}I + \rho_f\nu_f(\hat{\nabla}\hat{v}\hat{F}^{-1} + \hat{F}^{-T}\hat{\nabla}\hat{v}^T),$$

$$\hat{\sigma}|_{\hat{\Omega}_s} := \begin{cases} \hat{J}^{-1}\hat{F}(\lambda_s \mathrm{tr}\hat{E}\, I + 2\mu_s\hat{E})\hat{F}^T & (STVK), \\ -\hat{p}I + \mu_s(\hat{F}\hat{F}^T - I), & (INH). \end{cases}$$

Problem 7 (ALE formulation of stationary FSI problem). *Find* $\{\hat{v}, \hat{p}, \hat{u}\} \in \{\hat{v}^D + \hat{V}^0\} \times \hat{L} \times \{\hat{u}^D + \hat{V}^0\}$, *such that*

$$(\hat{\chi}_f\hat{J}\hat{\rho}\hat{F}^{-1}\hat{v}\cdot\hat{\nabla})\hat{v}, \hat{\psi}^v) + (\hat{J}\hat{\sigma}\hat{F}^{-T}, \hat{\nabla}\hat{\psi}^v) = (\hat{g}, \hat{\psi}^v)_{\hat{\Gamma}_N} + (\hat{\rho}\hat{J}\hat{f}, \hat{\psi}^v),$$
$$(\hat{\chi}_f \widehat{\mathrm{div}}(\hat{J}\hat{F}^{-1}\hat{v}), \hat{\psi}^p) + \hat{\alpha}_p\{(\hat{\chi}_s\hat{\nabla}\hat{p}, \hat{\nabla}\hat{\psi}^p) - (\partial_n\hat{p}, \hat{\psi}^p)_{\hat{\Gamma}_i}\} = 0, \quad (STVK)$$
$$(\hat{\chi}_f \widehat{\mathrm{div}}(\hat{J}\hat{F}^{-1}\hat{v}), \hat{\psi}^p) + (\hat{\chi}_s(\hat{J}-1), \hat{\psi}^p) = 0, \quad (INH)$$
$$-(\hat{\chi}_s\hat{v}, \hat{\psi}^u) + \hat{\alpha}_u\{(\hat{\chi}_f\hat{\nabla}\hat{u}, \hat{\nabla}\hat{\psi}^u) - (\hat{\partial}_n\hat{u}, \hat{\psi}^u)_{\hat{\Gamma}_i}\} = 0, \tag{5.5}$$

for all $\{\hat{\psi}^v, \hat{\psi}^p, \hat{\psi}^u\} \in \hat{V}^0 \times \hat{L} \times \hat{V}^0$, *where* $\hat{\alpha}_p, \hat{\alpha}_u$ *are small positive constants, and all other quantities are as defined in Problem 6.*

5.2. *The fully Eulerian formulation of the FSI problem*

The variational Eulerian formulation of the fluid problem is posed on the domain Ω_f, while the corresponding formulation of the structure problem is posed on the domain Ω_s. By construction the fluid-structure interaction interface Γ_i of both problems match. Now, both problems are combined into one complete problem on the combined domain $\Omega = \Omega_f \cup \Gamma_i \cup \Omega_s$.

Again, exactly as in the ALE formulation described above, continuity of the velocity across the fluid-structure interface Γ_i is strongly enforced by requiring one common continuous field for the velocity on Ω. The stress interface condition $\sigma_f n_f = \sigma_s n_f$ on Γ_i is still present in the form of a jump of the Cauchy stresses of both systems

$$([\sigma\cdot n], \psi^v)_{\Gamma_i} = (\sigma_f n_f, \psi^v)_{\Gamma_i} + (\sigma_s n_s, \psi^v)_{\Gamma_i}$$

on the right hand side. By omitting this boundary integral jump the (weak) continuity of the normal stress becomes an implicit condition of the combined variational formulation.

The remaining parts of the Neumann data g_f and g_s now form the Neumann boundary data on $\Gamma_N := \Gamma_{fN} \cup \Gamma_{sN}$ and are combined to g. Analogously, the Dirichlet boundary data v_f^D and v_s^D on parts of $\partial\Omega$ are merged into a suitable velocity field $v^D \in \mathcal{V}$. The right hand side functions f_f and f_s are combined to f. The Cauchy stress tensor is again defined areawise on the whole domain by

$$\sigma(x) := \begin{cases} \sigma_f(x), & x - u(x) \in \hat{\Omega}_f, \\ \sigma_s(x), & x - u(x) \in \hat{\Omega}_s \cup \hat{\Gamma}_i. \end{cases}$$

Analogously, we introduce the characteristic functions of the (unknown) subdomains Ω_f and Ω_s by

$$\chi_f(x) := \begin{cases} 1, & x - u(x) \in \hat{\Omega}_f, \\ 0, & x - u(x) \in \hat{\Omega}_s \cup \hat{\Gamma}_i, \end{cases} \qquad \chi_s := 1 - \chi_f.$$

The above definitions imply that we need to provide some kind of deformation u not only on the structure but also on the fluid domain. This is accomplished by the concept of the "Initial Position (IP) set" introduced below.

5.2.1. *The IP set approach*

We recall that for rewriting the structure equations in an Eulerian framework, we need the pressure $\hat{p}_s$, displacement $\hat{u}_s$, and its gradient $\hat{\nabla}\hat{u}_s$ expressed in the Eulerian sense, which are denoted by p_s, u_s, and ∇u_s, respectively. There holds

$$p_s(x) = \hat{p}_s(T(x)) = \hat{p}_s(\hat{x}), \quad u_s(x) = \hat{u}_s(T(x)) = \hat{u}_s(\hat{x}), \tag{5.6}$$

where T is the (inverse) displacement function of points in the deformed domain Ω_s back to points in the initial domain $\hat{\Omega}_s$,

$$\begin{aligned} \hat{T} : \hat{\Omega}_s \to \Omega_s, \quad & \hat{T}(\hat{x}) = \hat{x} + \hat{u}_s = x, \\ T : \Omega_s \to \hat{\Omega}_s, \quad & T(x) = x - u_s = \hat{x}. \end{aligned} \tag{5.7}$$

Since $\det\hat{\nabla}\hat{T} = \det\hat{F} \neq 0$ the displacements T and $\hat{T}$ are well defined. Further, for the deformation gradient, we found the relation $\hat{\nabla}\hat{u} = (I - \nabla u)^{-1} - I$.

The immediate difficulty with the above relations is that u_s is only implicitly determined by $\hat{u}_s$, since $T(x)$ also depends on u_s. This is unpractical, and we therefore need a direct way of determining the displacement $u(x)$ of a "material" point located at x with respect to its initial position at point $\hat{x}$. To achieve this, we introduce the so-called *"set of initial positions"* (IP set) $\varphi(\Omega, t)$ of all points of Ω at time t. If we look at a given "material" point at the position $x \in \Omega$ and the time $t \in (0, T]$, then the value $\varphi(x, t)$ will tell us what the initial position of this point was at time $t = 0$. These points are transported in the full domain with a certain velocity w. The convection velocity in the structure will be the structure velocity itself, $w_{|\Omega_s} = v_s$. If the fluid velocity were to be used for convection in the fluid domain, then the displacements there would eventually become very entangled. For this reason, we use an alternative velocity. With this notation, the mapping φ is determined by the following variational problem.

Problem 8 (Variational IP set problem). *Find $\varphi(\cdot, t) \in \varphi_0 + V^0$, $t \in I$, such that*

$$(\partial_t\varphi + w \cdot \nabla\varphi, \psi) = 0 \quad \forall\psi \in V^0. \tag{5.8}$$

where φ_0 is a suitable extension of the Dirichlet data along the boundaries,

$$\varphi(x, 0) = x, \quad x \in \Omega,$$

$$\varphi(x, t) = x, \quad \{x, t\} \in \partial\Omega \times (0, T].$$

This means that $\hat{x} + \hat{u}(\hat{x}, t) = x$, for any point with the initial position $\hat{x}$ and the position x later at time t. Since $\hat{x} = \varphi(\hat{x}, 0) = \varphi(x, t)$ and $\hat{u}(\hat{x}, t) = u(x, t)$ it follows that

$$x = \varphi(x, t) + u(x, t). \tag{5.9}$$

Using this in the IP set Eq. (5.8) yields

$$(\partial_t u - w + w \cdot \nabla u, \psi) = 0 \quad \forall\psi \in V^0. \tag{5.10}$$

The interface Γ_i will usually intersect mesh cells. Due to this, we need a reasonable continuation of the displacement and its gradient from the structure domain into the fluid domain. The value of u in the fluid domain will be determined by the choice of the convection velocity w. If we were to use the fluid velocity this would eventually lead to increasing entanglement, which would necessitate a continual reinitialization of the IP set. As an alternative, we use the harmonic continuation of the structure velocity v_s to the whole domain Ω, which is likewise denoted by w and satisfies

$$(\chi_s(w-v), \psi) + \alpha_w\{(\chi_f \nabla w, \nabla\psi) - (\partial_n w, \psi)_{\Gamma_i}\} = 0, \quad \forall \psi \in V^0, \tag{5.11}$$

where α_w is a small positive parameter. However, any other continuous extension of v_s satisfying physically reasonable boundary conditions would serve the same purpose. By this construction, the deflection u_f in the fluid domain becomes an artificial quantity without any real physical meaning, i.e., $d_t u_s = v_s$, but generally $d_t u_f \neq v_f$.

5.2.2. *Eulerian formulation of the FSI problem*

Now, we can combine the Eulerian formulations of the flow and the structure part of the problem into a complete variational formulation of the nonstationary as well as the stationary FSI problem in Eulerian framework. In the case of STVK material the (non-physical) pressure p_s in the structure subdomain is determined as harmonic extension of the flow pressure p_f.

Problem 9 (Eulerian formulation of FSI problem). *Find fields* $\{v, p, w, u\} \in \{v^D + \mathcal{V}^0\} \times \mathcal{L} \times \mathcal{V}^0 \times \mathcal{V}^0$, *such that* $v(0) = v^0$, $u(0) = u^0$, *and*

$$\begin{aligned}
(\rho(\partial_t v + v\cdot\nabla v), \psi) + (\sigma, \varepsilon(\psi)) &= (g_3, \psi)_{\partial\Omega} + (f_3, \psi), \\
(\operatorname{div} v, \chi) &= 0 \quad (INH), \\
(\chi_f \operatorname{div} v, \chi) + \alpha_p\{(\chi_s \nabla p, \nabla\chi) - (\partial_n p, \chi)_{\Gamma_i}\} &= 0 \quad (STVK), \\
(\partial_t u - w + w\cdot\nabla u, \xi) &= 0, \\
(\chi_s(w-v), \varphi) + \alpha_w\{(\chi_f \nabla w, \nabla\varphi) - (\partial_n w, \varphi)_{\Gamma_i}\} &= 0,
\end{aligned} \tag{5.12}$$

for all $\{\psi, \chi, \xi, \varphi\} \in V^0 \times L \times V^0 \times V^0$, *where* α_p, α_w *are small positive constants,* $F := I - \nabla u$, $J := \det F$, $E := \frac{1}{2}(F^{-T}F^{-1} - I)$,

$$\chi_f := \begin{cases} 1, & x - u \in \hat\Omega_f \backslash \hat\Gamma_i, \\ 0, & x - u \in \hat\Omega_s, \end{cases} \qquad \chi_s = 1 - \chi_f,$$

and

$$\sigma|_{\Omega_f} := -pI + 2\rho_f \nu_f \varepsilon(v),$$

$$\sigma|_{\Omega_s} := \begin{cases} -pI + \mu_s(F^{-1}F^{-T} - I) & (INH), \\ JF^{-1}(\lambda_s(\mathrm{tr}E)I + 2\mu_s E)F^{-T} & (STVK). \end{cases}$$

In this variational formulation the position of the fluid structure interface Γ_i is implicitly given by the displacement u and the characteristic function χ_s,

$$\Gamma_i(t) = \{x \in \Omega,\ x - u(x,t) \in \hat{\Gamma}_i\}. \tag{5.13}$$

Notice that the system (5.12) is *nonlinear* even if the two subproblems are linear, e.g., for a Stokes fluid interacting with a linear elastic structure.

In some situations the solution of an FSI problem may tend to a "steady state" as $t \to \infty$. For later purposes, we derive the set of equations determining such a steady state solution $\{v^*, p^*, w^*, u^*\} \in \{v^D + V^0\} \times L \times V^0 \times V^0$. The corresponding limits of the characteristic functions and subdomains are denoted by χ_f^*, χ_s^* and Ω_f^*, Ω_s^*, respectively. Further, the fluid velocity becomes constant in time, $v_f^* := \lim_{t\to\infty} v|_{\Omega_f}$, and the structure velocity vanishes, $v_s^* \equiv 0$, which in turn implies $w^* \equiv 0$. The steady state structure displacement u_s^* is likewise well defined, but the corresponding ("non-physical") fluid displacement is merely defined by $u_f^* = u_f^{\lim} := \lim_{t\to\infty} u|_{\Omega_f}$ and therefore depends on the chosen construction of $w|_{\Omega_f}$ as continuation of w_{Ω_s}. Actually, it could be defined by any suitable continuation of u_s^* to all of Ω, e.g., by harmonic continuation. On the other hand the steady state pressure p^* is to be determined from the limiting equations. Then, with suitable extensions u^D and v^D of the prescribed Dirichlet data on $\partial\Omega$, the FSI system (5.12) reduces to the following "stationary" form (for simplicity dropping the stars):

Problem 10 (Eulerian formulation of stationary FSI problem). *Find* $\{v, p, u\} \in \{v^D + V^0\} \times L \times \{u^D + V^0\}$, *such that*

$$(\rho v \cdot \nabla v, \psi) + (\sigma, \varepsilon(\psi)) = (g_3, \psi)_{\partial\Omega} + (f_3, \psi),$$

$$(\mathrm{div}\, v, \chi) = 0 \quad (INH),$$

$$(\chi_f \mathrm{div} v, \chi) + \alpha_p\{(\chi_s \nabla p, \nabla \chi) - (\partial_n p, \chi)_{\Gamma_i}\} = 0 \quad (STVK),$$
$$(\chi_f(u - u_f^{\lim}), \varphi) + (\chi_s v, \varphi) = 0, \tag{5.14}$$

for all $\{\psi, \chi, \varphi\} \in V^0 \times L \times V^0$, *where* α_p *is a small positive constant,* $F := I - \nabla u$, $J := \det F$, $E := \frac{1}{2}(F^{-T}F^{-1} - I)$, *and*

$$\sigma|_{\Omega_f} := -pI + 2\rho_f \nu_f \varepsilon(v),$$
$$\sigma|_{\Omega_s} := \begin{cases} -pI + \mu_s(F^{-1}F^{-T} - I) & (INH), \\ JF^{-1}(\lambda_s(\mathrm{tr}E)I + 2\mu_s E)F^{-T} & (STVK). \end{cases}$$

In the following Table 1, we summarize the two monolithical variational formulations of the FSI problem, the (arbitrary) Lagrangian–Eulerian (ALE) and the (fully) Eulerian formulation.

Table 1. Overview of variational formulations of the FSI problem for INH material: ALE (left) and Eulerian (right).

I) ALE formulation	II) Eulerian formulation
$\hat\Omega = \hat\Omega_f \cup \hat\Gamma_i \cup \hat\Omega_s$	$\Omega = \Omega_f \cup \Gamma_i \cup \Omega_s$
Find $\{\hat v, \hat p, \hat u\}$ for $\hat v(0)$, $\hat u(0)$ with	Find $\{v, p, w, u\}$ for $v(0)$, $u(0)$ with
$(\rho \hat J \partial_t \hat v + \hat\chi_f \rho \hat J(\partial_t \hat u - \hat v)\cdot \hat\nabla \hat v \hat F^{-1}, \hat\psi)$	$(\rho\partial_t v + \rho v \cdot \nabla v, \psi) + (\sigma, \varepsilon(\psi)) = f(\psi)$
$+(\hat J \hat\sigma \hat F^{-T}, \hat\varepsilon(\hat\psi)) = \hat f(\hat\psi)$	$(\mathrm{div} v, \chi) = 0$
$(\hat\chi_s(\hat J - 1) + \hat\chi_f \widehat{\mathrm{div}}(\hat J \hat v \cdot \hat F^{-T}), \hat\chi) = 0$	$(\chi_s(w - v), \varphi) + \alpha_w\{(\chi_f \nabla w, \nabla\varphi)$
$(\partial_t \hat u - \hat\chi_s \hat v, \hat\varphi) + \hat\alpha_u\{(\hat\chi_f \hat\nabla \hat u, \hat\nabla\hat\varphi)$	$-(\partial_n w, \varphi)_{\Gamma_i}\} = 0$
$-(\hat\partial_n \hat u, \hat\varphi^u)_{\hat\Gamma_i}\} = 0$	$(\partial_t u - w + w \cdot \nabla u, \xi) = 0$
(harmonically continued $\hat u_s$ into Ω_f)	(harmonically continued w_s into Ω_f)
for all test fields $\{\hat\psi, \hat\chi, \hat\varphi\}$	for all test fields $\{\psi, \chi, \xi, \varphi\}$, where
$\hat\chi_s(\hat x) := \begin{cases} 0, & \hat x \in \hat\Omega_f \\ 1, & \hat x \in \hat\Omega_s \cup \hat\Gamma_i \end{cases}$	$\chi_s(x) := \begin{cases} 0, & x - u \in \hat\Omega_f \\ 1, & x - u \in \hat\Omega_s \cup \hat\Gamma_i \end{cases}$
$\hat\chi_f := 1 - \hat\chi_s$	$\chi_s := 1 - \chi_f$
$\hat\sigma := \begin{cases} -\hat p_f + \mu_f(\hat\nabla \hat v_f \hat F^{-1} + \hat F^{-T}\hat\nabla \hat v_f^T) \\ -\hat p_s + \mu_s(\hat F \hat F^T - I) \text{ in } \hat\Omega_s \cup \hat\Gamma_i \end{cases}$	$\sigma := \begin{cases} -p_f I + 2\mu_f \varepsilon(v_f) \text{ in } \Omega_f \\ -p_s I + \mu_s(F^{-1}F^{-T} - I) \text{ in } \Omega_s \end{cases}$
$\hat F := I + \hat\nabla \hat u, \quad \hat J := \det \hat F$	$F := I - \nabla u$

6. Discretization

In this section, we detail the discretization in space and time of the FSI problem based on its different variational formulations. Our method of choice is the Galerkin finite element (FE) method with "conforming" finite elements. For a general introduction to the FE method, we refer to Carey and Oden,[15] Girault and Raviart,[27] Brenner and Scott,[12] and Braess.[11] First, we provide the framework for the finite element method. Then, we describe the complete variational formulations which are the basis of the Galerkin discretization. Since we are using a so-called "equal-order" approximation of all physical quantities, additional pressure stabilization has to be incorporated which is done here by the "local projection" technique of Becker and Braack.[4] The time discretization is by the first-order backward Euler scheme or the second-order Fractional-Step-θ scheme. We refer to Bristeau, Glowinski and Periaux,[13] Rannacher,[46,47] and Glowinski[28] for a detailed discussion of these time discretization schemes.

At each time step a nonlinear algebraic problem is solved using a Newton-like method. This relies on solving the linear defect-correction problem, which requires the evaluation of the corresponding Jacobi matrix. Due to the large size and the strongly nonlinear nature of the complete FSI problems in the ALE or the Eulerian frameworks, calculating the Jacobi matrix can be cumbersome. This difficulty is overcome following an approach that is also used in the method of "automatic differentiation", see Griewank.[29]

6.1. *Mesh notation and finite element spaces*

The spatial discretization is by a conforming finite element Galerkin method on meshes $\mathbb{T}_h$ consisting of cells denoted by K,

$$\bar{\Omega} = \bigcup_{i=1,\dots,N} \bar{K}_i,$$

which are (convex) quadrilaterals in 2D or hexahedrals in 3D. Such a decomposition $\mathbb{T}_h$ is referred to as "regular" if any cell edge is either a subset of the domain boundary components Γ_D, Γ_N, or a complete face or edge of another cell. However, to facilitate mesh refinement and coarsening, we allow the cells to have a certain number of nodes that are at the midpoint of sides or faces of neighboring cells. These “hanging nodes” do not carry degrees of freedom and the corresponding function values are determined by linear or bilinear interpolation of neighboring “regular” nodal points. For more details on this construction, we refer to Carey and Oden[15] or Bangerth and Rannacher.[2]

The mesh parameter h is a scalar cellwise constant function defined by $h|_K := h_K = \mathrm{diam}(K)$. We set $h_{\max} := \max_{K\in\mathbb{T}_h} h_K$. For a cell K, we denote by ρ_K the diameter of the maximal inscribed ball of K and by $\alpha_K^{\max}$ its maximum interior angle. To ensure proper approximation properties of the finite element spaces which are constructed based on the meshes $\mathbb{T}_h$, we require the following regularity condition to be fulfilled:

Mesh regularity condition: *Each cell $K \in \mathbb{T}_h$ is the image of the reference unit cube $\hat{K} = [0, 1]^d$ under some d-linear mapping $\sigma_K : \hat{K} \to K$. This mapping is uniquely described by the 2^d coordinate values of the corners of K, if the ordering of the corners is preserved, see Fig. 3. The Jacobian tensors σ_K' of these mappings are invertible and satisfy the uniform bounds*

$$\sup_{h>0}\max_{K\in\mathbb{T}_h} \|\sigma_K'\| \le c, \qquad \sup_{h>0}\max_{K\in\mathbb{T}_h} \|(\sigma_K')^{-1}\| \le c. \tag{6.1}$$

This condition is satisfied if the cells $K \in \mathbb{T}_h$ possess the usual structural properties of uniform “non-degeneracy”, “uniform shape”, and “uniform size property”.

To increase the number of cells in a decomposition $\mathbb{T}_h$, we employ “mesh refinement”, which consists of subdividing a cell into 2^d subcells. Cell subdivision is done by connecting the midpoints of opposing edges or faces of a cell. A refinement is “global” if this is done for every cell. An example of a regular mesh and two global refinements is shown in Fig. 4. Each of the resulting meshes after refinement is also regular. “Coarsening”

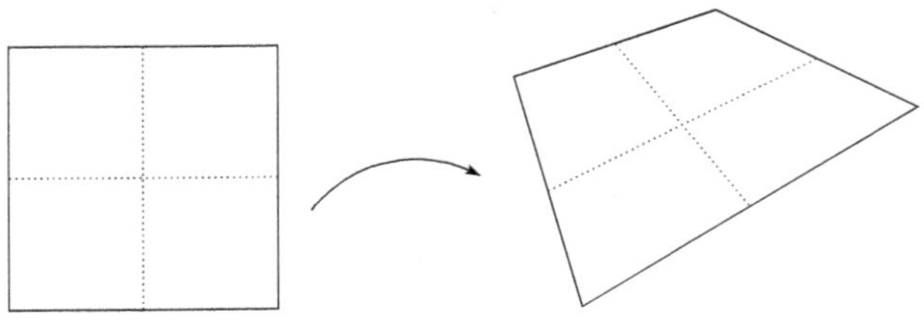

Fig. 3. Reference mapping $\sigma_K : \hat{K} \to K$.

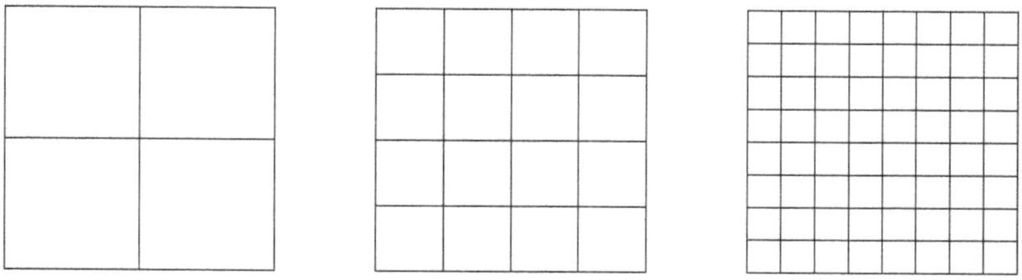

Fig. 4. A regular mesh after two cycles of global refinement.

of 2^d cells is possible if they were generated by prior refinement of some "parent cell". A group of 2^d such cells is referred to as a "(cell) patch".

In addition to "global" refinement, we will also use "local" refinement. This consists of only subdividing some cells in a given decomposition. Such refinement leads to cells nodes that are placed on the middle of the neighboring cells' edges or faces, i.e., to "hanging nodes", where only one hanging node is allowed per edge or face. In Fig. 5 local refinement is applied twice leading to hanging nodes indicated by "dots".

Sometimes, we will require that a decomposition $\mathbb{T}_h$ is organized in a patchwise manner. This means that $\mathbb{T}_h$ is the result of global refinement of the coarser decomposition $\mathbb{T}_{2h}$, as shown in Fig. 6.

Given a function space V the decomposition $\mathbb{T}_h$ and the cellwise space of polynomial functions $Q(K)$, we construct the corresponding finite element subspace $V_h \subset V$ by

$$V_h := \{\varphi \in V,\ \varphi|_K \in Q(K),\ K \in \mathbb{T}_h\}.$$

Each polynomial function space $Q(K)$ is actually defined on a "reference cell" $\hat{K} := (0, 1)^d$ as the reference function space $\hat{Q}(\hat{K})$. The function

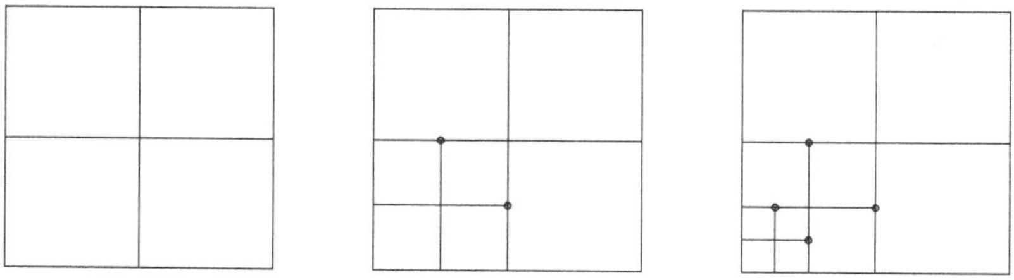

Fig. 5. A regular mesh after two cycles of local refinement.

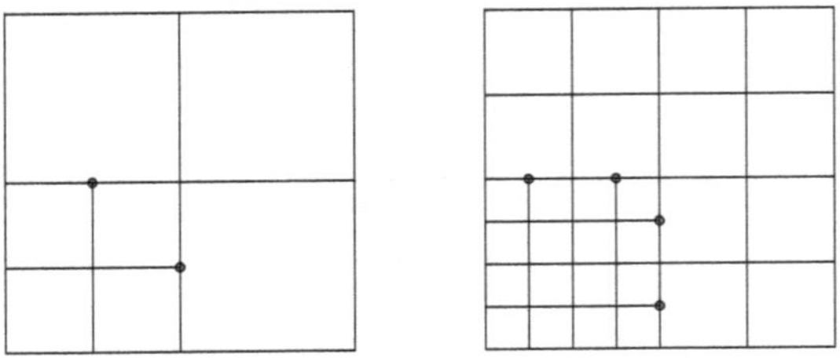

Fig. 6. A regular mesh after two cycles of patchwise local refinement.

space of polynomial degree $p \geq 0$ on $\hat{K}$ we denote as

$$\hat{Q}_p(\hat{K}) := \text{span}\{\hat{x}^\alpha,\ \hat{x} = (\hat{x}_1, \dots, \hat{x}_d) \in \hat{K},\ \alpha = (\alpha_1, \dots, \alpha_d),$$
$$\alpha_i \in \{0, \dots, p\}\},$$

with the usual multi-index notation. In the numerical tests presented below, only finite elements with $p = 1$ ("d-linear elements") are used. Therefore, we omit the degree p and simply refer to $\hat{Q}(\hat{K})$. The reference function space $\hat{Q}(\hat{K})$ is mapped to the corresponding cell K with the help of the d-linear mapping $\sigma_K : \hat{K} \to K$,

$$Q(K) = \left\{\varphi(x) = \hat{\varphi}\big(\sigma_K^{-1}(x)\big), \quad \hat{\varphi} \in \hat{Q}(\hat{K}), \quad x \in K\right\}.$$

Since, in this case, the reference function space and the mapping function space are the same, the resulting finite elements are referred to as "isoparametric".

6.2. *Galerkin formulation*

For the discretization of Problems 6 and 7 (ALE framework) or 9 and 10 (Eulerian framework) in space, we use equal-order Q_1 finite elements as described above for all unknowns where the corresponding finite element subspaces are denoted by

$$L_h \subset L, \quad V_h \subset V, \quad W_h \subset W := V \times V \times V \times V.$$

Within the present abstract setting the discretization in time is likewise thought as by a Galerkin method, such as the dG(r) ("discontinuous" Galerkin) or cG(r) ("continuous" Galerkin) method of degree $r \geq 0$. The dG(0) method is closely related to the first-order backward Euler scheme (indeed even identical to this method for autonomous problems) and the cG(1) method to the second-order Crank–Nicolson scheme. However, in the test computations described below, we have used the Galerkin method only in space but the finite difference schemes in time. The full space-time Galerkin framework is mainly introduced as basis for a systematic approach to residual-based *a posteriori* error estimation as described below. In the following, we write the discretization only for the FSI problem written in the Eulerian framework. In the ALE framework it looks quite similar.

At first, we introduce a compact form of the variational formulation of the FSI problem. For arguments $U = \{v, p, w, u\}$ and $\Psi = \{\psi^v, \psi^p, \psi^w, \psi^u\} \in \mathcal{W} := \mathcal{V} \times \mathcal{V} \times \mathcal{V} \times \mathcal{V}$, we introduce the space-time semilinear form

$$
\begin{aligned}
&A(U)(\Psi) \\
&\quad := \int_0^T \Big\{(\rho(\partial_t v + v \cdot \nabla v), \psi^v) + (\sigma(U), \varepsilon(\psi^v)) \\
&\qquad + \begin{cases} (\mathrm{div}\, v, \psi^p) & (INH) \\ (\chi_f \mathrm{div}\, v, \psi^p) + \alpha_p\{(\chi_s \nabla p, \nabla \psi^p) - (\partial_n p, \psi^p)_{\Gamma_i}\} & (STVK) \end{cases} \\
&\qquad - (g_3, \psi^v)_{\partial\Omega} - (f_3, \psi^v) + (\partial_t u - w + w \cdot \nabla u, \psi^u) \\
&\qquad + (\chi_s(w - v), \psi^w) + \alpha_w\{(\chi_f \nabla w, \nabla \psi^w) - (\partial_n w, \psi^w)_{\Gamma_i}\}\Big\}\, dt.
\end{aligned}
$$

With this notation, we can write the variational problem (9) in compact form.

Problem 11 (Compact Eulerian formulation of FSI problem). *Find $U \in U^D + \mathcal{W}^0$, such that*

$$A(U)(\Psi) = 0 \quad \forall \Psi \in \mathcal{W}^0, \tag{6.2}$$

where U^D is an appropriate extension of the Dirichlet boundary and initial data and the space $\mathcal{W}^0$ is defined by

$$
\begin{aligned}
\mathcal{W}^0 := \{\Psi = \{\psi^v, \psi^p, \psi^w, \psi^u\} \in \mathcal{V}^0 \times \mathcal{V}^0 \times \mathcal{V}^0 \times \mathcal{V}^0,& \\
\psi^u(0) = \psi^v(0) = 0\}.&
\end{aligned}
$$

The spatial discretization by "equal-order" finite elements for velocity and pressure needs stabilization in order to compensate for the missing "inf-sup stability". We use the so-called *"local projection stabilization"* (LPS) introduced by Becker and Braack.[4,5] An analogous approach is also employed for stabilizing the convection in the flow model as well as in the transport equation for the displacement u. Alternative methods of stabilization within Galerkin schemes use the "upwind/Petrov-Galerkin" approach of Hughes and Brooks[36] and Hughes, Franca and Balestra.[37]

We define the mesh-dependent bilinear form

$$(\varphi, \psi)_\delta := \sum_{K \in \mathbb{T}_h} \delta_K \, (\varphi, \psi)_K,$$

where the parameter δ_K is adaptively determined by

$$\delta_K := \frac{\alpha h_K^2}{\chi_f \rho_f \nu_f + \chi_s \mu_s + \beta \rho |v_h|_{\infty;K} h_K + \gamma |w_h|_{\infty;K} h_K}.$$

Further, we introduce the "fluctuation operator" $\pi_h : V_h \to V_{2h}$ on the finest mesh level $\mathbb{T}_h$ by $\pi_h := I - P_{2h}$, where $P_{2h} : V_h \to V_{2h}$ denotes the L^2-projection. The operator π_h measures the fluctuation of a function in V_h with respect to its projection into the next coarser space V_{2h}. With this notation, we define the stabilization form

$$S_\delta(U_h)(\Psi_h) := \int_0^T \{(\nabla \pi_h p_h, \nabla \pi_h \psi_h^p)_\delta + (\rho v_h \cdot \nabla \pi_h v_h, v_h \cdot \nabla \pi_h \psi_h^v)_\delta$$
$$+ (w_h \cdot \nabla \pi_h u_h, w_h \cdot \nabla \pi_h \psi_h^u)_\delta\} \, dt,$$

where the first term stabilizes the fluid pressure, the second one the INH structure pressure, the third one the transport in the flow model, and the fourth one the transport of the displacement u_h. The LP stabilization has the important property that it acts only on the diagonal terms of the coupled system and that it does not contain any second-order derivatives. However, it is only "weakly" consistent, as it does not vanish for the continuous solution, but it tends to zero with the right order as $h \to 0$. The choice of the numbers α, β, γ in the stabilization parameter δ_K is, based on practical experience, in our computations $\alpha = 1/2$, and $\beta = \gamma = 1/6$.

With this notation the stabilized Galerkin finite element approximation of problem (6.2) reads as follows.

Problem 12 (Spatial Galerkin approximation of FSI problem in Eulerian framework). *Find* $U_h \in U_h^D + \mathcal{W}_h^0$, *such that*

$$A_\delta(U_h)(\Psi_h) := A(U_h, \Psi_h) + S_\delta(U_h)(\Psi_h) = 0 \quad \forall \Psi_h \in \mathcal{W}_h^0, \tag{6.3}$$

where the "discrete" finite element space $\mathcal{W}_h^0$ *is defined analogously as its "continuous" counterpart* $\mathcal{W}^0$.

As on the spatially continuous level the existence of solutions to this semi-discrete problem is not guaranteed and has to be justified separately for each particular situation.

6.3. *Time discretization*

The discretization in time is by the so-called *"Fractional-Step-θ scheme"* in which each time step $t_{n-1} \to t_n$ is split into three substeps $t_{n-1} \to t_{n-1+\theta} \to t_{n-\theta} \to t_n$. For brevity, we formulate this time stepping method for an abstract differential-algebraic equation (DAE) of the form

$$\begin{bmatrix} M & 0 \\ 0 & 0 \end{bmatrix} \begin{bmatrix} \dot{v}(t) \\ \dot{p}(t) \end{bmatrix} + \begin{bmatrix} A(v(t)) & B \\ -B^T & C \end{bmatrix} \begin{bmatrix} v(t) \\ p(t) \end{bmatrix} = \begin{bmatrix} b(t) \\ c(t) \end{bmatrix}, \tag{6.4}$$

which resembles the operator form of the spatially discretized incompressible Navier–Stokes equations with pressure stabilization. With the parameters $\theta = 1 - \sqrt{2}/2 = 0.292893\ldots$, $\theta' = 1 - 2\theta$, $\alpha \in (1/2, 1]$, and $\beta = 1 - \alpha$, the fractional-step-θ scheme reads:

$$\begin{bmatrix} M + \alpha\theta k A^{n-1+\theta} & \theta k B \\ -B^T & C \end{bmatrix} \begin{bmatrix} v^{n-1+\theta} \\ p^{n-1+\theta} \end{bmatrix} = \begin{bmatrix} [M - \beta\theta k A^{n-1}]v^{n-1} + \theta k b^{n-1} \\ c^{n-1+\theta} \end{bmatrix}$$

$$\begin{bmatrix} M + \beta\theta' k A^{n-\theta} & \theta' k B \\ -B^T & C \end{bmatrix} \begin{bmatrix} v^{n-\theta} \\ p^{n-\theta} \end{bmatrix} = \begin{bmatrix} [M - \alpha\theta' k A^{n-1+\theta}]v^{n-1+\theta} + \theta' k b^{n-\theta} \\ c^{n-\theta} \end{bmatrix}$$

$$\begin{bmatrix} M + \alpha\theta k A^n & \theta k B \\ -B^T & C \end{bmatrix} \begin{bmatrix} v^n \\ p^n \end{bmatrix} = \begin{bmatrix} [M - \beta\theta k A^{n-\theta}]v^{n-\theta} + \theta k b_h^{n-\theta} \\ c^n \end{bmatrix},$$

where $A^{n-1+\theta} := A(x^{n-1+\theta})$, $b^{n-1} := b(t_{n-1})$, etc. This scheme is of second-order and has a similar work complexity as the well-known Crank–Nicolson scheme (case $\alpha = 1/2$). The fractional-step-θ scheme was originally proposed in form of an operator splitting scheme separating the two complications "nonlinearity" and "incompressibility" within each cycle $t_{n-1} \to t_{n-1+\theta} \to t_{n-\theta} \to t_n$. However, it has also very attractive features as a pure time-stepping method. Being *strongly* A-stable, for any choice of $\alpha \in (1/2, 1]$, it possesses the full smoothing property in the case of rough initial data, in contrast to the Crank–Nicolson scheme which is only conditionally smoothing (for $k \sim h^2$). Furthermore, it is less dissipative than most of the other second-order implicit schemes and therefore suitable for computing oscillatory solutions; for more details, we refer to Bristeau, Glowinski and Periaux,[13] Rannacher,[46,47] and Glowinski.[28]

For computing steady state solutions, we use a pseudo-time stepping techniques based on the simple (first-order) backward Euler scheme, which in the notation from before reads

$$\begin{bmatrix} M + kA^n & kB \\ -B^T & C \end{bmatrix} \begin{bmatrix} v^n \\ p^n \end{bmatrix} = \begin{bmatrix} Mv^{n-1} + kb_h^{n-1} \\ c^n \end{bmatrix}.$$

6.4. *Solution of the algebraic systems*

After time and space discretization, in each substep of the fractional-step-θ scheme (or any other fully implicit time-stepping scheme) a quasi-stationary nonlinear algebraic system has to be solved. This is done by a standard Newton-type method with adaptive step-length selection, in which most of the nonlinear terms (i.e., the transport terms, the structure stress terms, the ALE mapping terms) are correctly linearized. Only the stabilization terms and the terms involving the characteristic function χ_f, determining the position of the interface, are treated by a simple functional iteration. In all cases the iteration starts from the values at the preceding time level. The resulting linear subproblems are then solved by the "Generalized Minimal Residual (GMRES)" method with preconditioning by a geometric multigrid method with block-ILU smoothing. Since such an approach is rather standard nowadays, we omit its details and refer to some relevant literature, e.g., see Turek,[54] Rannacher,[46] or Hron and Turek.[33] For the implementational details of using the multigrid method on locally refined meshes, we refer to Becker and Braack.[3]

6.5. *Evaluation of directional derivatives*

The resulting linear operator is essentially (if time-stepping parts and factors stemming from the approximation of the temporal derivatives are neglected) the directional derivatives of the governing semilinear form of the variational formulation, i.e., the form $A(U)(\Psi)$ in the Eulerian framework,

$$A'(U)(\Phi, \Psi) := \frac{d}{d\varepsilon} A(U + \varepsilon\Phi)(\Psi)|_{\varepsilon=0}. \tag{6.5}$$

For only "weakly nonlinear" systems such as the original Navier–Stokes equations in Eulerian framework obtaining the directional derivative is a straight forward task and can be done analytically "by hand". For structure mechanical systems (for example based on the St. Vernant–Kirchhoff material law) though writing down the explicit directional derivative can become a combersome. For example in the Lagrangian case the scalar product $(\hat{J}\hat{F}^{-T}, \widehat{\nabla}\hat{\varphi}^v)$ is strongly nonlinear in $\hat{u}$,

$$\hat{J}\hat{\sigma}\hat{F}^{-T} = \hat{F}(\lambda_s \mathrm{tr}\hat{E}\, I + 2\mu_s \hat{E}), \quad \hat{F} = I + \widehat{\nabla}\hat{u}, \quad \hat{E} = \frac{1}{2}(\hat{F}^T\hat{F} - I).$$

However, in the Eulerian framework the corresponding scalar product takes the form $(\sigma, \nabla\varphi)$, which does not become any easier since the Cauchy stress tensor σ is based on the inverse of the "reverse deformation gradient" $I - \nabla u$,

$$\sigma = JF^{-1}(\lambda_s \mathrm{tr}E\, I + 2\mu_s E)F^{-T}, \quad F = I - \nabla u, \quad E = \frac{1}{2}(F^{-T}F^{-1} - I).$$

To alleviate this problem one may use a method that is the basis of "Automatic Differentiation" such as described in Rall[45] and Griewank.[29] The method is used to determine the derivative of a function at a given position. It is based on the technique of mechanically applying the basic rules of differentiation to the "serialized evaluation" of a function. This is achieved by breaking down the evaluation of the function for a given value into a sequence of basic elementary evaluations. Consequently, since evaluation is done in a sequence the resulting values from one evaluation are used in a later evaluation. To these elementary parts the rules of differentiation (i.e., the chain rule, the sum rule and the product rule) are applied (see Table 2).

The method of automatic differentiation lies between those of symbolic differentiation and the approximation of derivatives by divided differences. It is similar to symbolic differentiation in so far that the results are calculated by evaluating the same sequence of functions. It is thus just as accurate as symbolic differentiation. The difference is that, in contrast to symbolic differentiation, all "parsing" is done before compilation of the program, when the function evaluation is serialized and differentiation is applied to all levels of the serialization. This parsing before compilation is what gives the method a slight similarity to the method of divided differences. The full theory of automatic differentiation usually includes implementing the method in the form of a "precompiler" that completely relieves the user of applying the method and literally generates the derivatives in an automatic and efficient fashion. However, for the computational examples presented in this paper the method is used with differentiation done "by hand". In the first step, the "forward sweep", the function is broken down into a sequence of basic elementary evaluations. Each of these evaluations is stored in a variable. In the second step, the "reversed sweep", the rules of differentiation are applied. As a basic example, we present the calculation of the derivative of the function $f(x) := \sin(x \tanh(x)) \log(x - 1/x)$ at the position x_0. The details of the realization of this method of automatic differentiation for the FSI problem formulated in the ALE or the Eulerian framework can be found in Dunne.[22,23]

The only difference between the Eulerian and the ALE directional derivative are the differences of the directional derivatives concerning the displacement φ^u and $\hat{\varphi}^u$, respectively. In the Eulerian framework, we obtain "interface Dirac functions", whereas in the ALE framework the transformations acting on fluid equations are "derived". This problem is typically encountered in the field of shape optimization, see Sokolowski,

Table 2. Example of automatic differentiation.

forward sweep		reverse sweep
$f_1 := 1/x_0$		$f_1' := -1/x_0^2$
$f_2 := \log(x_0 - f_1)$		$f_2' := (1 - f_1')/(x_0 - f_1)$
$f_3 := \tanh(x_0)$	$\rightarrow$	$f_3' := 1 - \tanh^2(x_0)$
$f_4 := x_0 f_3$		$f_4' := f_3 + x_0 f_3'$
$f_5 := \sin(f_4)$		$f_5' := f_4' \cos(f_4)$
$f_6 := f_5 f_2$		$f_6' := f_5' f_2 + f_5 f_2'$
$f(x_0) := f_6$		$f'(x_0) := f_6'$

Zolésio[50] and Allaire, de Gournay, Jouve and Toader.[1] For strong solutions $U = \{u, v, p\}$ and $\hat{U} = \{\hat{u}, \hat{v}, \hat{p}\}$, with $\hat{U}(\hat{x}) = U(\hat{x} + \hat{u}(\hat{x}))$ for $\hat{x} \in \hat{\Omega}$, of the different formulations of the FSI problem the directional derivatives with respect to velocity and pressure are equal. For the details, we again refer to Dunne.[22,23]

The directional derivatives play a direct role in the setup of the (linearized) "dual problem" which occurs in the method for *a posteriori* error estimation and "goal-oriented" mesh adaptation used in our test computations. The differences due to the use of the ALE or the Eulerian framework will be investigated at simple model situations, below.

7. Mesh Adaptation

Now, we come to one of the main issues of this article, namely the automatic mesh adaptation within the finite element solution of the FSI problem. The computations shown in Secs. 8 and 9, below, have been done on three different types of meshes:

- globally refined meshes obtained using several steps of uniform refinement of a coarse initial mesh,
- locally refined meshes obtained using a purely geometry-based criterion by marking all cells for refinement which have certain prescribed distances from the fluid-structure interface,

- locally refined meshes obtained using a systematic residual-based criteria by marking all cells for refinement which have error indicators above a certain threshold.

The ultimate goal is to employ the so-called *"Dual Weighted Residual Method"* (DWR method) for the adaptive solution of FSI problems. This method has been developed in Becker and Rannacher[7,8] (see also Bangerth and Rannacher[2]) as an extension of the duality technique for *a posteriori* error estimation described in Eriksson, Estep, Hansbo and Johnson.[25] The DWR method provides a general framework for the derivation of "goal-oriented" *a posteriori* error estimates together with criteria of mesh adaptation for the Galerkin discretization of general linear and nonlinear variational problems, including optimization problems. It is based on a complete *variational* formulation of the problem, such as (6.2) for the FSI problem. In fact, this was one of the driving factors for deriving the Eulerian formulation underlying (6.2). In order to incorporate also the time discretization into this framework, we have to use a fully space-time Galerkin method, i.e., a standard finite element method in space combined with the dG(r) or cG(r) ("discontinuous" Galerkin or "continuous" Galerkin) method in time. The following discussion assumes such a space-time Galerkin discretization, though in our test computations, we have used the fractional-step-θ scheme which is a difference scheme. Accordingly, in this paper the DWR method is used only in its stationary form in computing either steady states or intermediate quasi-steady states within the time stepping process.

7.1. *The DWR method*

We begin with the description of the DWR method for the special case of an FSI problem governed by an abstract variational equation such as (6.2). We restrict us to a simplified version of the DWR method, which suffices for the present purposes. For a more elaborated version, which is particularly useful in the context of optimization problems, we refer to the literature stated above. For notational simplicity, we think the nonhomogeneous boundary and initial data U^D to be incorporated into a linear forcing term $F(\cdot)$, or to be exactly representable in the approximating space $\mathcal{W}_h$. Then, we seek

$U \in U^D + \mathcal{W}^0$ such that

$$A(U)(\Psi) = F(\Psi) \quad \forall \Psi \in \mathcal{W}^0. \tag{7.1}$$

The corresponding (stabilized) Galerkin approximation seeks $U_h \in U_h^D + \mathcal{W}_h^0$ such that

$$A(U_h)(\Psi_h) + S_\delta(U_h)(\Psi_h) = F(\Psi) \quad \forall \Psi_h \in \mathcal{W}_h^0. \tag{7.2}$$

Suppose that the goal of the computation is the evaluation of the value $J(U)$ for some functional $J(\cdot)$ (for simplicity assumed to be linear) which is defined on $\mathcal{W}$. We want to control the quality of the discretization in terms of the error

$$J(U - U_h) = J(U) - J(U_h).$$

To this end, we introduce the directional derivative $A'(U)(\Phi, \cdot)$ the existence of which is assumed. With the above notation, we introduce the bilinear form

$$L(U, U_h)(\Phi, \Psi) := \int_0^1 A'(U_h + s(U - U_h))(\Phi, \Psi)\, ds,$$

and formulate the "dual problem"

$$L(U, U_h)(\Phi, Z) = J(\Phi) \quad \forall \Phi \in \mathcal{W}^0. \tag{7.3}$$

In the present abstract setting the existence of a solution $Z \in \mathcal{W}^0$ of the dual problem (7.3) has to be assumed. Now, taking $\Phi = U - U_h \in \mathcal{W}^0$ in (7.3) and using the Galerkin orthogonality property

$$A(U)(\Psi_h) - A(U_h)(\Psi_h) = S_\delta(U_h)(\Psi_h), \quad \Psi \in \mathcal{W}_h^0,$$

yields the error representation

$$\begin{aligned} J(U - U_h) &= L(U, U_h)(U - U_h, Z) \\ &= \int_0^1 A'(U_h + s(U - U_h))(U - U_h, Z)\, ds \\ &= A(U)(Z) - A(U_h)(Z) \\ &= F(Z - \Psi_h) - A(U_h)(Z - \Psi_h) - S_\delta(U_h)(\Psi_h) \\ &=: \rho(U_h)(Z - \Psi_h) - S_\delta(U_h)(\Psi_h), \end{aligned}$$

where $\Psi_h \in \mathcal{W}^0$ is an arbitrary element, usually taken as the generic nodal interpolant $I_h Z \in \mathcal{W}_h^0$ of Z. For the evaluation of the terms on the right-hand side, we split the integrals in the residual term $\rho(U_h)(Z - \Psi_h)$ into their contributions from the single mesh cells $K \in \mathbb{T}_h$ and integrate by parts. This results in an estimate of the error $|J(U - U_h)|$ in terms of computable local residual terms $\rho_K(U_h)$ multiplied by certain weight factors $\omega_K(Z)$ which depend on the dual solution Z,

$$|J(U - U_h)| \leq \sum_{K \in \mathbb{T}_h} \rho_K(U_h)\,\omega_K(Z) + |S_\delta(U_h)(\Psi_h)|. \tag{7.4}$$

The explicit form of the terms in the sum on the right-hand side will be stated for a special situation below. The second term due to the regularization is assumed to be small and is therefore neglected.

Since the dual solution Z is unknown, the evaluation of the weights $\omega_K(Z)$ requires further approximation. First, we linearize by assuming

$$L(U, U_h)(\Phi, \Psi) \approx L(U_h, U_h)(\Phi, \Psi) = A'(U_h)(\Phi, \Psi),$$

and use the approximate "discrete" dual solution $Z_h \in \mathcal{W}_h^0$ defined by

$$A'(U_h)(\Phi, Z_h) = J(\Phi_h) \quad \forall \Phi_h \in \mathcal{W}_h^0. \tag{7.5}$$

From Z_h, we generate improved approximations to Z in a post-processing step by patchwise higher-order interpolation. For example in 2D on 2×2-patches of cells in $\mathbb{T}_h$ the 9 nodal values of the piecewise bilinear Z_h are used to construct a patchwise biquadratic function $\tilde{Z}$ as indicated in Fig. 7.

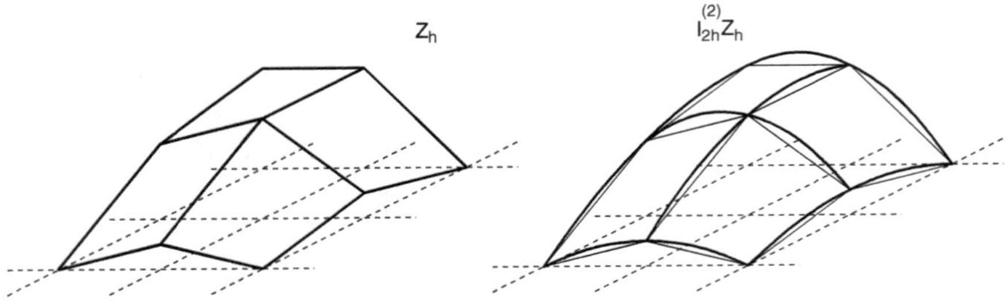

Fig. 7. Local postprocessing by patchwise "biquadratic" interpolation, $I_{2h}^{(2)} Z_h$, of the "bilinear" discrete solution Z_h in 2D.

This is then used to obtain the approximate error estimate

$$|J(U - U_h)| \approx \eta := \sum_{K \in \mathbb{T}_h} \rho_K(U_h)\, \omega_K(\tilde{Z}), \tag{7.6}$$

which is the basis of automatic mesh adaptation.

Remark 1. The dual solution Z has the features of a "generalized" Green function $G(K, K')$, as it describes the dependence of the target error quantity $J(U - U_h)$, which may be concentrated at some cell K, on local properties of the data, i.e., in this case the residuals $\rho_{K'}$ on cells K'; see Fig. 8.

Remark 2. The solvability of the primal and dual problems (7.1), (7.2) and (7.3), (7.5), respectively, is not for granted. This is a difficult task in view of the rather few existence results in the literature for general FSI problems. Further, the assumption of differentiability cause concerns in treating the FSI problems in the Eulerian framework since the dependence of the characteristic function $\chi_f(x - u)$ on the deflection u is generally not differentiable (only Lipschitzian). However, this non-differentiability can be resolved by the "Hadamard structure theorem", on the assumption that the interface between fluid and structure forms a lower dimensional manifold and the differentiation is done in a weak variational sense. In essence this has the same effect as discretizing along the interface and replacing the directional derivative by a mesh-size dependent difference quotient, a pragmatic approach that has proven itself in similar situations, e.g., for Hencky elasto-plasticity in Rannacher and Suttmeier.[48]

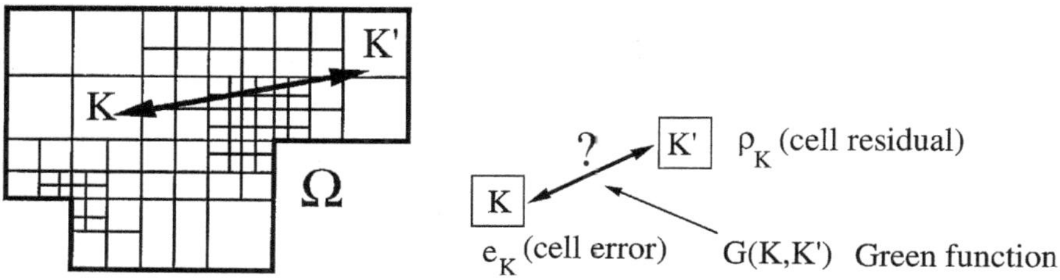

Fig. 8. Finite element mesh and scheme of error propagation.

For the *primal* problem the directional (Gâteaux) derivative of the complete FSI problem does not need to be exact, it only needs to be "good enough" for Newton iteration to ensure convergence, leading to a reduction of the residuals of the nonlinear system. Thus, for the primal problem the nonlinear system is used to measure the "quality" of the approximation. For the *dual* problem though things may initially seem less clear, since the dual problem is simply a *linear* problem directly based on the Gâteaux derivative. Of course, an immediate "measure of quality" of the discrete dual solution is the residual of the linear system. But there is no immediate measure for the quality of the discrete dual solution in relation to the *continuous* dual solution. This uncertainty stems from the highly nonlinear influence of the displacement u in the Gâteaux derivative. For the ALE framework this is seen in the transformed fluid equations. For the Eulerian framework this is seen in additional boundary Dirac integrals, which stem from the shape derivatives. This seemingly lack of clarity though is not typical to FSI problems. It is only more obvious in such problems since everything *visible* depends on the position of the interface. Generally though this uncertainty concerning the discrete dual solution is present in *all* nonlinear problems, since in such problems the Gâteaux derivatives depend on the primal solution and can only be approximated by using the discrete primal solutions.

In the case of FSI problems, we assume that the interface obtained on the current mesh is already in good agreement with the correct one, $\Gamma_{hi} \approx \Gamma_i$, and set up the dual problem formally with Γ_{hi} as a fixed interface. This approach has proven very successful in similar situations, e.g., for Hencky elasto-plasticity in Rannacher and Suttmeier.[48] In all test computations, we did not encounter difficulties in obtaining the discrete solutions. In fact the performance of the error estimator for a given goal functional was always good for both the ALE and the Eulerian framework.

A common measure of the accuracy of the error estimator is the *"effectivity index"* defined by

$$I_{\text{eff}} := \frac{\eta}{|J(U - U_h)|},$$

which is the overestimation factor of the error estimator. It should desirably be close to one. A second quality measure for the error estimator is how effective its results are as indicators for adaptive mesh refinement. The error indicators $\eta_K := \rho_K \omega_K$ are the cellwise contributions of the error estimator

$$\eta = \sum_{K \in \mathbb{T}_h} \eta_K.$$

Again, in all our test computations the error estimator performed well for both the ALE and the Eulerian frameworks.

7.2. *Mesh adaptation algorithm*

The approach we use for the adaptive refinement of the spatial mesh is straightforward. Particularly, for the refinement criteria there exist much more sophisticated versions, which are not used here for sake of simplicity. Let an error tolerance *TOL* be given. Then, on the basis of the (approximate) *a posteriori* error estimate (7.6), the mesh adaptation proceeds as follows:

1. Compute the primal solution U_h from Eq. (7.2) on the current mesh, starting from some initial state, e.g., that with zero deformation.
2. Compute the solution $\tilde{Z}_h$ of the approximate discrete dual problem (7.5).
3. Evaluate the cell-error indicators $\eta_K := \rho_K(U_h)\,\omega_K(\tilde{Z}_h)$.
4. If $\eta < TOL$ then accept U_h and evaluate $J(U_h)$, otherwise proceed to the next step.
5. Determine the 30% cells with largest and the 10% cells with smallest values of η_K. The cells of the first group are refined and those of the second group coarsened. Then, continue with Step 1. (Coarsening usually means canceling of an earlier refinement. Further refinement may be necessary to prevent the occurrence of too many hanging nodes. In two dimensions this strategy leads to about a doubling of the number of cells in each refinement cycle. By a similar strategy it can be achieved that the number of cells stays about constant during the adaptation process within a time stepping procedure.

Remark 3. The error representation (7.4) has been derived assuming the error functional $J(\cdot)$ as linear. In many applications nonlinear, most often quadratic, error functionals occur. An example is the spatial L^2-norm error

$$J(U_h) := \|(U - U_h)(T)\|,$$

at the end time T. For nonlinear (differentiable) error functionals the DWR approach can be extended to yield an error representation of the form (7.4); see Becker and Rannacher,[8] Bangerth and Rannacher.[2]

Remark 4. The DWR method can also be applied to optimization problems of the form

$$\min_{q \in Q} J(u, q)! \quad a(u, q)(\psi) = f(\psi) \quad \forall \psi \in \mathcal{V}^0,$$

for instance in the context of an FSI setting. To this end we introduce the Lagrangian functional $\mathcal{L}(u, q, \lambda) := J(u, q) + f(\lambda) - a(u, q)(\lambda)$, with the adjoint variable $\lambda \in \mathcal{V}$. Its stationary points $\{u, q, \lambda\}$ are possible solutions of the optimization problem. These are determined by the nonlinear variational equation (so-called KKT system)

$$\mathcal{L}'(u, q, \lambda)(\varphi, \chi, \psi) = 0 \quad \forall \{\varphi, \chi, \psi\} \in \mathcal{V}^0 \times Q \times \mathcal{V}^0,$$

which has saddle-point character. For the solutions $\{u_h, q_h, \lambda_h\}$ of the corresponding finite element Galerkin approximation, there are residual-based error estimates available similar to (7.6); see Becker and Rannacher,[8] Bangerth and Rannacher.[2]

7.3. *A stationary special case*

We will develop the explicit form of the error representation (7.4) and the approximate dual problem (7.5) for the *stationary* FSI model with an "incompressible neo-Hookean" (INH) solid. We assume the system as being driven only by Dirichlet boundary conditions, i.e., volume and surface forces are zero, $f_3 \equiv 0$ and $g_3 \equiv 0$. Let $\{v, p, w, u\} \in \{v^D + V^0\} \times L \times V^0 \times V^0$ be a steady state solution of the corresponding FSI model (5.12) determined by the system (5.14). In order to simplify the formulation of the corresponding dual problem, we omit higher order terms, e.g., in the Cauchy stress tensor for the structure. Since there is no movement in the structure domain, the mass conservation condition $\mathrm{div}\, v_s = 0$ is not practical for the sensitivity analysis. For this reason the conservation condition in the structure domain will be $\det(I - \nabla u) = 1$, from which we again omit the higher order terms by approximating $\det(I - \nabla u) - 1 \approx \mathrm{div} u = 0$. Then, the variational formulation of the stationary FSI problem further reduces to

$$A(U)(\Psi) = F(\Psi) \quad \forall \Psi \in W^0, \tag{7.7}$$

with the (time-independent) semilinear form

$$\begin{aligned} A(U)(\Psi) := (\rho v \cdot \nabla v, \psi^v) + (\sigma(U), \varepsilon(\Psi)) + (\chi_f \mathrm{div} v + \chi_s \mathrm{div} u, \psi^p) \\ + (\chi_f u + \chi_s v, \psi^u), \end{aligned}$$

and the linear functional $F(\Psi) := (\chi_f u_f, \psi^u)$. Here, the stress–strain relation is given by

$$\sigma(U) = \begin{cases} -pI + 2\rho_f \nu_f \varepsilon(v), & \text{in } \Omega_f, \\ -pI + 2\mu_s \varepsilon(u), & \text{in } \Omega_s. \end{cases}$$

where the (small-strain) approximation $FF^T - I \approx 2\varepsilon(\psi^u)$ has been used.

Suppose now that the discretization error is to be controlled with respect to some linear functional on W of the form $J(\Phi) = j^v(\varphi^v) + j^p(\varphi^p) + j^u(\varphi^u)$. In order to correctly set up the corresponding dual problem, we would have to differentiate the semi-linear form $A(U)(\Psi)$ with respect to all components of U. However, this is not directly possible since the unknown position of the interface Γ_i depends on the displacement function u in a non-differentiable way. Because of this difficulty, we will adopt a more heuristic approach which is rather common in solving problems with free (only implicitly determined) boundaries. We assume that the interface obtained on the current mesh is already in good agreement with the correct one, $\Gamma_{ih} \approx \Gamma_i$, and set up the dual problem formally with Γ_{ih} as a fixed interface.

Adopting these simplifications and dropping all stabilization terms leads us to the following approximate dual problem, in which we seek $Z = \{z^v, z^p, z^u\} \in W^0$, such that

$$\tilde{A}'(U_h)(\Phi, Z) = J(\Phi) \quad \forall \Phi = \{\varphi^v, \varphi^p, \varphi^u\} \in W^0, \tag{7.8}$$

with the bilinear form

$$\tilde{A}'(U_h)(\Phi, Z) = (\rho v_h \cdot \nabla\varphi^v, z^v) + (\rho\varphi^v \cdot \nabla v_h, z^v) + (\sigma'(U_h)\varepsilon(\Phi), \varepsilon(Z)) + (\chi_f \mathrm{div}\varphi^v + \chi_s \mathrm{div}\varphi^u, z^p) + (\chi_f \varphi^u + \chi_s \varphi^v, z^u),$$

where

$$\sigma'(U_h)\varepsilon(U) := \begin{cases} -pI + 2\rho_f \nu_f \varepsilon(v), & \text{in } \Omega_{f,h}, \\ -pI + 2\mu_s \varepsilon(u), & \text{in } \Omega_{s,h}. \end{cases}$$

Let Z be a solution of Eq. (7.8) and Z_h its finite element approximation. To evaluate the approximate error estimate

$$J(U - U_h) \approx F(Z - Z_h) - A(U_h)(Z - Z_h),$$

we introduce two modified submeshes

$$\mathbb{T}_{hs} := \{K \cap \Omega_{s,h},\ K \in \mathbb{T}_h\}, \quad \mathbb{T}_{hf} := \{K \cap \Omega_{f,h},\ K \in \mathbb{T}_h\},$$

and their union $\tilde{\mathbb{T}}_h := \mathbb{T}_{hs} \cup \mathbb{T}_{hf}$. The mesh $\tilde{\mathbb{T}}_h$ differs only from $\mathbb{T}_h$ in so far that the cells that contain the fluid-structure interface are subdivided into fluid domain part and structure domain part. Now, by cellwise integration by parts and rearranging boundary terms, we obtain

$$\begin{aligned} J(U - U_h) &= \sum_{K \in \tilde{\mathbb{T}}_h} \Big\{ (\mathrm{div}\sigma(U_h) - \rho v_h \cdot \nabla v_h, z^v - z_h^v)_K \\ &\quad - \left(\frac{1}{2}[\sigma(U_h) \cdot n], z^v - z_h^v\right)_{\partial K} - (\chi_f \mathrm{div} v_h + \chi_s \mathrm{div} u_h, z^p - z_h^p)_K \\ &\quad - (\chi_f (u_h - u_f) + \chi_s v_h, z^u - z_h^u)_K \Big\}, \end{aligned}$$

where $[\cdot]$ denotes the jump across intercell boundaries Γ. If Γ is part of the boundary $\partial\Omega$ the "jump" is assigned the value $[\sigma \cdot n] = 2\sigma \cdot n$. We note that in this error representation the "cell residuals" $\{\mathrm{div}\sigma(U_h) - \rho v_h \cdot \nabla v_h\}|_K$, $\{\chi_f \mathrm{div} v_h + \chi_s \mathrm{div} u_h\}|_K$, and $\{\chi_f(u_h - u_f) + \chi_s v_h\}|_K$ represent the degree of consistency of the approximate solution U_h, while the "edge term" $\frac{1}{2}[\sigma(U_h) \cdot n]$ measures its "discrete" smoothness. These residual terms are multiplied by the weights (sensitivity factors) $z^v - z_h^v$, $z^p - z_h^p$, and $z^u - z_h^u$, respectively. From this error representation, we can deduce the following approximate error estimate

$$|J(E)| \approx \sum_{K \in \tilde{\mathbb{T}}_h} \eta_K, \quad \eta_K := \sum_{i=1}^{4} \rho_K^{(i)} \omega_K^{(i)}, \tag{7.9}$$

with the residual terms and weights

$$\begin{aligned}
\rho_K^{(1)} &:= \|\mathrm{div}\sigma(U_h) - \rho v_h \cdot \nabla v_h\|_K, & \omega_K^{(1)} &:= \|Z - Z_h\|_K, \\
\rho_K^{(2)} &:= \frac{1}{2} h_K^{-1/2} \|[\sigma(U_h) \cdot n]\|_{\partial K}, & \omega_K^{(2)} &:= h_K^{1/2} \|z^v - z_h^v\|_{\partial K}, \\
\rho_K^{(3)} &:= \|\chi_f \mathrm{div} v_h + \chi_s \mathrm{div} u_h\|_K, & \omega_K^{(3)} &:= \|z^p - z_h^p\|_K, \\
\rho_K^{(4)} &:= \|\chi_f (u_h - u_f) + \chi_s v_h\|_K, & \omega_K^{(4)} &:= \|z^u - z_h^u\|_K.
\end{aligned}$$

The weights $\omega^{(i)}$ are approximated, for instance by post-processing the discrete dual solution Z_h as described above. Then, the cellwise error indicators η_K can be used for the mesh adaptation process.

7.4. *Numerical integration along the interface*

In the Eulerian framework, regardless of the refinement technique used, the interface line will be intersecting cells. In these interface cells equations change, e.g., the constitutive equations of the stress tensor. In the structure-structure interaction examples below only the material parameter of the structure changes. In the fluid-structure examples the constitutive equation of the stress tensor changes entirely. The primal approach for coping with the error at the interface is to increase the refinement. This is either done by employing zonal refinement along the whole interface or using sensitivity analysis as a guide for local refinement.

Of course the first cause for an error at the interface cells is when the discrete variables do not approximate the continuous values well enough. This error can only be resolved by cell refinement. If the error at the interface cells is in large parts only caused by quadrature errors, then refinement along the interface solely on this basis is expensive, since this increases the number of unknowns in the complete system. Additionally, even if the discrete variables do approximate the continuous values well, the quadrature error will still occur, due to the change in the model. Consider for example in Problem 10 the incompressibility condition for the fluid:

$$0 = \left(\chi_f \operatorname{div} v_h, \psi_h^p\right) = \sum_{K \in \mathbb{T}_h} \int_K \chi_f \operatorname{div} v_h \psi_h^p \, dx. \tag{7.10}$$

Generally, we will be using Gauß quadrature. This quadrature though is only good for smooth functions. For cells that are either completely in the fluid domain or in the structure domain the use of the Gauß quadrature is appropriate. But for interface calls, i.e., cells which are cut by the interface, this will lead to the cell integrals being wrongly weighted. In the context of Eq. (7.10) it will lead to the incompressibility condition either having a strong and undesirable influence on the structure velocity or on the other hand being influenced by the structure velocity. To reduce this error, we use an adaptive quadrature. On cells that are not cut by the interface, we continue using the Gauß rule. On the cells containing the interface, we use a more appropriate summed quadrature rule, which is based on the simple midpoint rule. This strategy has a very good effect on the quality of the approximation, and particularly on the mass conservation. For more details, we refer to Dunne.[22,23]

An additional source of discontinuous behavior is the exact evaluation of the characteristic functions. This stems from the way the basis functions on the cells couple, which may lead to a sudden on- and off-switching of the coupling between the nodes of interface cells and their neighbors, which in turn leads to sudden discontinuous behavior of node values. To alleviate this problem, we regularize the characteristic functions χ_f and $\chi_s := 1 - \chi_f$ like

$$\chi_{fh} := \frac{1}{2}(1 + \tanh(\alpha_\chi \varphi(x))), \qquad \chi_{sh} := 1 - \chi_{fh},$$

with a smoothing parameter α_χ and the signed distance function $\varphi(x) := (\chi_f - \chi_s)\operatorname{dist}(x, \Gamma_i)$. The smoothing parameter is chosen accordingly to the local mesh size h. We only use φ as a parameter to the tanh function, thus it is only necessary that it roughly approximate the distance. In the examples presented below material deformations at the interface are regular enough to allow the Eulerian distance function φ to be approximated by the reference domain distance $\varphi(x) \approx \hat{\varphi}(x - u)$ with $\hat{\varphi}(\hat{x}) := (\hat{\chi}_f - \hat{\chi}_s)\operatorname{dist}(\hat{x}, \hat{\Gamma}_i)$. This approach has similarity to the Volume-of-Fluid method (Hirt and

Nichols[32]), since both use a "fraction of equation" variable similar to our approximation of χ_f. For more details, we refer to Dunne.[22,23]

8. Numerical Test 1: Elastic Flow Cavity

For validating the numerical method based on the Eulerian framework, we use a simple stationary test example, the lid-driven cavity with an elastic bottom wall, as shown in Fig. 9. For simplicity, for modeling the fluid the linear Stokes equations are used and the material of the bottom wall is assumed to be linear neo-Hookean and incompressible. The structure material is taken as very soft such that a visible deformation of the fluid-structure interface can be expected. Then, the other material parameters are chosen such that flow and solid deformation velocity are small enough to allow for a stationary solution of the coupled linear systems. This solution is computed by a pseudo-time stepping method employing the implicit Euler scheme. A steady state is reached once the kinetic energy of the structure is below a prescribed small tolerance, here $\|v_s\|^2 \leq 10^{-8}$.

The cavity has a size of 2×2, and its elastic part has a height of 0.5. The material constants are $\rho_f = \rho_s = 1$, $\nu_f = 0.2$, and $\mu_s = 2.0$. At the top boundary Γ_{d1} the regularized tangential flow profile

$$v_0 = 0.5 \begin{cases} 4x, & x \in [0.0, 0.25], \\ 1, & x \in (0.25, 1.75), \\ 4(2-x), & x \in [1.75, 2.0], \end{cases}$$

is prescribed, in order to avoid problems due to pressure singularities.

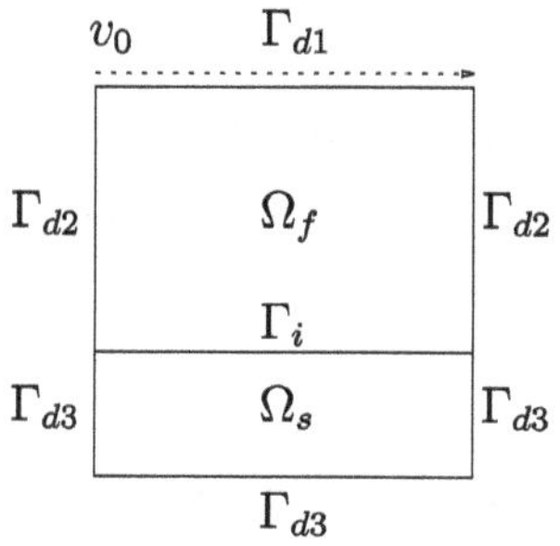

Fig. 9. Configuration of the "elastic" lid-driven cavity.

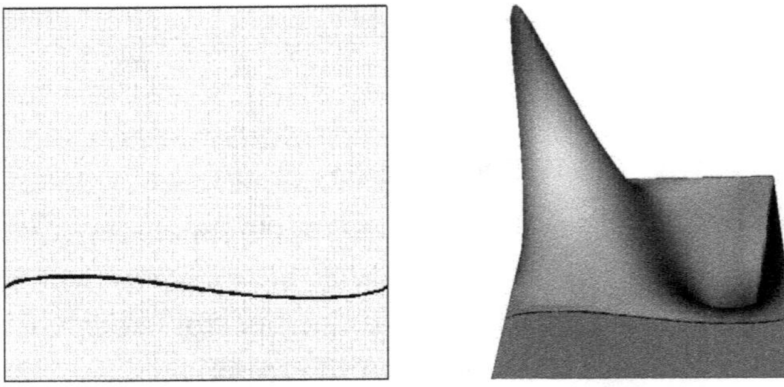

Fig. 10. Final position of interface (left) and vertical velocity (right).

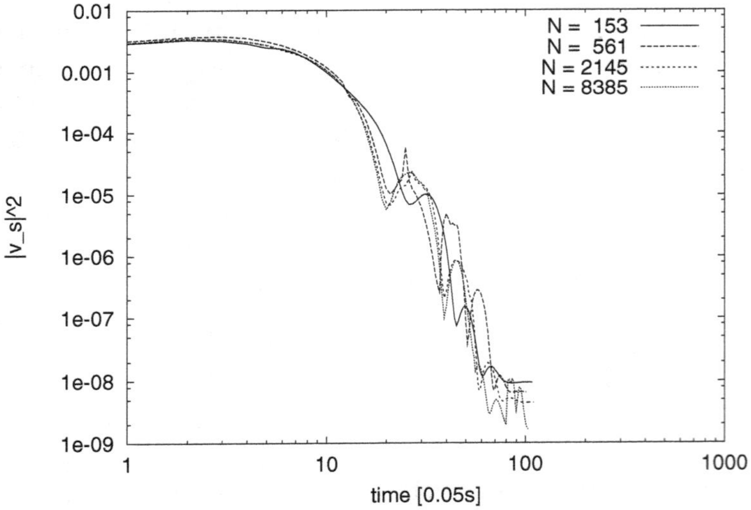

Fig. 11. Variation of $\|v_s\|^2$ in time for different numbers N of mesh cells.

8.1. *Computations on globally refined meshes*

Figure 10 shows the final steady state computed on globally uniform meshes. In Fig. 11, we monitor the development of $\|v_s\|^2$ during the pseudo-time stepping process depending on the number of cells of the mesh. As expected the kinetic energy tends to zero. The multiple "bumps" occur due to the way the elastic structure reaches its stationary state by "swinging" back and forth a few times. At the extreme point of each swing the kinetic energy has a local minimum. Figure 12 displays the mass error of the

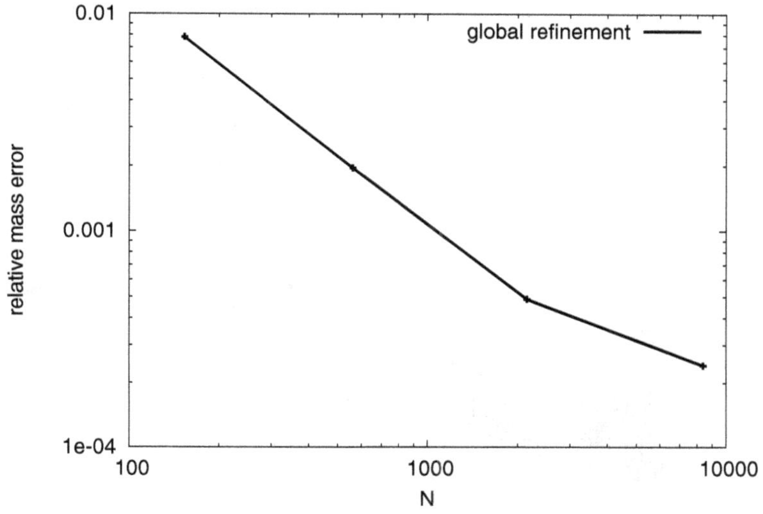

Fig. 12. Relative error of mass conservation in the steady state on globally and locally refined meshes.

structure at the stationary state and finds that it is actually of the expected order $O(h^2)$.

8.2. *Computations on locally adapted meshes*

Next, we apply the simplified stationary version of the DWR method as described in Sec. 7 for local mesh adaptation in the present test problem. For the "goal-oriented" *a posteriori* error estimation, we take the value of the pressure at the point $A = (0.5, 1.0)^T$ which is located in the flow region. To avoid sharp singularities in the corresponding dual solution, the associated functional is regularized to

$$J(u, p) = |K_A|^{-1} \int_{K_A} p\,dx \approx p(A),$$

where $K_A \in \mathbb{T}_h$ is a cell containing the point A. As a reference value of $p(A)$, we use the result obtained on a very fine uniform mesh.

Figure 13 shows a sequence of adapted meshes. As expected two effects can be seen. There is local refinement around the point of interest and since the position of the fluid-structure interface is a decisive factor for the pressure field, local refinement also occurs along the interface.

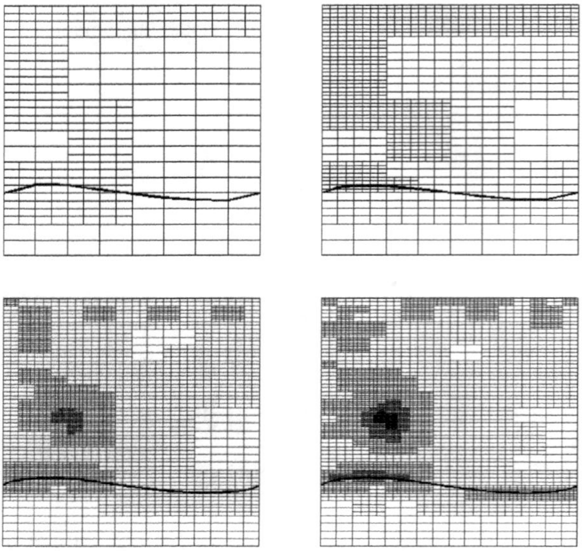

Fig. 13. Locally adapted meshes with $N = 335, 1031, 3309, 5123$ cells.

In Figs. 14 and 15 the resulting pressure error and the relative error in mass conservation is displayed as a function of the number of mesh cells.

It may seem surprising that in Fig. 15 there is no reduction of the mass error in the last iteration. This is due to the approach we are using here. After each step of mesh adaption a new primal solution is calculated, starting with the initial state of no deformation. The sensitivity analysis though does not take the initial state into account. Mesh adaption takes place around the final state of the interface, it does not reflect its initial state. An easy way of alleviating the mass error problem is to explicitly move a certain amount of local refinements with the interface from one time step to the next. Doing that though in this example would have made it unclear if the local refinement at the final interface position was due to the sensitivity analysis or the explicit movement of interface-bound refinement.

9. Numerical Test 2: FSI Benchmark FLUSTRUK-A

The second example is the FSI benchmark FLUSTRUK-A described in Ref. 34. A thin elastic bar immersed in an incompressible fluid develops self-induced time-periodic oscillations of different amplitude depending

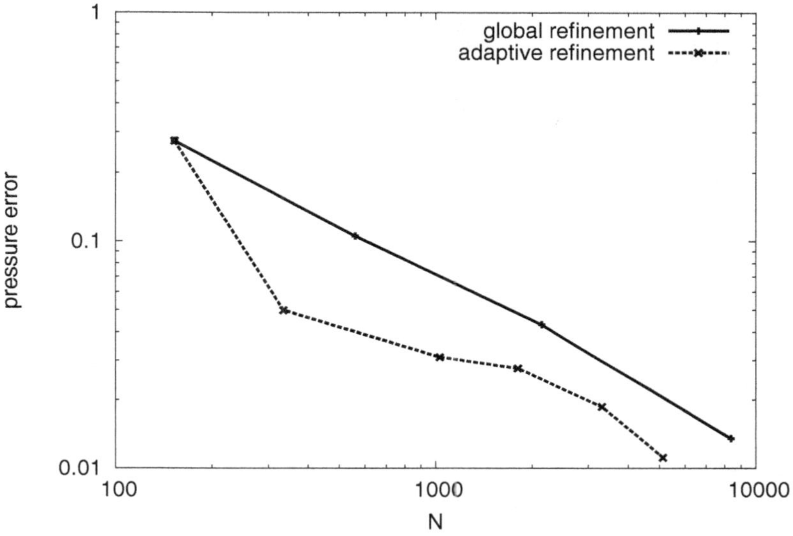

Fig. 14. Reduction of the pressure value error.

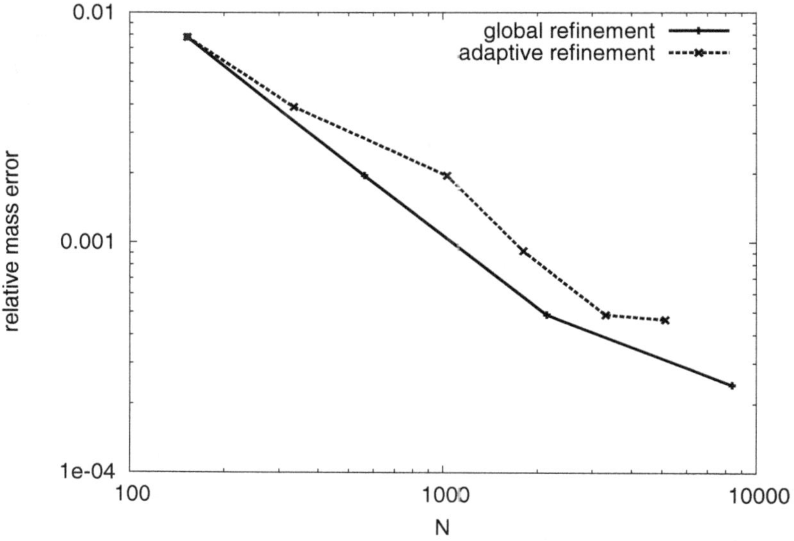

Fig. 15. Relative error of mass conservation.

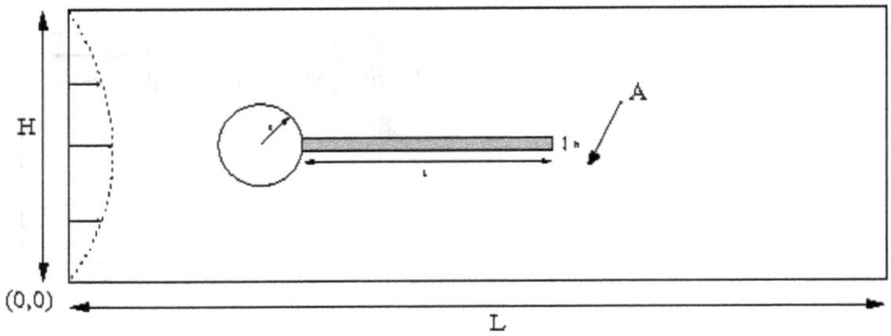

Fig. 16. Configuration of the FSI benchmark "FLUSTRUK-A".

on the material properties assumed. This benchmark has been defined to validate and compare the different computational approaches and software implementations for solving FSI problems. In order to have a fair comparison of our Eulerian-based method with the traditional Eulerian–Lagrangian approach, we have also implemented an ALE method for this benchmark problem. The configuration of this benchmark shown in Fig. 16 is based on the well-known CFD benchmark "Flow Around a Cylinder", see Turek and Schäfer.[55]

Configuration: The computational domain has length $L = 2.5$, height $H = 0.41$, and left bottom corner at $(0, 0)$. The center of the circle is positioned at $C = (0.2, 0.2)$ with radius $r = 0.05$. The elastic bar has length $l = 0.35$ and height $h = 0.02$. Its right lower end is positioned at $(0.6, 0.19)$ and its left end is clamped to the circle. Control points are $A(t)$ fixed at the trailing edge of the structure with $A(0) = (0.6, 0.20)$, and $B = (0.15, 0.2)$ fixed at the cylinder (stagnation point).

Boundary and initial conditions: The boundary conditions are as follows: Along the upper and lower boundary the usual "no-slip" condition is used for the velocity. At the (left) inlet a constant parabolic inflow profile,

$$v(0, y) = 1.5\,\bar{U}\,\frac{4y(H-y)}{H^2},$$

is prescribed which drives the flow, and at the (right) outlet zero-stress $\sigma \cdot n = 0$ is realized by using the "do-nothing" approach in the variational formulation.[31,46] This implicitly forces the pressure to have zero mean value at the outlet. The initial condition is zero flow velocity and structure displacement.

Material properties: The fluid is assumed as incompressible and Newtonian, the cylinder as fixed and rigid, and the structure as (compressible) St. Venant–Kirchhoff (STVK) type.

Discretization: The first set of computations is done on globally refined meshes for validating the proposed method and its software implementation. Then, for the same configuration adaptive meshes are used where the refinement criteria are either purely heuristic, i.e., based on the cell distance from the interface, or are based on a simplified stationary version of the DWR approach (at every tenth time step) as already used before for the cavity example. In all cases a uniform time-step size of $0.005s$ is used. The curved cylinder boundary is approximated to second-order by polygonal mesh boundaries as can be seen in Fig. 17.

The following four different test cases are considered:

- *Computational fluid dynamics test (CFD Test):* The structure is made very stiff, to the effect that we can compare the computed drag and lift coefficients with those obtained for a pure CFD test (with rigid structure).
- *Computational structure mechanics test (CSM Test):* The fluid is set to be initially in rest around the bar. The deformation of the bar under a vertical gravitational force is compared to the deformation of the same bar in a pure CSM test.
- *FSI tests:* Three configurations are treated corresponding to different inflow velocities and material stiffness parameters, and the Eulerian approach is compared to the standard ALE method.
- *FSI with large deflections:* The fluid is set to be initially in rest around the bar. The gravitational force on the bar is very large, causing a large

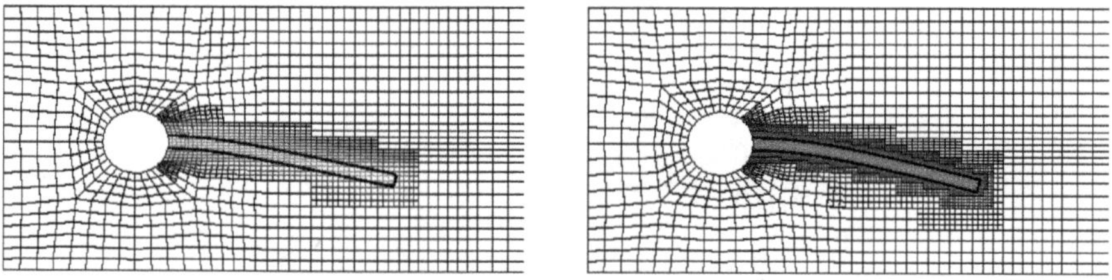

Fig. 17. CSM test: Stationary position of the control point A on heuristically refined meshes with $N = 1952$ and $N = 7604$ cells.

deformation of the bar and eventually it reaching and running up against the channel wall. This case is difficult for the ALE method but can easily be handled by the Eulerian approach.

9.1. *CFD test*

Here, the structure is set to be very stiff to the effect that we can compare derived drag and lift values with those obtained with a pure CFD approach. The forces are calculated based on the closed path S around the whole structure, cylinder and bar,

$$J(u, p) := \int_S \sigma_f \cdot n \, do. \tag{9.1}$$

The CFD test has been done with the parameters listed in Table 3.

For the chosen parameters there is a steady state solution. The reference values for the drag and lift forces are calculated using a pure CFD approach on globally refined meshes (see also Hron and Turek[34]). The results are shown in Table 4. Using the Eulerian FSI approach, we calculate the same forces again. As a method of mesh adaption, we use a heuristic approach as described above.

9.2. *CSM test*

Here, the inflow velocity is set to be zero and the fluid is initially at rest. A vertical gravitational force is applied, which causes the bar to slowly sink in the fluid-filled volume. Due to the viscous effect of the fluid the bar will eventually come to rest. The value of final displacement can be compared to the results calculated with a pure CSM approach in a Lagrangian

Table 3. Parameters for the CFD test.

Parameters	CFD test
$\rho_f[10^3 kg\, m^{-3}]$	1
$\nu_f[10^{-3} m^2 s^{-1}]$	1
ν_s	0.4
$\rho_s[10^6 kg\, m^{-3}]$	1
$\mu_s[10^{12} kg\, m^{-1} s^{-2}]$	1
$\bar{U}[m\, s^{-1}]$	1

Table 4. CFD test: Results of CFD computation on uniform meshes (left), and by the Eulerian FSI approach on heuristically adapted meshes (right).

N	dof	drag	lift	N	dof	drag	lift
1 278	3 834	145.75	10.042	1 300	9 100	122.66	12.68
4 892	14 676	133.91	10.239	2 334	16 338	126.13	11.71
19 128	57 384	136.00	10.373	8 828	61 796	132.17	11.93
75 632	226 896	136.54	10.366	9 204	64 428	131.77	10.53
300 768	902 304	136.67	10.369	36 680	256 760	134.47	10.45
∞	∞	136.70	10.530	∞	∞	136.70	10.530

Table 5. Parameters for the CSM test.

Parameters	CSM test
$\rho_f[10^3 kg\, m^{-3}]$	1
$\nu_f[10^{-3} m^2 s^{-1}]$	1
ν_s	0.4
$\rho_s[10^3 kg\, m^{-3}]$	1
$\mu_s[10^6 kg\, m^{-1} s^{-2}]$	0.5
$\bar{U}[m\, s^{-1}]$	0
$g[m\, s^{-2}]$	2

framework. The quantity of interest is the displacement of the point A at the middle of the trailing tip. The corresponding reference values are taken from Ref. 34. The CSM test has been done with the parameters listed in Table 5. Using the Eulerian FSI approach, we calculate the displacements with mesh adaption by the heuristic approach described above (see Table 6). The final stationary positions and the heuristically adapted meshes can be seen in Fig. 17.

Table 6. CSM test: Displacement of the control point *A* for three levels of heuristic mesh adaption.

N	dof	$u_x(A)$ $[10^{-3}m]$	$u_y(A)$ $[10^{-3}m]$
1952	13 664	−5.57	−59.3
3 672	25 704	−6.53	−63.4
7 604	53 228	−6.74	−64.6
∞	∞	−7.187	−66.10

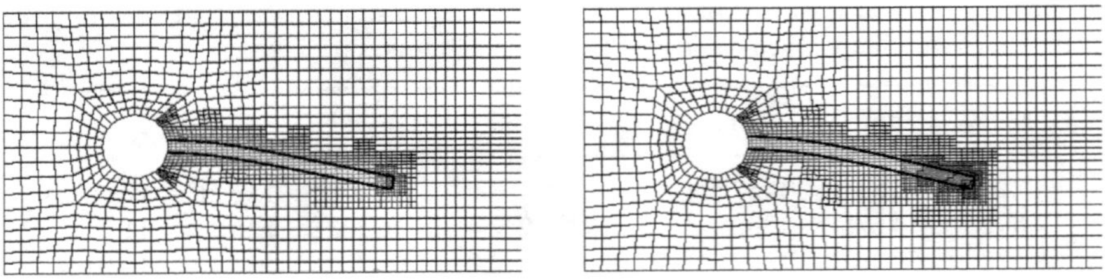

Fig. 18. CSM test: Stationary position of the bar computed on locally refined meshes (DWR method) with $N = 2{,}016$ and $N = 4{,}368$ cells.

Next, we apply the DWR method as described above to the CSM test case. In the dual problem, we use the Jacobi matrix of the model as presented in Sec. 6. In the first example the DWR method was always applied to the final stationary state. The results were used for mesh adaption. The generated mesh was then used with the initially unperturbed problem to determine a new final stationary state. In contrast to that approach, we now apply the DWR method at periodic intervals without restarting. To control the resulting mesh adaption at each interval, we try to keep the number of nodes N below a certain threshold N_d. This is achieved by reducing refinement and/or increasing coarsening at each interval. As an example we calculate the point-value of the component sum of $u(A)$ at the control point A. The position x_A is determined from $x_A - u(x_A) = A(0) = (0.6, 0.2)^T$. As a error control functional, we use a regularized delta function at x_A applied to $(e_1 + e_2)^T u$,

$$J(u) = |K_A|^{-1} \int_{K_A} (e_1 + e_2)^T u(x)\, dx,$$

where K_A is the cell in the Mesh $\mathbb{T}_h$ containing the point A. The results are shown in Table 7 and Fig. 18.

Table 7. CSM Test: Displacements of the control point A for three levels of locally refined meshes (DWR method).

N_d	N	dof	$u_x(A)$ $[10^{-3}m]$	$u_y(A)$ $[10^{-3}m]$
2 000	2 016	14 112	−5.73	−59.8
3 000	2 614	18 298	−6.54	−63.2
4 500	4 368	30 576	−6.88	−64.6
	∞	∞	−7.187	−66.10

Table 8. Parameter settings for the FSI test cases.

Parameter	FSI-2	FSI-2*	FSI-3	FSI-3*
Structure model	STVK	SZVK	STVK	INH
$\rho_f[10^3 kg\, m^{-3}]$	1	1	1	1
$\nu_f[10^{-3} m^2 s^{-1}]$	1	1	1	1
ν_s	0.4	0.4	0.4	0.5
$\rho_s[10^3 kg\, m^{-3}]$	10	20	1	1
$\mu_s[10^6 kg\, m^{-1} s^{-2}]$	0.5	0.5	2	2
$\bar{U}[m\, s^{-1}]$	1	0	2	2

9.3. *FSI tests*

Three test cases, FSI-2, FSI-3, and FSI-3*, are treated with different inflow velocities and material stiffness values as stated in Table 8. The parameters are chosen such that a visible transient behavior of the bar can be seen. The comparison values have been calculated using the ALE method on a very fine mesh. Using the Eulerian FSI approach, we calculate the displacements on three mesh levels, where the heuristic approach as described above is used for mesh refinement.

We begin with the FSI-2 and FSI-3 test cases. Some snapshots of the results of these simulations are shown in Figs. 19 and 21. The time-dependent behavior of the displacements for the tests are shown in Figs. 20 and 22. It can be seen that the time-periodic limit state is reached faster by the ALE method than by the Eulerian method. This phenomenon is observed also for the other test cases and therefore seems to be a characteristic feature of this particular version of the Eulerian method. In fact, we believe that it is the process of continuing the structure deformation into the fluid domain by harmonic extension (involving a sensible control parameter α_w), which causes the higher "stiffness" of the Eulerian model. However, this question needs further investigation.

The FSI-3* test case is used to illustrate some special features of the Eulerian solution approach. Figure 23 illustrates the treatment of corners in the structure by the IP set approach compared to the LS approach. In the LS method the interface is identified by all points for which $\varphi = 0$, while in the IP set method the interface is identified by all points which are on

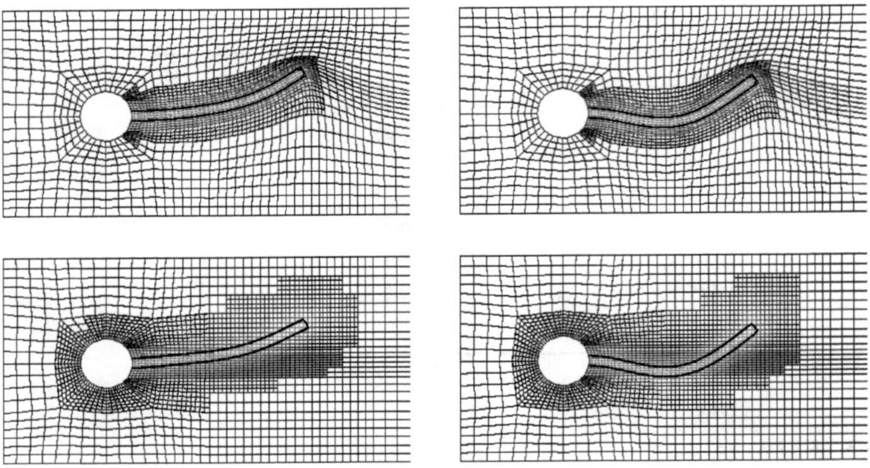

Fig. 19. FSI-2: Snapshots of results obtained by the ALE (top two) and by the Eulerian (bottom two) approaches.

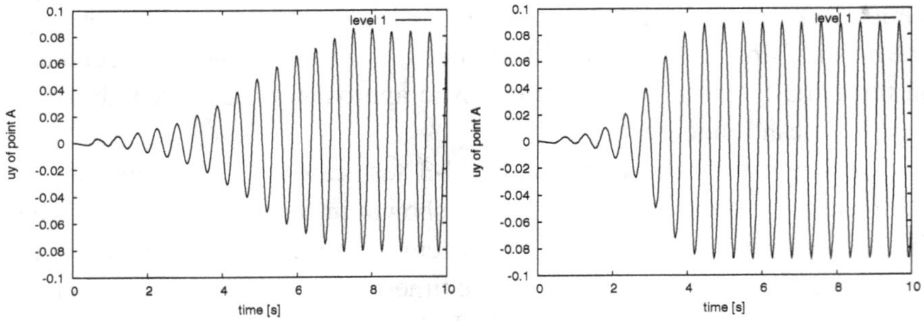

Fig. 20. FSI-2: Vertical displacement of the control point A, obtained by the Eulerian approach (left, $N = 2{,}082$ cells) with max. amplitude $2.226 \cdot 10^{-2}$ and frequency $1.92s^{-1}$, and by the ALE approach (right, $N = 2{,}784$ cells) with max. amplitude $2.68 \cdot 10^{-2}$ and frequency $1.953s^{-1}$.

one of the respective isoline segments belonging to the edges of the bar. The differences are visible in the cells that contain the corners.

Since in the Eulerian approach the structure deformations are not in a Lagrangian framework, it is not immediately clear, due to the coupling with the fluid, how well the mass of the structure is conserved in an Eulerian approach, especially in the course of an instationary simulation comprising hundreds of time steps. In Fig. 24, we display the bar's relative mass error

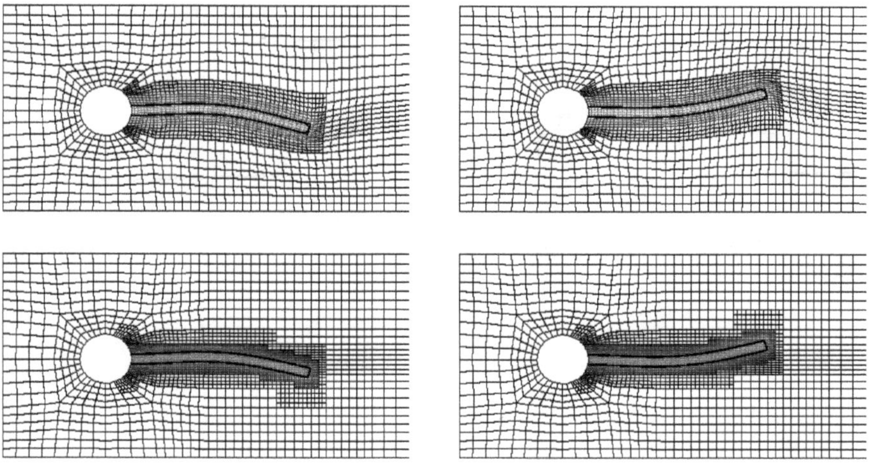

Fig. 21. FSI-3 test: Some snapshots of results obtained by the ALE (top two) and the Eulerian (bottom two) approaches.

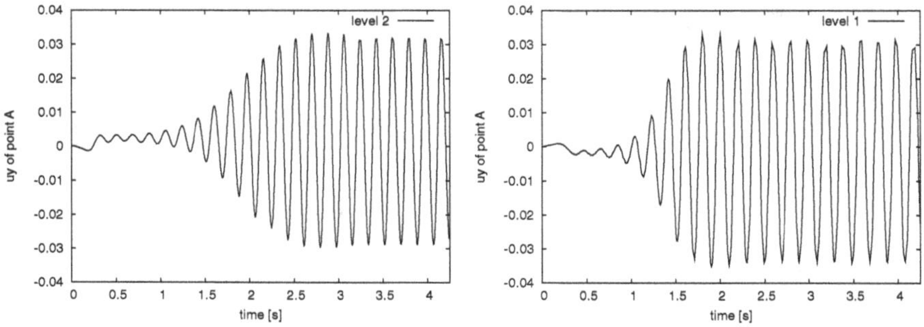

Fig. 22. FSI-3 test: Vertical displacement of the control point A, obtained by the Eulerian approach (left, $N = 3{,}876$ cells) with max. amplitude $6.01 \cdot 10^{-2}$ and frequency $5.48s^{-1}$, and by the ALE approach (right, $N = 2{,}082$ cells) with max. amplitude $6.37 \cdot 10^{-2}$ and frequency $5.04s^{-1}$.

as a function of time. Except for certain initial jitters, the relative error is less than 1%.

Finally, Fig. 25 illustrates the time dynamics of the structure and the adapted meshes over the time interval $[0, T]$. For both approaches, we obtain a periodic oscillation with maximum amplitudes and frequency, which are quite close to each other: 1.6e-2 versus 1.51e-2 and $6.86s^{-1}$ versus $6.70s^{-1}$.

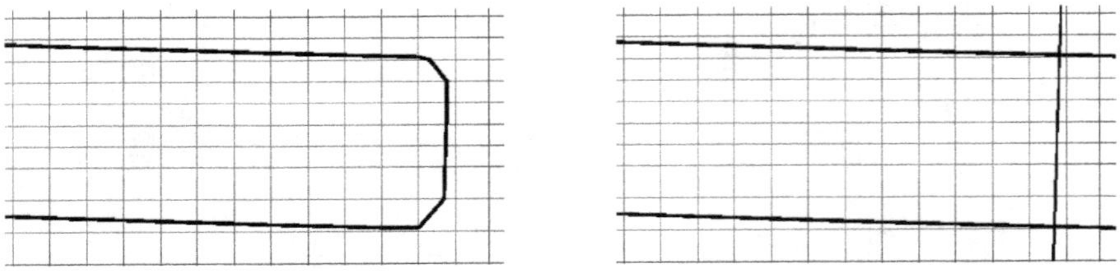

Fig. 23. FSI-3*: Treatment of corners by the LS method (left) and by the IP set method (right).

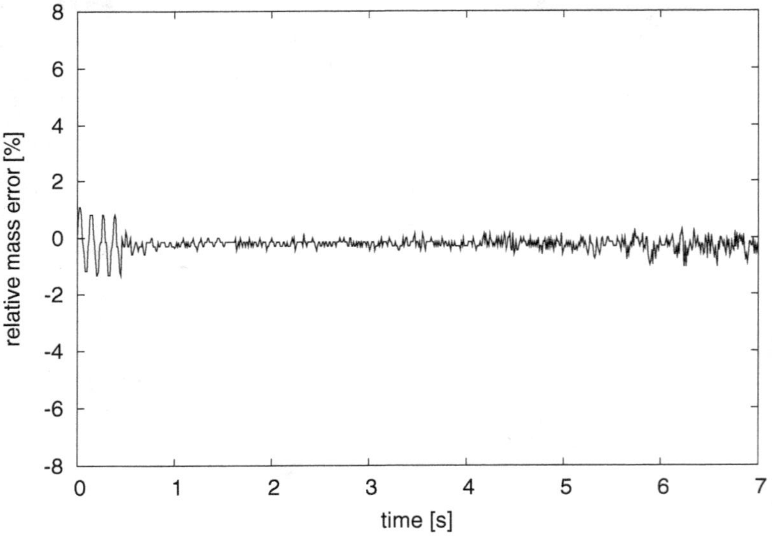

Fig. 24. FSI-3*: Relative mass error of the bar.

9.4. *FSI test with large deformations*

In the test case FSI-2* (see Table 8) the fluid is initially in rest and the bar is subjected to a vertical (gravitational) force. This causes the bar to bend downward until it touches the bottom wall. A sequence of snapshots of the transition to steady state obtained by the Eulerian approach for this problem is shown in Fig. 26 for zonal mesh refinement. The position of the trailing tip A is show in Fig. 27.

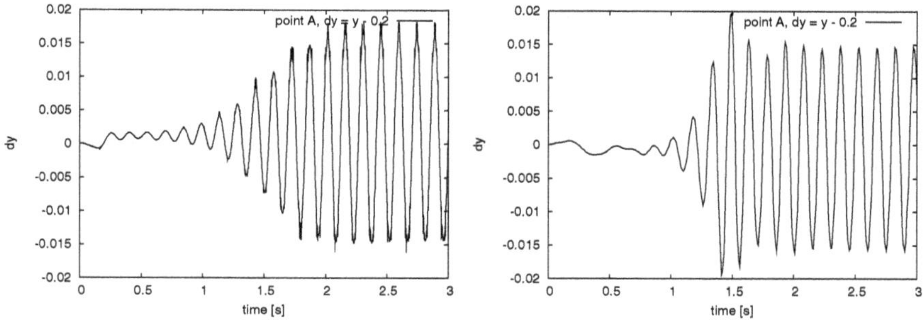

Fig. 25. FSI-3*: Vertical displacement of the control point A, obtained by the Eulerian approach (left) with max. amplitude $1.6 \cdot 10^{-2}$ and frequency $6.86s^{-1}$, and by the ALE approach (right) with max. amplitude $1.51 \cdot 10^{-2}$ and frequency $6.70s^{-1}$.

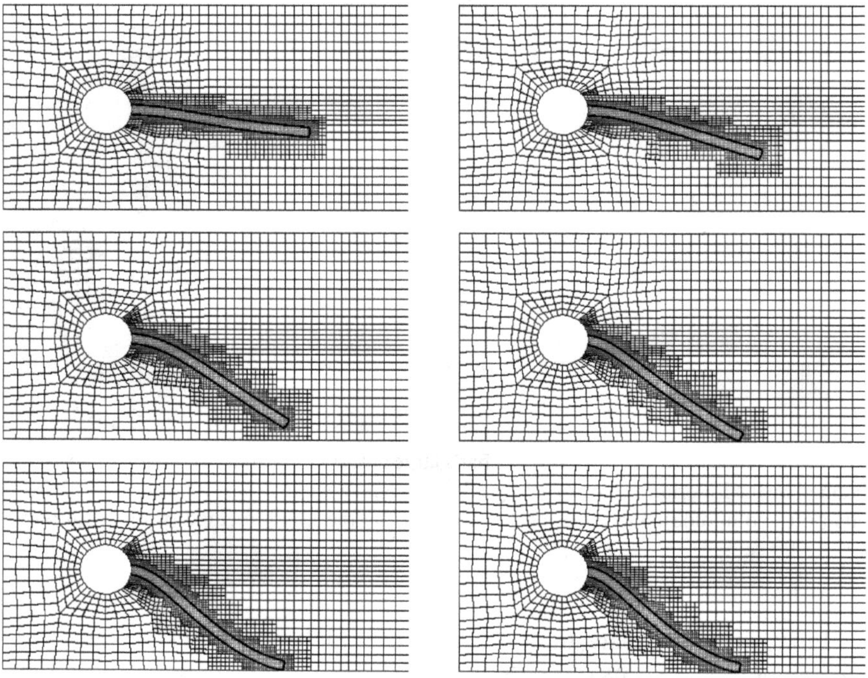

Fig. 26. FSI-2*: A sequence of snap-shots of the bar's large deformation under gravitational loading obtained by the Eulerian approach.

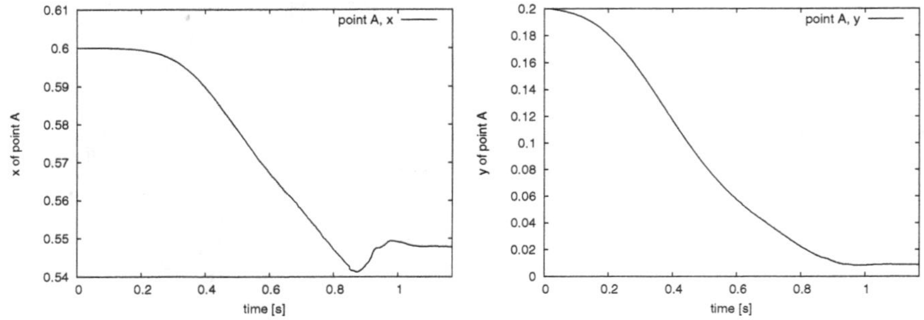

Fig. 27. FSI-2*: Position of the control point A during the deformation of the bar: x-coordinate (left) and y-coordinate (right).

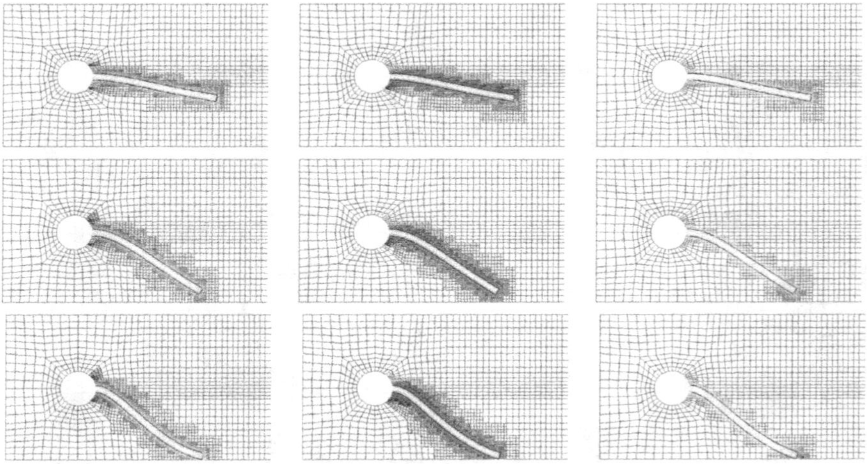

Fig. 28. A sequence of snap-shots of the bar's large deformation under gravitational loading obtained by the Eulerian approach using zonal refinement with $N \sim 3{,}000$ and $N \sim 12{,}000$ cells (left and midlde), and local mesh refinement by the DWR method (right) with only $N \sim 1{,}900$ cells.

Finally, we compare the efficiency of zonal refinement versus local refinement by the DWR method for this test case. Figure 28 shows corresponding sequences of snapshots, while the position of the trailing tip A is displayed in Fig. 29. We see that by sensitivity-driven local refinement within the DWR method on only $1{,}900$ cells, almost the same accuracy can be achieved as by zonal refinement on $12{,}300$ cells. The gain in CPU time

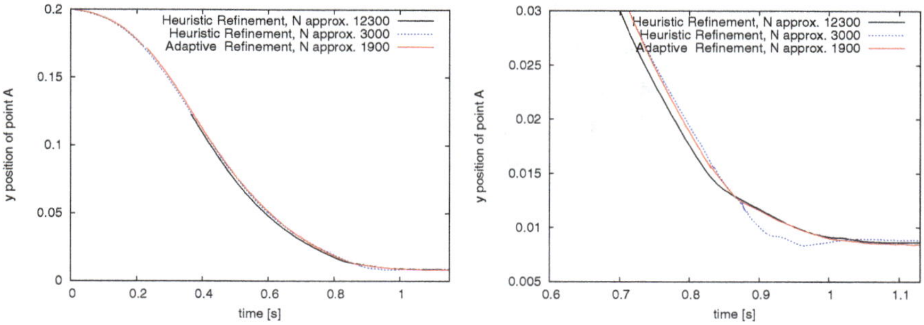

Fig. 29. Time varying position of the point $x_A(t)$ over the full time interval [0, 1.1] (left) and over a zoomed interval [0.6, 1.1] (right).

needed is almost 85% (about 30 h for the zonal versus about 4 h for the local refinement).

10. Summary and Future Development

In this paper we presented a fully Eulerian (Eulerian) variational formulation for "fluid-structure interaction" (FSI) problems. This approach uses the "initial position" set (IP set) method for interface capturing, which is similar to the "level set" (LS) method, but preserves sharp corners of the structure. The harmonic continuation of the structure velocity avoids the need of reinitialization of the IP set. This approach allows us to treat FSI problems with free bodies and large deformations. This is the main advantage of this method compared to interface tracking methods such as the arbitrary Lagrangian–Eulerian (ALE) method. At several examples the Eulerian approach turns out to yield results which are in good agreement with those obtained by the ALE approach. In order to have a "fair" comparison both methods have been implemented using the same numerical components and software library GASCOIGNE.[26] The method based on the Eulerian approach is inherently more expensive than the ALE method, by about a factor of two, but it allows to treat also large deformations and some kinds of topology changes.

As already mentioned above, theoretical results on existence of solution for fluid-structure interaction problems are rather sparse in the literature and can be found only for certain reduced systems and under very restrictive assumptions on the smallness of the data. The following list though is by far not comprehensive. Many results are available based on interaction of fluid with fixed rigid structures. In Desjardins and Esteban[18,19] the authors show that solutions exist for a finite number of rigid non-colliding structures embedded in the fluid. The considered fluids are incompressible as well as compressible isentropic fluids modeled by the Navier–Stokes equations. Previous work in this direction can be found in Desjardins.[17] Using an approach similar to that in Desjardins and Esteban[18,19] the authors of Desjardins, Esteban, Grandmont and LeTallec[20] prove the existence of weak solutions for an instationary fluid-elastic structure interaction model. This is achieved by "Leray's method", i.e., by finding weak solutions that satisfy bounds for the energy of the complete system. The authors model the elastic structure as a compressible linearized neo-Hookean material with a finite number of elastic modes. In LeTallec and Mani[42] the authors investigate an instationary linearized fluid-structure interaction problem for a viscous fluid and a thin elastic shell with small displacements. The authors simplify the problem by neglecting changes to the geometry. Based on these premises by using energy estimates they show that the problem is well posed, that a weak solution exists and that the discrete approximation, based on their discretization, converges to the continuous solution.

The monolithic variational formulation of the FSI problem provides the basis for the application of the "dual weighted residual" (DWR) method for "goal-oriented" *a posteriori* error estimation and mesh adaptation. In this method inherent sensitivities of the FSI problem are utilized by solving linear "dual" problems, similar as in the Euler–Lagrange approach to solving optimal control problems. The feasibility of the DWR method for FSI problems has, in a first step, been demonstrated for the computation of steady state solutions.

One conceptional disadvantage of the Eulerian approach to treating FSI problems is the need for a time-independent outer domain Ω, in which the FSI process takes place. This seems to prevent the use of this approach for simulating flow through blood vessels since here the time-varying vessel wall forms the "outer domain". This difficulty can be cured by embedding the whole vessel into an outer softer medium, as seems realistic from the biological point of view. The physical properties of this outer medium have to be provided by biomedical experience. The realization of this concept is the subject of ongoing research. The following topics have to be worked on in the future:

- The DWR method has to be applied also for nonstationary FSI problems, particularly for the simultaneous adaptation of spatial mesh and time step size.
- Another goal is the development of the Eulerian approach for three-dimensional FSI problems and to explore its potential for FSI problems with large deformations and topology changes, such as occurring in heart-valve simulations
- Application to optimal control problems with FSI

$$\min_{q \in Q} J(u, q)! \quad a(u, q)(\psi) = f(\psi) \quad \forall \psi \in \mathcal{V}^0.$$

In the context of the "all-at-once" approach (KKT System) goal-oriented error estimation is "relatively cheap".

Acknowledgment

This work has been supported by the Deutsche Forschungsgemeinschaft (DFG), Research Unit 493 "Fluid-Struktur Interaction: Modeling, Simulation, Optimization". This support is gratefully acknowledged.

References

1. G. Allaire, F. de Gournay, F. Jouve and A. Tonder, Structural optimization using topological and shape sensitivities via a level set method, *Cont. and Cybern.* **34** (2004): 59–80.
2. W. Bangerth and R. Rannacher, *Adaptive Finite Element Methods for Differential Equations* (Birkhäuser, 2003).
3. R. Becker and M. Braack, Multigrid techniques for finite elements on locally refined meshes, *Numer. Lin. Algebra Appl.* **7** (2000):363–379.
4. R. Becker and M. Braack, A finite element pressure gradient stabilization for the Stokes equations based on local projections, *Calcolo* **38** (2001):173–199.
5. R. Becker and M. Braack, A two-level stabilization scheme for the Navier–Stokes equations, *Proc. ENUMATH-03* (2003), pp. 123–130.
6. R. Becker, V. Heuveline and R. Rannacher, An optimal control approach to adaptivity in computational fluid mechanics, *Int. J. Numer. Meth. Fluids.* **40** (2002): 105–120.

7. R. Becker and R. Rannacher, Weighted *a-posteriori* error estimates in FE methods, Lecture ENUMATH-95, Paris, 18–22 September, 1995, in *Proc. ENUMATH-97* (eds), H.G. Bock, *et al.* (World Scientific Publ., Singapore, 1998), pp. 621–637.
8. R. Becker and R. Rannacher, An optimal control approach to error estimation and mesh adaption in finite element methods, in *Acta Numerica 2000*, ed. A. Iserles (Cambridge University Press, 2001), pp. 1–101.
9. R. Becker and R. Rannacher, A feed-back approach to error control in finite element methods: Basic analysis and examples, *East-West J. Numer. Math.* **4** (1996):237–264.
10. M. Braack and Th. Richter, Solutions of 3D Navier–Stokes benchmark problems with adaptive finite elements, *Comput. and Fluids* **35** (2006):372–392.
11. D. Braess, *Finite Elements* (Cambridge University Press, Cambridge, United Kingdom, 1997).
12. S. Brenner and R. L. Scott, *The Mathematical Theory of Finite Element Methods* (Springer, Berlin Heidelberg New York, 1994).
13. M. O. Bristeau, R. Glowinski and J. Periaux, Numerical methods for the Navier–Stokes equations, *Comput. Phys. Rep.* **6** (1987):73–187.
14. H.-J. Bungartz and M. Schäfer (eds), *Fluid-Structure Interaction: Modeling, Simulation, Optimization* (Springer, Berlin-Heidelberg, 2006).
15. G. Carey and J. Oden, *Finite Elements, Computational Aspects*, vol. 3. (Prentice-Hall, 1984).
16. Y. C. Chang, T. Y. Hou, B. Merriman and S. Osher, A level set formulation of Eulerian interface capturing methods for incompressible fluid flows, *J. Comput. Phys.* **123** (1996):449–464.
17. B. Desjardins, On weak solutions of the compressible isentropic Navier–Stokes equations, *Appl. Math. Lett.* **12** (1999):107–111.
18. B. Desjardins and M. J. Esteban, Existence of weak solutions for the motion of rigid bodies in a viscous fluid, *Arch. Ration. Mech. Anal.* **146** (1999):59–71.
19. B. Desjardins and M. J. Esteban, On weak solutions for fluid-rigid structure interaction: Compressible and incompressible models, *Commun. P.D.E.* **25** (2000):1399–1413.
20. B. Desjardins, M. J. Esteban, C. Grandmont and P. Le Tallec, Weak solutions for a fluid-elastic structure interaction model, *Rev. Math. Comput.* **14** (2001):523–538.
21. Th. Dunne, An Eulerian approach to fluid-structure interaction and goal-oriented mesh refinement, *Int. J. Numer. Meth. Fluids* **51** (2006):1017–1039.
22. Th. Dunne, Adaptive Finite Element Simulation of Fluid Structure Interaction Based on an Eulerian Formulation, Doctoral dissertation, Institute of Applied Mathematics, University of Heidelberg (2007).
23. Th. Dunne, *Adaptive Finite Element Simulation of Fluid Structure Interaction Based on an Eulerian Formulation*, Vieweg+Teubner Series Advances in Numerical Mathematics, to appear 2009.
24. Th. Dunne and R. Rannacher, Adaptive finite element simulation of fluid-structure interaction based on an Eulerian variational formulation, in *Fluid-Structure Interaction: Modeling, Simulation, Optimization* eds. H.-J. Bungartz and M. Schäfer (Springer, Berlin-Heidelberg, 2006) pp. 371–386.
25. K. Eriksson, D. Estep, P. Hansbo and C. Johnson, Introduction to adaptive methods for differential equations, in *Acta Numerica 1995*, ed. A. Iserles (Cambridge University Press, 1995), pp. 105–158.

26. Gascoigne, A C++ numerics library for scientific computing, Institute of Applied Mathematics, University of Heidelberg, url http://www.gascoigne.uni-hd.de/ .
27. V. Girault and P.-A. Raviart, *Finite Element Methods for the Navier–Stokes Equations* (Springer, Berlin-Heidelberg-New York, 1986).
28. R. Glowinski, Finite element methods for incompressible viscous flow, in *Handbook of Numerical Analysis, Volume IX: Numerical Methods for Fluids (Part 3)*, eds. P. G. Ciarlet and J. L. Lions (North-Holland, Amsterdam, 2003).
29. A. Griewank, *On Automatic Differentiation in Mathematical Programming: Recent Developments and Applications* (Kluwer Academic Publishers, 1989), pp. 83–108.
30. M. Heil, A. L. Hazel and J. Boyle, Solvers for large-displacement fluid-structure interaction problems: Segregated versus monolithic approaches, *Comput. Mech.* **43** (2008):91–101.
31. J. Heywood, R. Rannacher and S. Turek, Artificial boundaries and flux and pressure conditions for the incompressible Navier–Stokes equations, *Int. J. Numer. Math. Fluids* **22** (1992):325–352.
32. C. W. Hirt and B. D. Nichols, Volume of fluid (VOF) method for the dynamics of free boundaries, *J. Comput. Phys.* **39** (1981):201–225.
33. J. Hron and S. Turek, Proposal for numerical benchmarking of fluid-structure interaction between an elastic object and laminar incompressible flow, in *Fluid-Structure Interaction: Modelling, Simulation, Optimization*, eds. H.-J. Bungartz and M. Schäfer (Springer, Berlin-Heidelberg, 2006).
34. J. Hron and S. Turek, A monolithic FEM/multigrid solver for an ALE formulation of fluid-structure interaction with applications in biomechanics, in *Fluid-Structure Interaction: Modelling, Simulation, Optimization*, eds. H.-J. Bungartz and M. Schäfer, (Springer, Berlin-Heidelberg, 2006), pp. 146–170.
35. A. Huerta and W. K. Liu, Viscous flow with large free-surface motion, *Comput. Meth. Appl. Mech. Eng.* (1988).
36. T. J. R. Hughes and A. N. Brooks, Streamline upwind/Petrov–Galerkin formulations for convection dominated flows with particular emphasis on the incompressible Navier–Stokes equation. *Comput. Meth. Appl. Mech. Eng.* **32** (1982):199–259.
37. T. J. R. Hughes, L. P. Franc and M. Balestra, A new finite element formulation for computational fluid mechanics: V. Circumventing the Babuska–Brezzi condition: A stable Petrov–Galerkin formulation of the Stokes problem accommodating equal order interpolation, *Comput. Meth. Appl. Mech. Eng.* **59** (1986):85–99.
38. T. J. R. Hughes, W. K. Liu and T. K. Zimmermann, Lagrangian–Eulerian finite element formulations for incompressible viscous flows, *Comput. Meth. Appl. Mech. Eng.* **29** (1981):329–349.
39. S. Geller, J. Toelke and M. Krafczyk, Lattice–Boltzmann methods on quadtree-type grids for fluid-structure interaction, in *Fluid-Structure Interaction: Modelling, Simulation, Optimization*, eds. H.-J. Bungartz and M. Schäfer (Springer, Berlin-Heidelberg, 2006), pp. 270–293.
40. D. D. Joseph and Y. Y. Renardy, *Fundamentals of Two-Fluid Dynamics. Part I and II* (Springer, New York, 1993).
41. A. Legay, J. Chessa and T. Belytschko, An Eulerian–Lagrangian method for fluid-structure interaction based on level sets, *Computational Methods in Applied Mathematics* (2004).

42. P. Le Tallec and S. Mani, Numerical analysis of a linarized fluid-structure interaction problem, *Numer. Math.* **87** (2000):317–354.
43. C. Liu and N. J. Walkington, An Eulerian description of fluids containing visco-elastic particles, *Arch. Ration. Mech. Anal.* **159** (2001):229–252.
44. S. Osher and J. A. Sethian, Propagation of fronts with curvature based speed: Algorithms based on Hamilton-Jacobi formulations, *J. Comput. Phys.* **79** (1988):12.
45. L. B. Rall, *Automatic Differentiation — Techniques and Application* (Springer, New York, 1981).
46. R. Rannacher, Finite element methods for the incompressible Navier–Stokes equations, in *Fundamental Directions in Mathematical Fluid Mechanics*, eds. G. P. Galdi, J. Heywood, R. Rannacher (Birkhäuser, Basel-Boston-Berlin, 2000), pp. 191–293.
47. R. Rannacher, Incompressible Viscous Flow, in *Encyclopedia of Computational Mechanics*, eds. E. Stein, *et al.* (John Wiley, Chichester, 2004).
48. R. Rannacher and F.-T. Suttmeier, Error estimation and adaptive mesh design for FE models in elasto-plasticity, in *Error-Controlled Adaptive FEMs in Solid Mechanics*, ed. E. Stein (John Wiley, Chichster, 2002), pp. 5–52.
49. J. A. Sethian, *Level Set Methods and Fast Marching Methods* (Cambridge University Press, 1999).
50. J. Sokolowski and J. P. Zolesio, *Introduction to Shape Optimization: Shape Sensitivity Analysis*, (Springer, Berlin, 1992).
51. T. E. Tezduyar, M. Behr and J. Liou, A new strategy for finite element flow computations involving moving boundaries and interfaces — the deforming-spatial-domain/space-time procedures: I. The concept and preliminary tests, *Computer Methods in Applied Mechanics and Engineering* (1992).
52. T. E. Tezduyar, M. Behr and J. Liou, A new strategy for finite element flow computations involving moving boundaries and interfaces-the deforming-spatial-domain/space-time procedures: II. Computation of free-surface flows, two-liquid ows and ows with drifting cylinders, *Computer Methods in Applied Mechanics and Engineering* (1992).
53. T. E. Tezduyar, S. Sathe, K. Stein and L. Aureli, Modeling of fluid-structure interaction with the space-time technique, in *Fluid-Structure Interaction: Modelling, Simulation, Optimization*, eds. H.-J. Bungartz and M. Schäfer (Springer, Berlin-Heidelberg, 2006), pp. 50–81,
54. S. Turek, *Efficient Solvers for Incompressible Flow Problems: An Algorithmic and Computational Approach* (Springer, Heidelberg-Berlin-New York, 1999).
55. S. Turek and M. Schäfer, Benchmark computations of laminar flow around a cylinder, in Flow Simulation with High-Performance Computers II', Vol. 52 of *Notes on Numerical Fluid Mechanics*, ed. E. H. Hirschel (Vieweg, 1996).
56. J. Vierendeels, Implicit coupling of partial fluid-structure interaction solvers using reduced-order models, in *Fluid-Structure Interaction: Modelling, Simulation, Optimization*, eds. H.-J. Bungartz and M. Schäfer (Springer, Berlin-Heidelberg, 2006), pp. 1–18.
57. VISUSIMPLE, An open source interactive visualization utility for scientific computing, Institute of Applied Mathematics, University of Heidelberg, http://visusimple.uni-hd.de/ .

58. W. A. Wall, Fluid-structure interaction with stabilized finite elements, doctoral dissertation, Report No. 31 (1999), Institute of Structural Mechanics, University of Stuttgart.
59. W. A. Wall, A. Gerstenberger, P. Gamnitzer, C. Forster and E. Ramm, Large deformation fluid-structure interaction: Advances in ALE methods and new fixed grid approaches, in *Fluid-Structure Interaction: Modelling, Simulation, Optimization*, eds. H.-J. Bungartz and M. Schäfer (Springer, Berlin-Heidelberg, 2006), pp. 195–232.

CHAPTER 2

EXTERIOR FLOWS AT LOW REYNOLDS NUMBERS: CONCEPTS, SOLUTIONS, AND APPLICATIONS

Vincent Heuveline

Karlsruher Institut für Technologie
Institut für Angewandte und Numerische Mathematik IV
AG Numerische Simulation, Optimierung und Hochleistungsrechnen
Vincent.Heuveline@kit.edu

Peter Wittwer

Département de Physique Théorique Université de Genève, Switzerland
peter.wittwer@unige.ch

We discuss fluid structure interaction for exterior flows, for Reynolds numbers ranging from about one to several thousand. New applications demand a better quantitative understanding of the details of such flows, and this has stimulated a revival of interest in this topic. Astonishingly, in spite of the apparent simplicity of low Reynolds number flows, their precise prediction turns out to be numerically demanding even with today's computers. On the basis of simple examples we review the progress that has been made over the recent years in the analysis of such problems through a combined use of techniques from asymptotic analysis, symbolic computation, and computational fluid dynamics, and discuss open problems which one should be able to solve with the techniques presented here.

1. Introduction

We discuss fluid structure interaction at low Reynolds numbers. Fluids filling up all of space and flowing past spheres, plates, and cylinders of various cross sections are much studied examples of such situations. Examples of more complicated arrangements are the motion of bubbles rising in a liquid close to a wall and the sedimentation of particles that undergo collisions. In all these cases it turns out to be of great practical importance to be able to determine the forces that the fluid exerts on the structure with good precision. The vertical speed of bubbles rising near a wall depends for example on the drag, and the exact distance from the wall at which the bubbles rise requires one to find the position relative to the wall where the transverse force is zero. This is an example of a computation that turns out to be very delicate, since at low Reynolds numbers the transverse forces typically are orders of magnitude smaller than the forces along the flow. Such questions can therefore only be answered with the help of high precision computations, which, if done by brute force, are excessively costly even with today's computers. Luckily there are better ways: in what follows we review the progress that has been made over recent years in the analysis of such problems through a combination of techniques from asymptotic analysis, symbolic computation, and computational fluid dynamics. We focus this review on the theoretical framework of the analysis. The goal is to present the general ideas of the theory in a self-contained way since it is the basis also for ongoing research. The different sections are written at various levels of mathematical rigor. Section 2 contains a pedagogical introduction to the analysis of the large time asymptotics of solutions of parabolic and elliptic partial differential equations as it has been developed over the last ten years. The analysis is based on simple examples but contains detailed proofs and serves as a basis for the other sections. In particular it provides easy access to the results in Sec. 3. The section also contains an introduction to the function spaces that have been used successfully for the study of many more difficult problems, including the stationary Navier–Stokes equations. In Sec. 3 we explain how to apply the techniques of Sec. 2 in order to obtain precise analytic results for the downstream asymptotics of stationary and time periodic solutions of the Navier–Stokes equations. We give a detailed review of existing results and state conjectures which we expect to be proved using the same techniques. Section 4 describes on a much more heuristic level the connection between the results of Sec. 3, i.e., the downstream asymptotic behavior of a solution for a given problem, and the asymptotic behavior in other directions. The section also contains a proof of the connection between certain invariant quantities at large distance and the forces that act on the structures. In

Sec. 5 we formulate artificial boundary conditions for flows in two and three dimensions. These artificial boundary conditions are based on the results in Secs. 3 and 4. This section is essentially self-contained and a reader who is mainly interested in implementing the boundary conditions can directly start reading there. In Sec. 6 we discuss the numerical scheme that has been used for solving the Navier–Stokes equations with the artificial boundary conditions in some explicit cases. Section 7 contains a review of some numerical results. Section 8, finally, contains an extensive bibliography of related work on low Reynolds number flows. Throughout the article we also discuss open problems. Whenever possible we have formulated these problems in terms of concrete conjectures which one should be able to prove with the techniques of Secs. 2 and 3. We hope that these conjectures will stimulate further research in this direction and will serve as a starting point for important further development.

1.1. *Applications*

In many practical applications Reynolds numbers are extremely large and the corresponding flows turbulent, and this is the reason why such flows are most intensively studied. In contrast to these cases we will concentrate here on the regime of flows that are either stationary or time periodic. Surprisingly, in spite of this apparent simplicity when compared with turbulent flows, there is little reliable quantitative information available for such cases. This is not without reason since, as indicated above, the precise prediction of the forces turns out to be computationally demanding. Indeed, linearized theories (Stokes, Oseen) provide a good quantitative description (forces determined within an error of one percent, say) only for Reynolds numbers less than one (see Batchelor (1967)), whereas approximation schemes based on some version of boundary layer theory work well only for Reynolds numbers in excess of some fifty thousand (see Carmichael (1981)). For the intermediate regime where neither the viscous nor the inertial forces dominate, the full Navier–Stokes equations need to be solved. However, when truncating an infinite exterior domain for numerical purposes to a finite sub-domain one is confronted with the problem of finding appropriate boundary conditions on the outer boundary of this sub-domain in order to mimic the boundary conditions at infinity. See for example Heywood *et al.* (1992), Sergej *et al.* (2001) and Nazarov and

Specovius-Neugebauer (2003). It turns out that for the Reynolds numbers under consideration any such naïve choice modifies the hydrodynamic forces significantly unless excessively large computational domains are used (several hundred times the size of the structures). Based on the asymptotic work reviewed in subsequent chapters we present in Sec. 5 boundary conditions that are simple to implement and allow a significant reduction of the size of the computational domain. Namely, we provide explicit expressions of vector fields that describe the solutions at large distances. These expressions depend on the forces of interest, and these forces can therefore be determined in a self-consistent way as part of the solution process. When compared with other schemes the size of the computational domains that is needed to determine the forces with a given precision are drastically reduced. This leads in turn to an overall gain in computational efficiency of typically several orders of magnitude. We finally note that the topic which is of particular interest here is the study of wakes since, as we will see, the asymptotic behavior of low Reynolds number flows is entirely encoded in the wakes. See also Jaxquim *et al.* (2003) and Afanasyev and Yakov (2005). In the remainder of this section we provide a bibliographic survey for two main areas of applications: the problem of "flight at low Reynolds numbers" and various problems related to the micro-physics relevant for "climate modeling". The first case includes besides the study of flight as such, the problem of swimming at low Reynolds numbers and questions related to flow control. The second case regroups the bibliography for questions related to the free fall of small particles like ice crystals in clouds and the sedimentation of particles in the oceans that may in addition undergo collisions.

1.2. *Flight at low Reynolds numbers*

The wish to construct aircraft that can fly at low Reynolds numbers is not new. Important publications which are concerned with the design of low speed airfoils and which are still of relevance today are due to Werle (1974), Eppler and Somers (1979), Eppler and Somers (1981), Lissaman (1983), Eppler and Somers (1985) and Ladson *et al.* (1996). When these papers were written, Reynolds numbers of the order of fifty to hundred thousand were considered small, but nowadays the interest focuses on flows with Reynolds number in the range from as low as some hundred to several thousand. Quite recently a new experimental facility has been built with the specific goal of measuring flows for this range of Reynolds numbers. See Hanf (2004). Earlier studies are due to Mueller (1999). See also Suhariyono *et al.* (2006). The goal of these experimental works and of the corresponding theoretical studies is the engineering of so-called micro-air vehicles (MAV). There is an extensive recent literature on the subject. See in particular Pelletier and Mueller (2000), Abdulrahim and Cocquyt (2000), Mueller (2001), and Mueller and DeLaurier (2003). The size of such micro-air vehicles is close to the size of large insects and, even though the discussion of subsequent sections mainly concerns the case of so-called fixed wing vehicles, it is a natural question to study the so-called flapping wing technology which leads in the regime of Reynolds numbers under consideration to time periodic flows. Such questions are extensively studied in Mueller (2001). Interesting information concerning these questions can also be obtained by studying directly insect flight. See for example Dickinson *et al.* (1999), Wang (2000) and Michelson and Naqvi (2003a).

A typical application of micro-air vehicles are reconnaissance missions, but they are also a much awaited tool for field studies in climate research, where they will allow cheap in situ measurement of various parameters of the atmosphere at variable height. Such measurements are considered an important input for future improvements of climate models. See the next section for more details. Other applications of low Reynolds number flight include the possibility to fly at very low speed at high altitudes, again either for reconnaissance missions or for the study of the physics of high lying clouds for climate research purposes. See for example the work by Greer *et al.* (1999). At high altitudes Reynolds numbers are small because of the low density of the atmosphere, but at the same time the Mach number is small for the same reason. Subsonic flight at such altitudes is therefore only possible at Reynolds numbers which are again in the range of interest of this review. There is also an increasing interest in flight in the atmosphere on Mars, which leads again to low Reynolds numbers, see Michelson and Naqvi (2003b). Finally there are many questions which are related to, but go beyond what we will be able to discuss here. To mention just a few further examples there is a whole body of research concerning the design of airfoils with high critical Mach number, see Kropinski (1997), questions related to flow control, see Bewley (2001), and questions related to the swimming of certain marine animals like small molluscs, see for example Childress and Dudley (2004) and Avron *et al.* (2004).

1.3. *Climate modeling*

There are several examples of low Reynolds number flows that play a role in climate modeling and weather prediction. Of particular importance is the need to predict the terminal velocity of ice crystals and rain drops falling within clouds and in the high atmosphere, as well as the speed of sedimentation of particles and the motion of small bubbles in the oceans. For the case of ice crystals undergoing free fall in the atmosphere the knowledge of their terminal velocity together with a knowledge of the speed of the upwind allows to predict the size of the ice crystals that populate the clouds. This in turn influences the albedo value of these clouds which is an important parameter in climate modeling. See for example Weinstein (1969), Heymsfield (1972), Baker (1997), and Heymsfield and Iaquinta (1999). A related question is the determination of the terminal speed of raindrops undergoing free fall, see Gunn and Kinzer (1949), Beard and Pruppacher (1969), Foote and du Toit (1969), Sostarecz and Belmonte (2003) and Necasova (2004). Such questions are the reason why even the flow around a sphere as the simplest example of a falling particle is still a subject of interest today in spite of its long history, see Le Clair *et al.* (1970) and Brown and Lawler (2003). Other cases of low Reynolds number flows that are relevant for climate modeling are the description of bubbly flows, see Esmaeeli and Tryggvason (1999) and Esmaeeli and Tryggvason (2000) as well as the prediction of the speed of sedimentation of particles in the ocean, see Yuan and Li (2006). For dense populations of ice crystals the collision between the crystals also needs to be described, see for example Wang and Ji (2000), Lamura *et al.* (2001) or Ripoll *et al.* (2004). Further related questions are discussed in Johnson and Wu (1979) and Vaidya (2006).

2. Large Time Asymptotics

The goal of this section is to provide a simple and self contained introduction to the so-called renormalization group technique which has been developed over the past ten years as a tool to prove existence and to analyze the long time behavior of solutions of nonlinear parabolic and elliptic partial differential equations. An extensive bibliography of the corresponding body of work can be found in Sec. 2.10. The same ideas can be applied to some extent to hyperbolic equations, but this case will not be discussed here.

Basically, in all cases, the goal will be to show that the dominant asymptotic behavior of the solution of a given nonlinear problem is given by the solution of an appropriate linear problem. The strength of the method is its robustness with respect to the addition of general nonlinear terms (universality). In some cases the appropriate linear problem will be easy to identify, but in other cases already the identification of the correct linear problem is in itself an important step. We will also give an introduction to the techniques needed for this task.

As we will see, the renormalization group technique comes in several flavors. There are two continuous versions and one discrete version. The discrete version is the most robust one inasmuch the addition of nonlinear terms is concerned. One of the continuous versions can be understood as a special case of the discrete version but involving only one iterative step. It is this version that is most useful for elliptic problems. The second continuous version is more ambitious in its aim. It does not only provide existence and an asymptotic analysis of the solutions but offers a geometric description of the results through the construction of attractive invariant manifolds. This version is very appealing because of the elegance of the mathematical description of the results, but it is technically more involved and less robust in its applications.

2.1. *Introduction*

In what follows we consider the Cauchy problem for the one dimensional heat equation

$$\begin{aligned} \dot{u}(x,t) &= u''(x,t), \\ u(x,t_0) &= u_0(x), \qquad u_0 \in L^1(\mathbf{R},dx). \end{aligned} \tag{1}$$

where $x \in \mathbf{R}$, $t \geq t_0$, and where by definition $\dot{u}(x,t) \equiv \partial_t u(x,t)$ and $u'(x,t) \equiv \partial_x u(x,t)$. For the moment we set $t_0 = 0$, but later we will usually choose $t_0 = 1$. The main question that we want to answer is: what can one say about the limit $\lim_{t\to\infty} u(x,t)$ "as a function of" u_0? The solution

u of Eq. (1) is

$$u(x,t) = \frac{1}{\sqrt{4\pi t}} \int_{\mathbf{R}} e^{-\frac{(x-y)^2}{4t}} u_0(y)\, dy. \tag{2}$$

Note that for $t > 0$ u is smooth as a function of x, even for non-smooth initial conditions $u_0 \in L^1(\mathbf{R}, dx)$. For what follows, the regularity of the initial conditions u_0 will only play a minor role, but the behavior of u_0 at infinity will be essential. From Eq. (2) we immediately get the following two basic inequalities

$$\|u_t\|_{L^1} \leq \|u_0\|_{L^1}\,, \tag{3}$$

$$\|u_t\|_{L^\infty} \leq \frac{1}{\sqrt{4\pi t}} \|u_0\|_{L^1}\,, \tag{4}$$

where, by definition, $u_t(x) = u(x,t)$. Inequality (4) implies that u_t converges pointwise to zero at least like $t^{-1/2}$, but if we integrate u_t with respect to x over the whole space we find that

$$\int_{\mathbf{R}} u_t(x)\, dx = \int_{\mathbf{R}} u_0(x)\, dx = a_0, \tag{5}$$

i.e., the constant a_0 is an invariant of the time evolution. These two observations motivate to introduce the scaled functions $\tilde{u}$,

$$\tilde{u}(x,t) = \sqrt{t}u(\sqrt{t}x, t). \tag{6}$$

For the function $\tilde{u}_t$, $\tilde{u}_t(x) = \tilde{u}(x,t)$, we still have the invariant quantity a_0,

$$\int_{\mathbf{R}} \tilde{u}_t(x)\, dx = a_0, \tag{7}$$

but now we have instead of Eqs. (3) and (4) the inequalities

$$\|\tilde{u}_t\|_{L^1} \leq \|u_0\|_{L^1}, \tag{8}$$

$$\|\tilde{u}_t\|_{L^\infty} \leq \frac{1}{\sqrt{4\pi}} \|u_0\|_{L^1}, \tag{9}$$

i.e., $\tilde{u}_t$ does not anymore converge pointwise to zero like u_t did. Indeed, the following proposition shows that the function $\tilde{u}_t$, has an interesting long time behavior.

Proposition 1 (universality). *Let* $u_0 \in L^1(\mathbf{R}, dx)$, *let* $\tilde{u}_t$ *and* a_0 *be as defined above, and let* $\tilde{u}_{\text{as}}$ *be defined by*

$$\tilde{u}_{\text{as}}(x) = \frac{a_0}{\sqrt{4\pi}} e^{-\frac{x^2}{4}}. \tag{10}$$

Then

$$\lim_{t\to\infty} \|\tilde{u}_t - \tilde{u}_{\text{as}}\|_{L^\infty} = 0. \tag{11}$$

Remark 2. The fact that $\tilde{u}_{\text{as}}$ depends on u_0 only through the quantity a_0 is what we mean when we say that the limit is "universal". The number a_0 labels the so-called "universality classes", i.e., all initial conditions with the same value of a_0 have the same limit.

Remark 3. Since u is smooth Eq. (11) implies in particular that for all $x \in \mathbf{R}$, $\tilde{u}_t(x) \underset{t\to\infty}{\longrightarrow} \tilde{u}_{\text{as}}(x)$, i.e., we have pointwise convergence of $\tilde{u}_t$ to the limit $\tilde{u}_{\text{as}}$.

Remark 4. Let, for $t > 0$, $x \in \mathbf{R}$,

$$u_{\text{as}}(x,t) = \frac{1}{\sqrt{t}} \tilde{u}_{\text{as}}\left(\frac{x}{\sqrt{t}}\right) = \frac{a_0}{\sqrt{4\pi t}} e^{-\frac{x^2}{4t}}.$$

Since, for any fixed $t \geq 1$,

$$\begin{aligned}\sup_{x\in\mathbf{R}} |\tilde{u}_t(x) - \tilde{u}_{\text{as}}(x)| &= \sup_{x\in\mathbf{R}} |\tilde{u}_t(x/\sqrt{t}) - \tilde{u}_{\text{as}}(x/\sqrt{t})| \\ &= \sqrt{t} \sup_{x\in\mathbf{R}} |(u - u_{\text{as}})(x,t)|,\end{aligned}$$

we can write instead of Eq. (11) equivalently

$$\lim_{t\to\infty} \sqrt{t}\left(\sup_{x\in\mathbf{R}} |(u - u_{\text{as}})(x,t)|\right) = 0. \tag{12}$$

For all of Sec. 2 we will use the more compact notation in Eq. (11). But starting with Sec. 3, when we discuss the down stream behavior of the Navier–Stokes equation, we will rather use the notation in Eq. (12), since the asymptote will have a more complicated structure involving two different scalings.

We now give a proof of Proposition 1. The interest of this proof is that it will carry over to the nonlinear case with very few modifications. The main tool is the Fourier transform (see for example Stein and Weiss (1975) or Titchmarsh (1937)). Let

$$\hat{u}_0(k) = \mathcal{F}(u_0)(k) = \int_{\mathbf{R}} e^{ikx} u_0(x)\, dx. \tag{13}$$

By the Riemann–Lebesgue lemma $\hat{u}_0 \in C_\infty(\mathbf{R})$, the space of continuous functions that converge to zero at infinity. For the invariant quantity a_0 we find

$$a_0 = \hat{u}_0(0).$$

From Eq. (13) we immediately get the inequality (a special case of the Hausdorff–Young inequality)

$$\|\hat{u}_0\|_{L^\infty} \leq \|u_0\|_{L^1}. \tag{14}$$

We will make extensive use of this inequality in what follows. Let

$$\hat{u}(k, t) = \int_{\mathbf{R}} e^{ikx} u(x, t)\, dx. \tag{15}$$

In Fourier space, Eq. (1) becomes

$$\begin{aligned} \frac{d}{dt}\hat{u}(k, t) &= -k^2 \hat{u}(k, t), \\ \hat{u}(k, 0) &= \hat{u}_0(k), \end{aligned} \tag{16}$$

which has the solution

$$\hat{u}(k, t) = e^{-k^2 t} \hat{u}_0(k). \tag{17}$$

Note that all nonzero frequencies k are exponentially damped by the evolution Eq. (17). Only a rescaled region near $k = 0$ survives at large times, which is the reason for the scaling $\sqrt{t}$ introduced above. Namely, from Eq. (15) we find for the Fourier transform $\widehat{\tilde{u}}$ of the scaled function $\tilde{u}$ the expression

$$\begin{aligned} \widehat{\tilde{u}}(k, t) &= \int_{\mathbf{R}} e^{ikx} \tilde{u}(x, t)\, dx = \sqrt{t} \int_{\mathbf{R}} e^{ikx} u(\sqrt{t}x, t)\, dx \\ &= \int_{\mathbf{R}} e^{i\frac{k}{\sqrt{t}}x} u(x, t)\, dx = \hat{u}\left(\frac{k}{\sqrt{t}}, t\right). \end{aligned} \tag{18}$$

We therefore define in Fourier space the rescaled function $\widetilde{\hat{u}}$ by

$$\widetilde{\hat{u}}(k,t) = \widehat{\tilde{u}}(k,t) = \hat{u}\left(\frac{k}{\sqrt{t}}, t\right). \tag{19}$$

From Eq. (17) we find for $\widetilde{\hat{u}}$,

$$\widetilde{\hat{u}}(k,t) = e^{-k^2}\hat{u}_0\left(\frac{k}{\sqrt{t}}\right), \tag{20}$$

and for the Fourier transform $\widehat{\tilde{u}}_{\mathrm{as}}$ of $\tilde{u}_{\mathrm{as}}$ we have

$$\widehat{\tilde{u}}_{\mathrm{as}}(k) = a_0 e^{-k^2}, \tag{21}$$

and we set $\widetilde{\hat{u}}_{\mathrm{as}} = \widehat{\tilde{u}}_{\mathrm{as}}$. Note that $\widetilde{\hat{u}}_{\mathrm{as}} \in L^1(\mathbf{R}, dk)$. Now let $t \geq 1$ and let $\widetilde{\hat{u}}_t(k) = \widetilde{\hat{u}}(k,t)$. First, since $\hat{u}_0$ is a continuous function, we find that $\widetilde{\hat{u}}_t$ converges pointwise to $\widetilde{\hat{u}}_{\mathrm{as}}(k)$, i.e., for all $k \in \mathbf{R}$,

$$\widetilde{\hat{u}}_t(k) \underset{t\to\infty}{\longrightarrow} \widetilde{\hat{u}}_{\mathrm{as}}(k). \tag{22}$$

Second, the family of functions $\widetilde{\hat{u}}_t$ is uniformly bounded by a function which is in $L^1(\mathbf{R}, dk)$. Namely, using Eq. (14) we find that

$$|\widetilde{\hat{u}}_t(k)| \leq \|u_0\|_{L^1}\, e^{-k^2}. \tag{23}$$

From Eqs. (22) and (23) it follows from the Lebesgue dominated convergence theorem that

$$\lim_{t\to\infty} \|\widetilde{\hat{u}}_t - \widetilde{\hat{u}}_{\mathrm{as}}\|_{L^1} = 0, \tag{24}$$

and finally, using the Hausdorff–Young inequality for the inverse Fourier transform, we see that Eq. (24) implies Eq. (11). This completes the proof of Proposition 1.

2.2. *Other function spaces: a counter example*

In later subsections we will be confronted with initial conditions u_0 that are not in $L^1(\mathbf{R}, dx)$. It is therefore instructive to investigate what can be said about the large time behavior in such cases. So assume for a moment that $u_0 \in L^p(\mathbf{R}, dx)$, for some $1 < p \leq 2$, and define for given $x \in \mathbf{R}$ and $t \geq 1$ the function $g_{x,t}$ by

$$g_{x,t}(y) \equiv e^{-\frac{(x-y)^2}{4t}}.$$

Using Hölder's inequality we find from Eq. (2) that u_t satisfies the following pointwise bound,

$$\begin{aligned}|u_t(x)| &\leq \frac{1}{\sqrt{4\pi t}} \|g_{x,t} u_0\|_{L^1} \leq \|g_{x,t}\|_{L^q} \|u_0\|_{L^p} \\ &\leq \frac{1}{\sqrt{4\pi t}} \left(\frac{4\pi t}{q}\right)^{1/q} \|u_0\|_{L^p}, \end{aligned} \tag{25}$$

where $1/q + 1/p = 1$, i.e., $q = p/(p-1)$. This shows that we should not in general expect to find Eq. (11) for initial conditions that are not in $L^1(\mathbf{R}, dx)$. The following counter example (an adaptation of an example given in Collet and Eckmann (1992b)), shows that it is indeed not obvious how to enlarge the function space of initial conditions without losing Proposition 1.

Proposition 5. *There is an initial condition u_0, with $u_0 \in L^p(\mathbf{R}, dx)$ for all $p > 1$, and an increasing sequence of times T_n, $T_n < T_{n+1}$ with $\lim_{n\to\infty} T_n = \infty$, such that*

$$\lim_{n\to\infty} (-1)^n \tilde{u}(0, T_n) = 1.$$

The proof of Proposition 5 is by explicit construction of an appropriate initial condition u_0. The function u_0 is identically zero for $x < 0$. For $x > 0$ it is a sequence of more and more spread apart positive and negative peaks of compact support. The surface below the first peak is equal to one and below subsequent peaks minus or plus two. The positions of the peaks are such that at adequately constructed times the convolution in Eq. (2) is essentially given by the sum of the areas of the first n peaks. The details of the construction are given in Appendix 2.8.

2.3. *Power counting, asymptotic expansions*

We next discuss the question of what can be said about the limit when t goes to infinity in the case when $a_0 = \int_{\mathbf{R}} u_0(x)\, dx = 0$ and, a related question, what can be said about higher order corrections to the asymptotics.

2.3.1. *The case of compact support*

Consider an initial condition of compact support, i.e., $u_0 \in L^1(\mathbf{R}, dx) \cap C_0(\mathbf{R})$. In this case the Fourier transform is an entire analytic function. Namely, let $I = [-L, L]$ be a finite interval containing the support of u_0. Then

$$\hat{u}_0(k) = \int_I e^{ikx} u_0(x)\, dx = \sum_{m \geq 0} a_m (ik)^m, \tag{26}$$

where

$$a_m = \frac{1}{m!} \int_I x^m u_0(x)\, dx. \tag{27}$$

Since we have the bound

$$\left| \int_I x^m u_0(x)\, dx \right| \leq L^m \|u_0\|_1, \tag{28}$$

we find that the sum in Eq. (26) is absolutely convergent for all $k \in \mathbf{C}$. Using this representation we find that

$$\widetilde{\hat{u}}(k, t) = \sum_{m \geq 0} t^{-m/2} a_m (ik)^m e^{-k^2}. \tag{29}$$

In direct space we therefore have

$$\tilde{u}(x, t) = \frac{1}{\sqrt{4\pi}} \sum_{m \geq 0} t^{-m/2} a_m (-1)^m \partial_x^m e^{-\frac{x^2}{4}}. \tag{30}$$

Note that

$$(-1)^m \partial_x^m e^{-\frac{x^2}{4}} = e^{-\frac{x^2}{4}} H_m\left(\frac{x}{2}\right),$$

with H_m the mth Hermite Polynomial. We conclude that, for initial conditions with compact support, the term proportional to a_0 (which corresponds to the limit in Proposition 1) is nothing else than the first term of a more general expansion of the solution in inverse powers of $\sqrt{t}$. In particular, if $a_0 = 0$ but $a_1 \neq 0$, then the dominant term in the asymptotics will be the one with amplitude a_1, and the decay of the solution will be proportional to t^{-1} instead of $t^{-1/2}$. For initial conditions of non-compact support there still exists an asymptotic expansion for the solution, but the number of terms in this expansion is limited by the decay of the initial condition at infinity, since in particular the numbers a_m in Eq. (27) need to exist. This is the content of the next subsection.

2.3.2. *Weighted L^p spaces*

To illustrate somewhat further the dependence of the results as a function of the decay of the initial condition we consider now for $n = 0, 1, \ldots$ the weighted spaces $L^1(\mathbf{R}, (1+x^2)^{n/2}dx)$. Similar results can be obtained in weighted L^2 spaces. See Wayne (1997) and the appendix in Gallay and Wayne (2002b).

Proposition 6 (asymptotic expansion). *Let $t \mapsto u_t$ be the solution of the heat equation with initial condition $u_0 \in L^1(\mathbf{R}, (1+x^2)^{n/2}dx)$. Let, for $m = 0, \ldots, n$,*

$$\tilde{u}_{\text{as}}^m(x) = \frac{a_m}{\sqrt{4\pi}}(-1)^m e^{-\frac{x^2}{4}} H_m\left(\frac{x}{2}\right),$$

with

$$a_m = \frac{1}{m!}\int_{\mathbf{R}} x^m u_0(x)\,dx,$$

and let $\tilde{u}_t(x) = \sqrt{t}u_t(\sqrt{t}x)$. Then,

$$\lim_{t\to\infty} t^{n/2}\left\|\tilde{u}_t - \sum_{m=0}^{n} t^{-m/2}\tilde{u}_{\text{as}}^m\right\|_{L^\infty} = 0. \tag{31}$$

Note that for $n = 0$ the Proposition 6 reduces to Proposition 1.

Since $u_0 \in L^1(\mathbf{R}, (1+x^2)^{n/2}dx)$ the functions $x \mapsto x^m u(x)$ are in $L^1(\mathbf{R}, dx)$ for $m = 1, \ldots, n$, and therefore, by the Riemann–Lebesgue lemma, the Fourier transform $\hat{u}_0$ of u_0 is n times continuously differentiable. For the mth derivative of $\hat{u}_0$ we have

$$\hat{u}_0^{(m)}(k) = \int_{\mathbf{R}} e^{ikx}(ix)^m u_0(x)\,dx, \tag{32}$$

and therefore

$$a_m = \frac{1}{m!}\lim_{k\to 0}(-i)^m \hat{u}_0^{(m)}(k).$$

Let $t \geq 1$. Expanding in a Taylor series we get

$$\begin{aligned} R_n(k,t) &= t^{n/2}\left(\hat{u}_0\left(\frac{k}{\sqrt{t}}\right) - \sum_{m=0}^{n-1} t^{-m/2} a_m(-i)^m k^m\right) \\ &= t^{n/2}\int_0^{k/\sqrt{t}} dk_1 \int_0^{k_1} dk_2 \ldots \int_0^{k_{n-1}} dk_n\, \hat{u}_0^{(n)}(k_n) \\ &= \int_0^{k} dk_1 \int_0^{k_1} dk_2 \ldots \int_0^{k_{n-1}} dk_n\, \hat{u}_0^{(n)}\left(\frac{k_n}{\sqrt{t}}\right), \end{aligned} \tag{33}$$

from which we get that pointwise for $k \in \mathbf{R}$

$$\lim_{t\to\infty} R_n(k,t)e^{-k^2} = \frac{k^n}{n!}\hat{u}_0^{(n)}(0)e^{-k^2} = a_n(-ik)^n e^{-k^2}. \tag{34}$$

Furthermore, since by Eq. (32)

$$\left|\hat{u}_0^{(n)}\left(\frac{k}{\sqrt{t}}\right)\right| \leq \int_{\mathbf{R}} |x|^n |u_0(x)|\, dx = \text{const.} < \infty,$$

we get from Eq. (33) that

$$|R_n(k,t)e^{-k^2}| \leq \text{const.} |k|^n e^{-k^2}, \tag{35}$$

i.e., the function $k \mapsto R_n(k,t)e^{-k^2}$ is bounded uniformly in $t \geq 1$ by a function in $L^1(\mathbf{R}, dk)$. From Eqs. (34) and (35) we conclude by the Lebesgue dominated convergence theorem that

$$\lim_{t\to\infty} t^{n/2}\left\|\hat{u}_0\left(\frac{k}{\sqrt{t}}\right) - \sum_{m=0}^{n} t^{-m/2} a_m(-i)^m k^m\right\|_{L^1} = 0. \tag{36}$$

From Eq. (36) the Proposition (31) now follows as in the proof of Proposition 1 by using the Hausdorff–Young inequality for the inverse Fourier transform. This completes the proof of Proposition 6.

2.3.3. *Power counting*

The above discussion motivates to measure the "size" of various functions in inverse powers of $\sqrt{t}$. This will be important when we discuss the nonlinear case on a formal level. In what follows we mean by power counting a formal reference to the above results. Namely, we will summarize these results by saying that, asymptotically as $t \to \infty$, $u \sim t^{-1/2}$, and similarly that $\dot{u} \sim t^{-3/2}$, $u' \sim t^{-1}$ and $u'' \sim t^{-3/2}$.

2.3.4. *Higher order terms revisited: formal and asymptotic expansions*

Another way to proceed in order to prove the existence of the leading and subleading order terms is to set $u(x,t) = 1/\sqrt{t}\, f(x/\sqrt{t}) + u_1(x,t)$ (we choose $t_0 = 1$ here) and to plug this expression into the equation. In a first step one sets $u_1 \equiv 0$ and gets an ordinary differential equation for f,

$$f''(z) + \frac{1}{2} z f'(z) + \frac{1}{2} f(z) = 0,$$

which has in particular the solution $f(z) = \tilde{u}_{\mathrm{as}}(z)$, with $\tilde{u}_{\mathrm{as}}$ as defined in Eq. (10). With this function f one gets for u_1 again the heat equation but with the initial condition $u_1(x,0) = u_0(x) - \tilde{u}_{\mathrm{as}}(x)$. By definition of $\tilde{u}_{\mathrm{as}}$ one has that $\int_{\mathbf{R}} u_1(x,0) = 0$. Provided the initial condition u_0 decays sufficiently rapidly, higher order terms can be computed in a similar way. Finally, the remainder is estimated by solving the resulting equation in the appropriate function space. This again leads to the construction of asymptotic expansions. This procedure will be the method of choice in the case of nonlinear problems.

2.4. *Function spaces*

Based on the discussions above we now introduce function spaces directly for the Fourier transforms. These spaces have proved to be well adapted for studying the large time behavior of parabolic and elliptic problems. Related function spaces are used for the case of the Navier–Stokes equations. We set $t_0 = 1$ from now on.

Definition 7. Let $\alpha \geq 0$. Then, we define $\mathcal{A}_\alpha$ to be the Banach space of continuous, complex valued functions $\hat{f}: \mathbf{R} \to \mathbf{C}$, for which the norm $\|\hat{f}\|_\alpha$ defined by

$$\|\hat{f}\|_\alpha = \sup_{k \in \mathbf{R}} (1 + |k|^\alpha) |\hat{f}(k)| \tag{37}$$

is finite.

Remark 8. For $\alpha > 1$ a function $\hat{f} \in \mathcal{A}_\alpha$ is in $L^q(\mathbf{R}, dk)$ for all $q \geq 1$, and its inverse Fourier transform $f = \mathcal{F}^{-1}(\hat{f})$ is therefore in particular in $C_\infty \cap L^p(\mathbf{R}, dx)$ for $2 \leq p \leq \infty$.

For functions $\hat{f}$ of $k \in \mathbf{R}$ and $t \geq 1$ we write either $\hat{f}(k, t)$ or $\hat{f}_t(k)$.

Definition 9. Let $\alpha, \beta \geq 0$ and let $\hat{f}$ be a continuous map from $[1, \infty)$ to $\mathcal{A}_\alpha$. Let

$$\widetilde{\hat{f}}_t(k) = \hat{f}\left(\frac{k}{t^{1/2}}, t\right). \tag{38}$$

Then, we define $\mathcal{B}_{\alpha,\beta}$ to be the Banach space of all such maps for which the norm $\|\hat{f}\|_{\alpha,\beta}$ defined by

$$\|\hat{f}\|_{\alpha,\beta} = \sup_{t \geq 1} t^{\beta} \|\widetilde{\hat{f}}_t\|_\alpha \tag{39}$$

is finite.

Note that for all $\hat{f} \in \mathcal{B}_{\alpha,\beta}$ we have for all $k \in \mathbf{R}$ and $t \geq 1$ the bound

$$|\hat{f}(k, t)| \leq \frac{\|\hat{f}\|_{\alpha,\beta}}{t^\beta} \mu_\alpha(k, t), \tag{40}$$

where

$$\mu_\alpha(k, t) = \frac{1}{1 + (|k|\sqrt{t})^\alpha}. \tag{41}$$

Similarly, we find that if a continuous function f satisfies for some constant c the bound

$$|\hat{f}(k, t)| \leq \frac{c}{t^\beta} \mu_\alpha(k, t), \tag{42}$$

uniformly in $k \in \mathbf{R}$ and $t \geq 1$, then $\hat{f} \in \mathcal{B}_{\alpha,\beta}$, and $\|\hat{f}\|_{\alpha,\beta} \leq c$.

2.5. *The renormalization group*

Let again $t_0 = 1$. Let $\tau > 0$ (typically $\tau \gg 0$) and let $\mathcal{R}_\tau$ be the map that associates to the initial condition u_0 at $t = t_0 = 1$ the rescaled solution $\tilde{u}_t$, at $t = e^\tau$. Explicitly we have in Fourier space

$$\hat{u}(k, t) = e^{-k^2(t-1)} \hat{u}_0(k),$$

and therefore

$$\widetilde{\hat{u}}(k,t) = e^{-k^2(1-1/t)}\hat{u}_0\left(\frac{k}{\sqrt{t}}\right),$$

so that

$$\mathcal{R}_\tau(\hat{u}_0)(k) = e^{-k^2(1-e^{-\tau})}\hat{u}_0(ke^{-\tau/2}).$$

Note that it follows from the above that $\mathcal{R}_\tau$ is well defined as a map from $\mathcal{A}_\alpha$ to $\mathcal{A}_\alpha$, for all $\alpha \geq 0$. Now let $\hat{u}_1 = \mathcal{R}_{\tau_1}(\hat{u}_0)$ and $\hat{u}_2 = \mathcal{R}_{\tau_2}(\hat{u}_1)$. For the composition $\mathcal{R}_{\tau_2}\mathcal{R}_{\tau_1}$ of the two maps we have

$$\begin{aligned}
(\mathcal{R}_{\tau_2}\mathcal{R}_{\tau_1})(\hat{u}_0)(k) &= \mathcal{R}_{\tau_2}(\hat{u}_1)(k) \\
&= e^{-k^2(1-e^{-\tau_2})}\hat{u}_1(ke^{-\tau_2/2}) \\
&= e^{-k^2(1-e^{-\tau_2})}\left(e^{-k_1^2(1-e^{-\tau_1})}\hat{u}_0(k_1e^{-\tau_1/2})\right)|_{k_1=ke^{-\tau_2/2}} \\
&= e^{-k^2(1-e^{-\tau_2})}e^{-k^2(1-e^{-\tau_1})e^{-\tau_2}}\hat{u}_0(ke^{-\tau_2/2}e^{-\tau_1/2}) \\
&= e^{-k^2(1-e^{-(\tau_1+\tau_2)})}\hat{u}_0(ke^{-(\tau_1+\tau_2)}) \\
&= \mathcal{R}_{\tau_1+\tau_2}(\hat{u}_0)(k).
\end{aligned}$$

This means that $\mathcal{R}_\tau$ has a semi-group structure. This is the so-called Renormalization group which has been enormously successful as a frame of mind, as a way of structuring and organizing the proofs. Note that by construction $\mathcal{R}_\tau(u_{\text{as}}) = u_{\text{as}}$, i.e., the asymptotic behavior discussed above is recovered here in terms of a fixed point of $\mathcal{R}_\tau$. This is the so-called trivial (Gaussian) fixed point. The rescaled solution $\tilde{u}$ is a (forward) orbit of the Renormalization group, and the above results can be interpreted as saying that the fixed-point u_{as} is attractive.

In the so-called discrete version of the Renormalization group one first constructs a metric space (typically a ball in a Banach space) which, for some large but finite τ, is contracted into itself by $\mathcal{R}_\tau$, which implies the existence of a solution for some large but finite time. Once the existence of the solution is known, enough additional information on this solution is then obtained so that the procedure can be iterated using the semi-group properties, i.e., one analyzes $\mathcal{R}_{n\tau} = \mathcal{R}_\tau^n$ (n-fold composition) as $n \in N$ goes to infinity. This is the original technique introduced (for the nonlinear case) by Bricmont and Kupiainen (1994a), Bricmont *et al.* (1994), Bricmont and Kupiainen (1994b) and Bricmont and Kupiainen (1995), based in part on earlier work by Collet and Eckmann (1992b), Collet and Eckmann (1992a), Collet *et al.* (1992), Eckmann and Gallay (1993) and Eckmann and Wayne (1994). See also Pao (1993) and Bona *et al.* (1994). For higher order asymptotics see for example Bona *et al.* (1995).

In the simple continuous version of the RG one proves for a (small) set of initial conditions existence and bounds on $\mathcal{R}_\tau$ that are uniform in $\tau \geq 0$. The (attractive) fixed-points of $\mathcal{R}_\tau$ can then be obtained simply by taking the limit $\tau \to \infty$. This is the procedure that we follow below. It has been introduced in Gallay (1994), based on Gallay (1993) and is closest to the standard functional analytic techniques.

The second continuous version of the RG is constructed differently. Instead of studying $\mathcal{R}_\tau$, one sets $u(x,t) = v(\sqrt{t}x, \log(t))$ and then analyzes the semigroup obtained by solving the evolution equation for v rather than the one obtained by solving the equation for u. This leads to a connection with the idea of invariant manifolds and the theory of finite dimensional dynamical systems. See Wayne (1997) for an introduction and Gallay and Wayne (2002b) for an important application.

2.6. *Technical lemmas*

This section contains the main technical lemmas used in subsequent sections. The propositions are specific to the heat equation but the methods of proof and the basic ideas are independent of the particular case under consideration. It is therefore instructive to give the details of these proofs for this simple case. But, in order to proceed with the general discussion, we have relegated these details to Appendix 2.9 at the end of this section. Here, we only state the results.

The first proposition shows that the function spaces $\mathcal{B}_{\alpha\beta}$ are well adapted for the description of the time evolution generated by the heat equation.

Lemma 10. *Let α, β, $\gamma \geq 0$ with $\alpha + 2\beta \geq \gamma$. Then*

$$e^{-k^2(t-1)} \left(\frac{t-1}{t}\right)^{\beta} |k|^{\gamma} \mu_{\alpha}(k,1) \leq \frac{\text{const.}}{t^{\min\{\beta,\gamma/2\}}} \mu_{\alpha+2\beta-\gamma}(k,t),$$

uniformly in $t \geq 1$, $k \in \mathbf{R}$, and with μ_α as defined in Eq. (41).

The second proposition shows that the scaling built into the function spaces $\mathcal{B}_{\alpha\beta}$ naturally permits to extract the optimal time decay of nonlinear terms. Note that local nonlinear terms in direct space (products of u, u', and u'', say) become convolution products in Fourier space.

Lemma 11. *Let α, $\alpha' > 1$. Then we have*

$$\int_{\mathbf{R}} \mu_\alpha(k-k',t)\mu_{\alpha'}(k',t)dk' \leq \frac{\text{const.}}{\sqrt{t}}\mu_{\min\{\alpha,\alpha'\}}(k,t), \tag{43}$$

uniformly in $t \geq 1$ and $k \in \mathbf{R}$.

The third proposition exhibits properties of the semigroup generated by the heat equation. In particular we have a maximal regularity result in $\mathcal{B}_{\alpha\beta}$ (see Eq. (47)), which again shows that the spaces $\mathcal{B}_{\alpha\beta}$ provide a natural functional setting. See for example the book by Lunardi (2003) for a discussion of optimal regularity in parabolic problems.

Lemma 12. *Let $\beta > 1$. For $\alpha \geq 0$, we have the bound,*

$$\int_1^t e^{-k^2(t-s)}\frac{1}{s^\beta}\mu_\alpha(k,s)\,ds \leq \text{const.}\ \mu_\alpha(k,t), \tag{44}$$

$$\int_1^t \left(e^{-k^2(t-s)} - e^{-k^2(t-1)}\right)\frac{1}{s^\beta}\mu_\alpha(k,s)\,ds \leq \frac{\text{const.}}{t^{\beta-1}}\ \mu_\alpha(k,t), \tag{45}$$

for $\alpha \geq 1$ we have the bound,

$$\int_1^t e^{-k^2(t-s)}\frac{1}{s^\beta}\mu_{\alpha-1}(k,s)\,ds \leq \text{const.}\ \mu_\alpha(k,t), \tag{46}$$

and for $\alpha \geq 2$ we have the bound (maximal regularity),

$$\int_1^t e^{-k^2(t-s)}\frac{1}{s^\beta}\mu_{\alpha-2}(k,s)\,ds \leq \text{const.}\ \mu_\alpha(k,t), \tag{47}$$

uniformly in $t \geq 1$ and $k \in \mathbf{R}$.

2.7. *Nonlinear problems*

We now discuss the simplest non-linear cases. Consider the equation

$$\dot{u}(x,t) = u''(x,t) - u(x,t)^p, \quad p = 1,2,3\ldots \tag{48}$$
$$u(x,1) = u_0(x), \quad u_0 = \mathcal{F}^{-1}(\hat{u}_0), \text{ with } \hat{u}_0 \in \mathcal{A}_\alpha, \alpha > 1.$$

Here $\mathcal{F}^{-1}$ denotes the inverse Fourier transform, i.e.,

$$\mathcal{F}^{-1}(\hat{u}_0)(x) = \frac{1}{2\pi} \int_{\mathbf{R}} e^{-ikx} \hat{u}_0(k)\, dk.$$

The strength of the renormalization group method is that it makes no direct reference to the exact form of the nonlinearity. The same technique that we use now first for studying the nonlinearity u^p can then also be used to study a general nonlinearity of the form $F(u, u', u'')$, with F an arbitrary (nonlinear) function that is jointly analytic in its arguments near the origin.

On a formal level, if we assume that the linear heat equation is the relevant problem for the description of the large time asymptotics of the solution of Eq. (48), i.e., that the nonlinear term becomes negligible at large times t, then we have the following power counting: $u \sim t^{-1/2}$, and $\dot{u}$, and $u'' \sim t^{-3/2}$, and $u^p \sim (t^{-1/2})^p = t^{-p/2}$. We therefore find that u^p is indeed negligible when compared with $\dot{u}$ and u'', provided $p > 3$, and we get the following formal classification:

(i) $p > 3$, the nonlinearity is "irrelevant" (the trivial fixed point is stable and the linear heat equation is the relevant linear problem at large times).

(ii) $p = 3$, the nonlinearity is "marginal" (the trivial fixed point is marginally stable and, depending on the nonlinearity, the linear heat equation is or is not the relevant linear problem).

(iii) $p = 1, 2$, the nonlinearity is "relevant" (the trivial fixed-point is unstable, the linear heat equation is not the relevant linear problem).

In what follows we will mainly discuss the point (i) of this formal classification, since this is the relevant case for the analysis of the Navier–Stokes equation. For the readers interested in the analysis of the cases (ii) and (iii) we refer to Bricmont *et al.* (1994), Bricmont *et al.* (1996), Bricmont and Kupiainen (1996b) and Uecker (2006).

2.7.1. *The case of irrelevant perturbations, I*

For Eq. (48) we have the following analog of Proposition 1.

Proposition 13 (universality). *Let* $p > 3$ *and* $t \geq 1$ *and let* $t \mapsto \hat{u}_t$ *be the solution of Eq.* (48) *with initial condition* $\hat{u}_0 \in \mathcal{A}_\alpha$ *with* $\alpha > 1$. *Let*

$u(x,t) = \mathcal{F}^{-1}(\hat{u}_t)(x)$, and let

$$\tilde{u}_t(x) = \sqrt{t}u(\sqrt{t}x, t).$$

Define furthermore the function u_{as} by

$$\tilde{u}_{\text{as},p}(x) = \frac{a_{\text{as},p}}{\sqrt{4\pi}} e^{-\frac{x^2}{4}},$$

where

$$a_{\text{as},p} = \hat{u}_0(0) + \int_1^\infty \hat{u}_s^{*p}(0)\, ds. \tag{49}$$

Then,

$$\lim_{t\to\infty} \|\tilde{u}_t - \tilde{u}_{\text{as},p}\|_{L^\infty} = 0. \tag{50}$$

Remark 14. In contrast to Proposition 1, where because of the existence of an invariant quantity the amplitude of the limit could be computed directly from the initial condition, the amplitude (49) involves the solution. There are still universality classes of initial conditions (two initial conditions are equivalent if they have the same limit), but in contrast to the linear case studied above we have to solve the equation in order to know if two initial conditions belong to the same class or not.

In order to prove Proposition 13 we proceed in two steps. First we prove the existence of a solution, then we analyze its long time behavior. So let $p > 3$ and $t \geq 1$ and let $\hat{u}_0 \in \mathcal{A}_\alpha$ with $\alpha > 1$. We construct solutions of the Eq. (48) by solving in Fourier space the integral equation

$$\hat{u}(k,t) = e^{-k^2(t-1)}\hat{u}_0(k) + \int_1^t e^{-k^2(t-s)}\, \hat{q}(k,s)\, ds, \tag{51}$$

with

$$\hat{q}(k,t) = \hat{u}_t^{*p}(k), \tag{52}$$

with $*$ the convolution product with respect to the variable k. If $\hat{u}$ solves Eqs. (51) and (52), then the inverse Fourier transform,

$$u_t(x) = \mathcal{F}^{-1}(\hat{u}_t)(x) = \frac{1}{2\pi}\int_{\mathbf{R}} e^{-ikx}\hat{u}_t(k)\, dk,$$

solves Eq. (48). To prove the existence of a solution to Eqs. (51) and (52) for a given initial conditions $\hat{u}_0$ we will apply the contraction mapping principle to the map $\mathcal{N} = \mathcal{M}\mathcal{L}$, where $\mathcal{L}\colon \hat{q} \mapsto_{(51)} \hat{u}$, is the map that associates to $\hat{q}$ the function $\hat{u}$ using Eq. (51), and where $\mathcal{M}\colon \hat{u} \mapsto_{(52)} \hat{q}$, is the map that associates to $\hat{u}$ the function $\hat{q}$ using Eq. (52). Note that for $p > 3$ we have that $\beta = (p-1)/2 > 1$.

Proposition 15. *Let* $0 < \varepsilon_0$, $\alpha > 1$ *and let* $\beta = (p-1)/2$. *Let* $\hat{u}_0 \in \mathcal{A}_\alpha$ *with* $\|\hat{u}_0\|_\alpha = \varepsilon_0$ *and let* $\mathcal{U}_{\alpha,\beta}(\varepsilon_0) = \{\hat{q} \in \mathcal{B}_{\alpha,\beta} | \ \|\hat{q}\|_{\alpha,\beta} < \varepsilon_0\}$. *Then,*

(i) $\mathcal{L}$ *is well defined as a map from* $\mathcal{B}_{\alpha,\beta}$ *to* $\mathcal{B}_{\alpha,0}$.
(ii) $\mathcal{M}$ *is well defined as a map from* $\mathcal{B}_{\alpha,0}$ *to* $\mathcal{B}_{\alpha,\beta}$.
(iii) $\mathcal{N}$ *is well defined as a map from* $\mathcal{B}_{\alpha,\beta}$ *to* $\mathcal{B}_{\alpha,\beta}$.
(iv) $\mathcal{N}(\mathcal{U}_{\alpha,\beta}(\varepsilon_0)) \subset \mathcal{U}_{\alpha,\beta}(\varepsilon_0)$ *for* ε_0 *small enough.*
(v) *For* ε_0 *small enough,* $\|\mathcal{N}(\hat{q}_1) - \mathcal{N}(\hat{q}_2)\|_{\alpha,\beta} \leq \frac{1}{2}\|\hat{q}_1 - \hat{q}_2\|_{\alpha,\beta}$, *for all* $\hat{q}_1, \hat{q}_2 \in \mathcal{U}_{\alpha,\beta}(\varepsilon_0)$.
(vi) $\mathcal{N}$ *has a unique fixed point in* $\mathcal{U}_{\alpha,\beta}(\varepsilon_0)$ *for* ε_0 *small enough.*

Since $\mathcal{N} = \mathcal{M}\mathcal{L}$, (*iii*) follows from (*i*) and (*ii*). (*vi*) follows from (*iv*) and (*v*) using the contraction mapping principle. We now prove (*i*). Throughout all proofs we denote by ε a constant multiple of ε_0. This constant may depend on p and α and may be different from instance to instance. We first show that the function $\hat{u}_L$,

$$\hat{u}_L(k,t) = e^{-k^2(t-1)}\hat{u}_0(k),$$

is in $\mathcal{B}_{\alpha,0}$. Since $\hat{u}_0 \in \mathcal{A}_\alpha$ we have by definition that

$$|\hat{u}_0(k)| \leq \|\hat{u}_0\|_\alpha \, \mu_\alpha(k,1) \leq \varepsilon_0 \mu_\alpha(k,1),$$

and therefore we find using Lemma 10 that $\hat{u}_L(k,t) \leq \varepsilon\mu_\alpha(k,t)$. Therefore $u_L \in \mathcal{B}_{\alpha,0}$ and $\|\hat{u}_L\|_{\alpha,0} \leq \varepsilon$. Next let

$$\hat{u}_N(k,t) = \int_1^t e^{-k^2(t-s)} \, \hat{q}(k,s)\, ds.$$

Since $\hat{q} \in \mathcal{B}_{\alpha,\beta}$ we have that

$$|\hat{q}(k,t)| \leq \|\hat{q}\|_{\alpha,\beta} t^{-\beta}\mu_\alpha(k,t) \leq \varepsilon_0 t^{-\beta}\mu_a(k,t),$$

and therefore $\hat{u}_N \in \mathcal{B}_{\alpha,0}$ by Eq. (44) of Lemma 12, and $\|u_L\|_{\alpha,0} \leq \varepsilon$. Since $\hat{u} = \hat{u}_L + \hat{u}_N$ it now follows that $\hat{u} \in \mathcal{B}_{\alpha,0}$ as claimed. Furthermore

$\|\hat{u}\|_{\alpha 0} \leq \varepsilon$ by the triangle inequality. We now prove (ii). Let $\hat{u} \in \mathcal{B}_{\alpha,0}$. Applying Lemma 11 to the $p-1$ convolutions we get, since $\beta = \frac{p-1}{2}$, that $\mathcal{M}(\hat{u}) \in \mathcal{B}_{\alpha,\beta}$. Furthermore, if $\|\hat{u}\| \leq \text{const.}\varepsilon_0$, then

$$\|\mathcal{M}(\hat{u})\|_{\alpha,\beta} \leq \text{const.}\varepsilon_0^p.$$

Therefore, we find that for $\hat{q} \in \mathcal{U}_{\alpha,\beta}(\varepsilon_0)$,

$$\|\mathcal{N}(\hat{q})\|_{\alpha,\beta} \leq \text{const.}\varepsilon_0^p.$$

Now since $\text{const.}\varepsilon_0^p < \varepsilon_0$ for ε_0 small enough we find (iv). To prove (v) we consider, for $i = 1, 2$, the image $\hat{u}_i = \mathcal{L}(\hat{q}_i)$ of $\hat{q}_i \in \mathcal{U}_{\alpha,\beta}(\varepsilon_0)$. We have already shown that $\|\hat{u}_i\|_{\alpha,0} \leq \varepsilon$, $i = 1, 2$, and using exactly the same techniques one shows that

$$\|\hat{u}_1 - \hat{u}_2\|_{\alpha,0} \leq \text{const.}\|\hat{q}_1 - \hat{q}_2\|_{\alpha,\beta}.$$

Finally, since $\hat{u}_1^{*p} - \hat{u}_2^{*p} = P(\hat{u}_1, \hat{u}_2) * (\hat{u}_1 - \hat{u}_2)$ for a certain polynomial P of degree $p-1$ (all multiplications are convolution products here), we find that

$$\begin{aligned}\|\mathcal{N}(\hat{q}_1) - \mathcal{N}(\hat{q}_2)\|_{\alpha,\beta} &\leq \text{const.}\varepsilon_0^{p-1}\|\hat{q}_1 - \hat{q}_2\|_{\alpha,\beta} \\ &\leq \frac{1}{2}\|\hat{q}_1 - \hat{q}_2\|_{\alpha,\beta},\end{aligned}$$

for ε_0 small enough. This completes the proof of Proposition 15.

We can now prove the Proposition 13. The proof is essentially as in the linear case. Let

$$\hat{U}_t(k) = e^{-k^2(t-1)}\left(\hat{u}_0(k) + \int_1^t \hat{u}_s^{*p}(k)\, ds\right),$$

and let

$$\widetilde{\hat{U}}_t(k) = U_t\left(\frac{k}{t^{1/2}}\right).$$

First we note that for any fixed $k \in \mathbf{R}$,

$$\widetilde{\hat{u}}_{\text{as},p}(k) := \lim_{t\to\infty} \widetilde{\hat{U}}_t(k) = a_{\text{as},p}e^{-k^2} = \widehat{\tilde{u}}_{\text{as},p},$$

and that uniformly in $k \in \mathbf{R}$, and $t \geq 2$,

$$|\widetilde{\hat{U}}_t(k)| \leq \text{const.}\mu_\alpha(k, 1).$$

Therefore it follows from the Lebesgue dominated convergence theorem that

$$\lim_{t\to\infty} \|\widetilde{\hat{U}}_t - \widetilde{\hat{u}}_{\mathrm{as},p}\|_{L^1} = 0.$$

Applying the Hausdorff–Young inequality to the inverse Fourier transform we find for $\tilde{U}_t = \mathcal{F}^{-1}(\widetilde{\hat{U}}_t)$ that

$$\lim_{t\to\infty} \|\tilde{U}_t - \tilde{u}_{\mathrm{as},p}\|_{L^\infty} = 0.$$

Inequality (50) now follows using the triangle inequality, provided

$$\lim_{t\to\infty} \|\tilde{u}_t - \tilde{U}_t\|_{L^\infty} = 0. \tag{53}$$

For the difference of $\hat{U}_t$ and $\hat{u}_t$ we have that

$$|\hat{U}_t(k) - \hat{u}_t(k)| \le \varepsilon \int_1^t \left(e^{-k^2(t-s)} - e^{-k^2(t-1)}\right) \mu_\alpha(k,s)\frac{ds}{s^\beta},$$

and therefore we find using Eq. (45) in Lemma 12 that

$$|\hat{U}_t(k) - \hat{u}_t(k)| \le \frac{\varepsilon}{t^{\beta-1}}\mu_\alpha(k,t),$$

from which it follows, since $\alpha > 1$, that

$$\|\widetilde{\hat{u}}_t - \widetilde{\hat{U}}_t\|_{L^1} \le \varepsilon \int_{\mathbf{R}} \frac{1}{t^{\beta-1}}\mu_\alpha(k,1)\,dk \le \frac{\varepsilon}{t^{\beta-1}}.$$

Therefore $\lim_{t\to\infty} \|\widetilde{\hat{u}}_t - \widetilde{\hat{U}}_t\|_{L^1} = 0$, which implies Eq. (53) by the Hausdorff–Young inequality. This completes the proof of Proposition 13.

2.7.2. *The case of irrelevant perturbations, II*

We now discuss the case of the nonlinearity $q = uu'$, which is relevant for the Navier–Stokes equations. Namely, we consider the equation

$$\dot{u}(x,t) = u''(x,t) - u(x,t)u'(x,t). \tag{54}$$

According to the preceding power counting we find that $q \sim t^{-1/2}t^{-1} = t^{-3/2}$, i.e., the nonlinearity $q = uu'$ is *a priori*, a marginal perturbation of the heat equation. However, since $uu' = \frac{1}{2}(u^2)'$, and in contrast to the preceding nonlinearity, the present nonlinearity does not destroy the invariance property of the linear heat equation, i.e., the quantity $a_0 =$

$\int_R u_0(x)\,dx$ is preserved by the time evolution. This allows in turn to restrict the equation to the subspace of functions for which $a_0 = 0$. In this subspace the power counting for the linear heat equation is $u \sim t^{-1}$, $u' \sim t^{-3/2}$ and u'', $\dot{u} \sim t^{-2}$, and therefore $q = uu' \sim t^{-5/2}$ which is now an irrelevant perturbation. It is this case which is important for the analysis of the Navier–Stokes equations.

Proposition 16 (universality). *Let $t \mapsto \hat{u}_t$ be the solution of Eq. (54) for an initial condition $\hat{u}_0$ of the form $\hat{u}_0(k) = ik\hat{v}_0(k)$, with $\hat{v}_0 \in \mathcal{A}_\alpha$ and $\alpha > 3$. Let $u(x,t) = \mathcal{F}^{-1}(\hat{u}_t)(x)$, and let*

$$\tilde{u}_t(x) = t\,u(\sqrt{t}x, t).$$

Define furthermore the function $u_{\rm as}$ by

$$\tilde{u}_{\rm as}(x) = \frac{a_{\rm as}}{\sqrt{4\pi}}\frac{x}{2}e^{-x^2/4}, \tag{55}$$

where

$$a_{\rm as} = \hat{v}_0(0) + \int_0^\infty \hat{u}_s^{*2}(0)\,ds. \tag{56}$$

Then,

$$\lim_{t\to\infty} \|\tilde{u}_t - \tilde{u}_{\rm as}\|_{L^\infty} = 0. \tag{57}$$

In order to prove Proposition 16 we proceed exactly as in the proof of Proposition 13. First we prove the existence of a solution, then we analyze its long time behavior. We again construct a solution of the Eq. (54) by solving in Fourier space the corresponding integral equation, which for the present case is

$$\hat{u}_t(k) = ike^{-k^2(t-1)}\hat{v}_0(k) + ik\int_1^t e^{-k^2(t-s)}\,\hat{q}(k,s)\,ds, \tag{58}$$

with

$$\hat{q}(k,t) = \hat{u}_t^{*2}(k). \tag{59}$$

If $\hat{u}_t$ solves Eqs. (58) and (59), then the inverse Fourier transform $u_t(x) = \mathcal{F}^{-1}(\hat{u}_t)(x)$ solves Eq. (54). To prove the existence of a solution to Eqs. (58) and (59) for a given initial condition $\hat{v}_0$ we apply the contraction mapping principle to the map $\mathcal{N} = \mathcal{M}\mathcal{L}$, where $\mathcal{L}\colon \hat{q} \mapsto \hat{u}$ is the map that associates to $\hat{q}$ the function $\hat{u}$ using Eq. (58), and where $\mathcal{M}\colon \hat{u} \mapsto \hat{q}$, is the map that associates to $\hat{u}$ the function $\hat{q}$ using Eq. (59).

Proposition 17. *Let* $0 < \varepsilon_0$, $\alpha > 2$ *and let* $\beta = 3/2$. *Let* $\hat{v}_0 \in \mathcal{A}_\alpha$ *with* $\|\hat{v}_0\|_\alpha = \varepsilon_0$ *and let* $\mathcal{U}_{\alpha-1,\beta}(\varepsilon_0) = \{\hat{q} \in \mathcal{B}_{\alpha-1,\beta} |\ \|\hat{q}\|_{\alpha-1,\beta} < \varepsilon_0\}$. *Then,*

(i) $\mathcal{L}$ *is well defined as a map from* $\mathcal{B}_{\alpha-1,\beta}$ *to* $\mathcal{B}_{\alpha-1,1/2}$.
(ii) $\mathcal{M}$ *is well defined as a map from* $\mathcal{B}_{\alpha-1,1/2}$ *to* $\mathcal{B}_{\alpha-1,\beta}$.
(iii) $\mathcal{N}$ *is well defined as a map from* $\mathcal{B}_{\alpha-1,\beta}$ *to* $\mathcal{B}_{\alpha-1,\beta}$.
(iv) $\mathcal{N}(\mathcal{U}_{\alpha-1,\beta}(\varepsilon_0)) \subset \mathcal{U}_{\alpha-1,\beta}(\varepsilon_0)$ *for* ε_0 *small enough.*
(v) *For* ε_0 *small enough,* $\|\mathcal{N}(\hat{q}_1) - \mathcal{N}(\hat{q}_2)\|_{\alpha-1,\beta} \leq \frac{1}{2}\|\hat{q}_1 - \hat{q}_2\|_{\alpha-1,\beta}$, *for all* $\hat{q}_1, \hat{q}_2 \in \mathcal{U}_{\alpha-1,\beta}(\varepsilon_0)$.
(vi) $\mathcal{N}$ *has a unique fixed point in* $\mathcal{U}_{\alpha-1,\beta}(\varepsilon_0)$ *for* ε_0 *small enough.*

Since $\mathcal{N} = \mathcal{M}\mathcal{L}$, (*iii*) follows from (*i*) and (*ii*). (*vi*) follows from (*iv*) and (*v*) using the contraction mapping principle. We now prove (*i*). Let $u(k,t) = u_t(k)$. We first show that the function $\hat{u}_L$,

$$\hat{u}_L(k,t) = ike^{-k^2(t-1)}\hat{v}_0(k),$$

is in $\mathcal{B}_{\alpha-1,1/2}$. Since $\hat{v}_0 \in \mathcal{A}_\alpha$ we have by definition that

$$|\hat{v}_0(k)| \leq \|\hat{v}_0\|_\alpha\, \mu_\alpha(k,1) \leq \varepsilon_0 \mu_\alpha(k,1).$$

Therefore, $|ik\hat{v}_0(k)| \leq \varepsilon_0 |k| \mu_\alpha(k,1)$, and we find using Lemma 10 that $\hat{u}_L(k,t) \leq \varepsilon \mu_{\alpha-1}(k,t)/t^{1/2}$. Therefore $u_L \in \mathcal{B}_{\alpha-1,1/2}$ and $\|\hat{u}_L\|_{\alpha-1,1/2} \leq \varepsilon$. Next let

$$\hat{u}_N(k,t) = ik \int_1^t e^{-k^2(t-s)}\, \hat{q}(k,s)\, ds.$$

Since $\hat{q} \in \mathcal{B}_{\alpha-1,\beta}$ we have that

$$|\hat{q}(k,t)| \leq \|\hat{q}\|_{\alpha-1,\beta} t^{-\beta} \mu_{\alpha-2}(k,t) \leq \varepsilon_0 t^{-\beta} \mu_{a-2}(k,t).$$

Therefore we find by Eq. (46) of Lemma 12, and using that

$$|k|\mu_a(k,t) \leq \frac{\text{const.}}{t^{1/2}} \mu_{a-1}(k,t),$$

that $u_N \in \mathcal{B}_{\alpha-1,1/2}$ with $\|u_N\|_{\alpha-1,1/2} \leq \varepsilon$. Since $\hat{u} = \hat{u}_L + \hat{u}_N$ it now follows using the triangle inequality that $\hat{u} \in \mathcal{B}_{\alpha-1,1/2}$ and that

$\|\hat{u}\|_{\alpha-1,1/2} \leq \varepsilon$. We now prove (ii). Let $\hat{u} \in \mathcal{B}_{\alpha-1,1/2}$. Applying Lemma 11 we find that $\hat{q} = \mathcal{M}(\hat{u}) \in \mathcal{B}_{\alpha-1,\beta}$. Furthermore, if $\|\hat{u}\|_{\alpha-1,1/2} \leq \text{const}.\varepsilon_0$, then $\|\mathcal{M}(\hat{u})\|_{\alpha-1,\beta} \leq \text{const}.\varepsilon_0^2$. Therefore, $\|\mathcal{N}(\hat{q})\|_{\alpha,\beta} \leq \text{const}.\varepsilon_0^2$ for $\hat{q} \in \mathcal{U}_{\alpha-1,\beta}(\varepsilon_0)$. Since $\text{const}.\varepsilon_0^2 < \varepsilon_0$ for ε_0 small enough we find (iv). To prove (v) we consider the image of $\hat{q}_i$, $i = 1, 2$ in $\mathcal{U}_{\alpha-1,\beta}(\varepsilon_0)$. Let $\hat{u}_i = \mathcal{L}(\hat{q}_i)$, $i = 1, 2$. We have already shown that $\|\hat{u}_i\|_{\alpha-1,1/2} \leq \varepsilon$, $i = 1, 2$, and using exactly the same techniques one shows that

$$\|\hat{u}_1 - \hat{u}_2\|_{\alpha-1,1/2} \leq \text{const}.\|\hat{q}_1 - \hat{q}_2\|_{\alpha-1,\beta}.$$

Finally, since $\hat{u}_1^{*2} - \hat{u}_2^{*2} = (\hat{u}_1 + \hat{u}_2) * (\hat{u}_1 - \hat{u}_2)$, we find that

$$\begin{aligned}\|\mathcal{N}(\hat{q}_1) - \mathcal{N}(\hat{q}_2)\|_{\alpha-1,\beta} &\leq \text{const}.\varepsilon_0\|\hat{q}_1 - \hat{q}_2\|_{\alpha-1,\beta} \\ &\leq \frac{1}{2}\|\hat{q}_1 - \hat{q}_2\|_{\alpha-1,\beta},\end{aligned}$$

for ε_0 small enough. This completes the proof of Proposition 17.

We can now prove Proposition 16. Let

$$\hat{U}_t(k) = ike^{-k^2(t-1)}\left(\hat{v}_0(k) + \int_1^t \hat{u}_s^{*2}(k)\, ds\right),$$

and let

$$\widetilde{\hat{U}}_t(k) = t^{1/2}U_t\left(\frac{k}{t^{1/2}}\right).$$

First we note that for any fixed $k \in \mathbf{R}$,

$$\widetilde{\hat{u}}_{\text{as}}(k) := \lim_{t\to\infty} \widetilde{\hat{U}}_t(k) = a_{\text{as}}ike^{-k^2} = \widetilde{\hat{u}}_{\text{as}},$$

with $\tilde{u}_{\text{as}}$ as given in Eq. (55), and that uniformly in $k \in \mathbf{R}$, and $t > 2$,

$$|\widetilde{\hat{U}}_t(k)| \leq \text{const}.\mu_{\alpha-1}(k, 1).$$

Therefore it follows from the Lebesgue dominated convergence theorem that

$$\lim_{t\to\infty} \|\widetilde{\hat{U}}_t - \widetilde{\hat{u}}_{\text{as}}\|_{L^1} = 0.$$

Applying the Hausdorff–Young inequality to the inverse Fourier transform we find for $\tilde{U}_t = \mathcal{F}^{-1}(\widetilde{\hat{U}_t})$ that

$$\lim_{t\to\infty} \|\tilde{U}_t - \tilde{u}_{\rm as}\|_{L^\infty} = 0.$$

Inequality (57) now follows using the triangle inequality, provided

$$\lim_{t\to\infty} \|\tilde{u}_t - \tilde{U}_t\|_{L^\infty} = 0. \tag{60}$$

For the difference of $\hat{U}_t$ and $\hat{u}_t$ we have that

$$|\hat{U}_t(k) - \hat{u}_t(k)| \leq \varepsilon|k| \int_1^t \left(e^{-k^2(t-s)} - e^{-k^2(t-1)}\right) \mu_{\alpha-1}(k,s)\frac{ds}{s^\beta},$$

and therefore we find using Eq. (45) in Lemma 12 that

$$|\hat{U}_t(k) - \hat{u}_t(k)| \leq \frac{\varepsilon|k|}{t^{\beta-1}}\mu_{\alpha-1}(k,t) \leq \frac{\varepsilon}{t^{\beta-1/2}}\mu_{\alpha-2}(k,t),$$

from which it follows that

$$\|\widetilde{\hat{u}_t} - \widetilde{\hat{U}_t}\|_{L^1} \leq \varepsilon \int_{\mathbf{R}} \frac{1}{t^{\beta-1}}\mu_{\alpha-2}(k,1)\,dk \leq \frac{\varepsilon}{t^{\beta-1}}.$$

Therefore $\lim_{t\to\infty} \|\widetilde{\hat{u}_t} - \widetilde{\hat{U}_t}\|_{L^1} = 0$, which implies Eq. (60) by the Hausdorff–Young inequality. This completes the proof of Proposition 16.

2.7.3. *The case of marginal perturbations*

According to the power counting scheme there are two critical nonlinear terms, namely $q = u^3$ and $q = uu'$. We discuss these cases here only briefly and refer to the article of Bricmont *et al.* (1994) for details.

For the case $q = u^3$ there are so-called logarithmic corrections to scaling, i.e., the solution still converges to the Gaussian limit, however not like $1/\sqrt{t}$ but somewhat more rapidly, namely like $1/\sqrt{t\log(t)}$. In order to discuss this case the class of functions spaces $\mathcal{B}_{\alpha\beta}$ has therefore to be generalized to allow for a scaling behavior different from power laws. In addition, it is in this case not possible anymore to analyze the map $\mathcal{L}$ by separating it into a part $\hat{u}_L$ and a part $\hat{u}_N$, since the dominant terms in u_N and u_L compensate each other. But, except for these points, the proof is as above.

The case of the nonlinearity $q = uu'$ is the equation that we have discussed in the preceding subsection for initial conditions satisfying $a_0 = \int_{\mathbf{R}} u_0(x)\,dx = 0$. The same equation but for the case where $a_0 \neq 0$ is known as the dissipative Burgers equation. It has a one parameter family of nontrivial solutions and also front solutions ($a_0 = \infty$ for front solutions), and it has many nontrivial applications. We do not discuss this case here but rather refer the reader again to the papers Bricmont *et al.* (1994) and Bricmont *et al.* (1996) and to Uecker (2006) for a very interesting nontrivial application.

2.7.4. *The case of relevant perturbations*

The case of relevant perturbations is discussed in Bricmont and Kupiainen (1996b). In order to analyze, for example, the equation

$$\dot{u}(x,t) = u''(x,t) - u(x,t)^2 \tag{61}$$

one first has to identify the relevant linear problem. Indeed, there are interesting scale invariant solutions of Eq. (61), but they converge to zero like $1/t$ and not like $1/\sqrt{t}$. One therefore sets $u(x,t) = f(x/\sqrt{t})/t + u_1(x,t)$, and plugs this Ansatz into Eq. (61). For $u_1 = 0$ one gets a nonlinear ordinary differential equation for f. This equation has two solutions, one that decays like a modified Gaussian at infinity and one decaying like $1/x^2$ at infinity. Both of these functions can be taken as a starting point for an asymptotic analysis, and for appropriate initial conditions it can be shown that u_1 converges to zero faster than $1/t$.

2.7.5. *The case of irrelevant perturbations, III*

For completeness we present here also the analysis of the non-linear heat equation for the case where the nonlinearity $q = u^p$ in Eq. (48) is replaced by $q = uu''$. Formally, $q \sim t^{-1/2}t^{-3/2} = t^{-2}$, and therefore $q = uu''$ is an irrelevant perturbation of the heat equation. However, because of the second derivative its analysis requires the refined inequality (47) of Lemma 12 (maximal regularity).

Proposition 18. *Let $\alpha > 3$ and let $t \mapsto \hat{u}_t$ be the solution of Eq.* (48) *with initial condition $\hat{u}_0 \in \mathcal{A}_\alpha$. Let $u(x,t) = \mathcal{F}^{-1}(\hat{u}_t)(x)$, and*

$$\tilde{u}_t(x) = \sqrt{t}\,u(\sqrt{t}x, t).$$

Define furthermore the function $u_{\rm as}$ *by*

$$\tilde{u}_{\rm as}(x) = \frac{a_{\rm as}}{\sqrt{4\pi}} e^{-x^2/4}, \tag{62}$$

where

$$a_{\rm as} = \hat{u}_0(0) + \int_1^\infty \hat{q}(0,s)\,ds, \tag{63}$$

with

$$\hat{q}(k,t) = (\hat{u}_t * \hat{v}_t)(k), \tag{64}$$

and

$$\hat{v}_t(k) = k^2 \hat{u}_t(k). \tag{65}$$

Then,

$$\lim_{t\to\infty} \|\tilde{u}_t - \tilde{u}_{\rm as}\|_{L^\infty} = 0. \tag{66}$$

Again, we construct a solution by solving the equations in Fourier space. The only difference is that we have to require that $\hat{u}_0 \in \mathcal{A}_\alpha$ with $\alpha \geq 3$, since the nonlinearity involves in direct space second derivatives. For $\alpha \geq 3$ we can then use Eq. (47) rather than Eq. (44) of Proposition 12, which allows us to recover the decay in k. Namely, let $t \geq 1$ and let $\hat{u}_0 \in \mathcal{A}_\alpha$ with $\alpha > 3$. Then, the equation to be solved is

$$\hat{u}_t(k) = e^{-k^2(t-1)}\hat{u}_0(k) + \int_1^t e^{-k^2(t-s)}\,\hat{q}(k,s)\,ds, \tag{67}$$

with $\hat{q}$ as defined in Eqs. (64) and (65). To prove the existence of a solution to Eq. (67), we proceed as in the proof of Proposition 15. For a given initial condition $\hat{u}_0$ we apply the contraction mapping principle to the map $\mathcal{N} = \mathcal{M}\mathcal{L}$, where $\mathcal{L}\colon \hat{q} \mapsto \hat{u}$, is the map that associates to $\hat{q}$ the function $\hat{u}$ using Eq. (67), and where $\mathcal{M}\colon \hat{u} \mapsto \hat{q}$, is the map that associates to $\hat{u}$ the function $\hat{q}$ using Eqs. (64) and (65).

Proposition 19. *Let* $0 < \varepsilon_0$, $\alpha > 3$ *and let* $\beta = 3/2$. *Let* $\hat{u}_0 \in \mathcal{A}_\alpha$ *with* $\|\hat{u}_0\|_\alpha = \varepsilon_0$ *and let* $\mathcal{U}_{\alpha,\beta}(\varepsilon_0) = \{\hat{q} \in \mathcal{B}_{\alpha-2,\beta} | \; \|\hat{q}\|_{\alpha,\beta} < \varepsilon_0\}$. *Then,*

(i) $\mathcal{L}$ *is well defined as a map from* $\mathcal{B}_{\alpha-2,\beta}$ *to* $\mathcal{B}_{\alpha,0}$.
(ii) $\mathcal{M}$ *is well defined as a map from* $\mathcal{B}_{\alpha,0}$ *to* $\mathcal{B}_{\alpha-2,\beta}$.

(iii) $\mathcal{N}$ *is well defined as a map from* $\mathcal{B}_{\alpha-2,\beta}$ *to* $\mathcal{B}_{\alpha-2,\beta}$.
(iv) $\mathcal{N}(\mathcal{U}_{\alpha-2,\beta}(\varepsilon_0)) \subset \mathcal{U}_{\alpha-2,\beta}(\varepsilon_0)$ *for* ε_0 *small enough.*
(v) *For* ε_0 *small enough,* $\|\mathcal{N}(\hat{q}_1) - \mathcal{N}(\hat{q}_2)\|_{\alpha-2,\beta} \leq \frac{1}{2}\|\hat{q}_1 - \hat{q}_2\|_{\alpha-2,\beta}$, *for all* $\hat{q}_1, \hat{q}_2 \in \mathcal{U}_{\alpha-2,\beta}(\varepsilon_0)$.
(vi) $\mathcal{N}$ *has a unique fixed point in* $\mathcal{U}_{\alpha-2,\beta}(\varepsilon_0)$ *for* ε_0 *small enough.*

The proof is again essentially identical to the proof of Proposition 15. (iii) again follows from (i) and (ii), and (vi) follows from (iv) and (v). We now prove (i). We have already shown that the functions $\hat{u}_L(k,t) = e^{-k^2(t-1)}\hat{u}_0(k)$ are in $\mathcal{B}_{\alpha,0}$ and that $\|\hat{u}_L\|_{\alpha,0} \leq \varepsilon$. Next let

$$\hat{u}_N(k,t) = \int_1^t e^{-k^2(t-s)}\,\hat{q}(k,s)\,ds.$$

Since $\hat{q} \in \mathcal{B}_{\alpha-2,\beta}$ we have that

$$|\hat{q}(k,t)| \leq \|\hat{q}\|_{\alpha-2,\beta} t^{-\beta}\mu_{\alpha-2}(k,t) \leq \varepsilon_0 t^{-\beta}\mu_{\alpha-2}(k,t),$$

and therefore $\hat{u}_N \in \mathcal{B}_{\alpha,0}$ by Eq. (47) of Lemma 12, and $\|u_N\|_{\alpha,0} \leq \varepsilon$. Since $\hat{u} = \hat{u}_L + \hat{u}_N$ it now follows that $\hat{u} \in \mathcal{B}_{\alpha,0}$ as claimed. Furthermore $\|\hat{u}\|_{\alpha 0} \leq \varepsilon$. We now prove (ii). Let $\hat{u} \in \mathcal{B}_{\alpha,0}$ with $\|\hat{u}\|_{\alpha,0} < \varepsilon$, and let $\hat{v}(k,t) = \hat{v}_t(k)$ be as defined in Eq. (65). Then,

$$|\hat{v}(k,t)| \leq \varepsilon|k|^2\mu_{\alpha,0}(k,t) \leq \frac{\varepsilon}{t}\mu_{\alpha-2}(k,t),$$

and therefore $\hat{v} \in \mathcal{B}_{\alpha-2,1}$ and $\|\hat{v}\| \leq \varepsilon$. Applying Lemma 11 to the convolution in Eq. (64) we get that $\mathcal{M}(\hat{u}) \in \mathcal{B}_{\alpha-2,\beta}$ for $\hat{u} \in \mathcal{B}_{\alpha,0}$. Furthermore, if $\|\hat{u}\| \leq \text{const.}\varepsilon_0$, then $\|\mathcal{M}(\hat{u})\|_{\alpha-2,\beta} \leq \text{const.}\varepsilon_0^2$. Therefore we find that, for $\hat{q} \in \mathcal{U}_{\alpha-2,\beta}(\varepsilon_0)$, $\|\mathcal{N}(\hat{q})\|_{\alpha-2,\beta} \leq \text{const.}\varepsilon_0^2$. Now since $\text{const.}\varepsilon_0^2 < \varepsilon_0$ for ε_0 small enough we find (iv). To prove (v) we consider, for $i = 1, 2$, the image $\hat{u}_i = \mathcal{L}(\hat{q}_i)$ of $\hat{q}_i \in \mathcal{U}_{\alpha-2,\beta}(\varepsilon_0)$. We have already shown that $\|\hat{u}_i\|_{\alpha,0} \leq \varepsilon$, $i = 1, 2$, and using exactly the same techniques one shows that

$$\|\hat{u}_1 - \hat{u}_2\|_{\alpha,0} \leq \text{const.}\|\hat{q}_1 - \hat{q}_2\|_{\alpha-2,\beta},$$

$$\|\hat{v}_1 - \hat{v}_2\|_{\alpha-2,1} \leq \text{const.}\|\hat{q}_1 - \hat{q}_2\|_{\alpha-2,\beta},$$

where $v_i(k,t) = k^2 u_i(k,t)$, $i = 1, 2$. Using that

$$\begin{aligned}\hat{q}_1 - \hat{q}_2 &= u_1 * v_1 - u_2 * v_2 \\ &= (u_1 - u_2) * v_1 + u_2 * (v_1 - v_2),\end{aligned}$$

we find that, for ε_0 small enough,

$$\begin{aligned}\|\mathcal{N}(\hat{q}_1) - \mathcal{N}(\hat{q}_2)\|_{\alpha-2,\beta} &\leq \text{const.}\varepsilon_0 \|\hat{q}_1 - \hat{q}_2\|_{\alpha-2,\beta} \\ &\leq \frac{1}{2}\|\hat{q}_1 - \hat{q}_2\|_{\alpha-2,\beta}.\end{aligned}$$

This completes the proof of Proposition 19.

We now prove Proposition 18. Let

$$\hat{U}_t(k) = e^{-k^2(t-1)}\left(\hat{u}_0(k) + \int_1^t \hat{q}(k,s)\, ds\right),$$

with $\hat{q}$ as in Eq. (64) above, and let $\widetilde{\hat{U}}_t(k) = U_t\left(k/t^{1/2}\right)$. For fixed $k \in \mathbf{R}$ we have that $\widetilde{\hat{u}}_{\text{as}}(k) := \lim_{t\to\infty} \widetilde{\hat{U}}_t(k) = a_{0,p} e^{-k^2} = \widehat{\tilde{u}}_{\text{as}}$, with $\tilde{u}_{\text{as}}$ given by Eq. (62), and furthermore that $|\widetilde{\hat{U}}_t(k)| \leq \text{const.}\mu_{\alpha-2}(k,1)$, uniformly in $k \in \mathbf{R}$, $t \geq 2$. Therefore it follows from the Lebesgue dominated convergence theorem that $\lim_{t\to\infty} \|\widetilde{\hat{U}}_t - \widetilde{\hat{u}}_{\text{as}}\|_{L^1} = 0$. Applying the Hausdorff–Young inequality to the inverse Fourier transform we find for $\tilde{U}_t = \mathcal{F}^{-1}(\widetilde{\hat{U}}_t)$ that $\lim_{t\to\infty} \|\tilde{U}_t - \tilde{u}_{\text{as}}\|_{L^\infty} = 0$. It remains to be shown that

$$\lim_{t\to\infty} \|\tilde{u}_t - \tilde{U}_t\|_{L^\infty} = 0. \tag{68}$$

Inequality (66) then follows using the triangle inequality. Since $\alpha > 3$ we can proceed for the difference of $\hat{U}_t$ and $\hat{u}_t$ as in the proof of Proposition 13. Namely,

$$\begin{aligned}|\hat{U}_t(k) - \hat{u}_t(k)| &\leq \varepsilon \int_1^t \left(e^{-k^2(t-s)} - e^{-k^2(t-1)}\right) \mu_{\alpha-2}(k,s) \frac{ds}{s^\beta} \\ &\leq \frac{\varepsilon}{t^{\beta-1}} \mu_{\alpha-2}(k,t),\end{aligned}$$

and therefore we get that

$$\|\widetilde{\hat{u}}_t - \widetilde{\hat{U}}_t\|_{L^1} \leq \varepsilon \int_{\mathbf{R}} \frac{1}{t^{\beta-1}} \mu_{\alpha-2}(k,1)\, dk \leq \frac{\varepsilon}{t^{\beta-1}}.$$

Therefore, $\lim_{t\to\infty} \|\widetilde{\hat{u}}_t - \widetilde{\hat{U}}_t\|_{L^1} = 0$, and Eq. (68) now follows by using the Hausdorff–Young inequality. This completes the proof of Proposition 18.

2.8. *Appendix I: Construction of a counter example*

In what follows we give the details of the construction of an initial condition which satisfies Proposition 5. The example is based on Collet and Eckmann (1992b). We set, for $x > 0$, $u_0 = \sqrt{4\pi} f'$, where f is the function that we now construct. Let for $0 \le x \le 1$

$$h(x) = c_2 e^{-1/x} e^{-1/(1-x)}, \tag{69}$$

with

$$c_2 = \left(\int_0^1 e^{-1/x} e^{-1/(1-x)} dx \right)^{-1} = 142.25\ldots, \tag{70}$$

so that $\int_0^1 h(x)\,dx = 1$. The function h is infinitely differentiable on $[0, 1]$, and satisfies $0 \le h(x) \le h(1/2) = h_{\max} = 2.6\ldots$. Now let, for $0 \le x \le 1$,

$$H(a,b)(x) = a + (b-a) \int_0^x h(y)\,dy. \tag{71}$$

By construction the function $H(a, b)$ interpolates between a and b on the interval $[0, 1]$. For the mth derivative $H^{(m)}$ of H we have for all $m \ge 1$, $H^{(m)} = (b-a)h^{(m-1)}$, and therefore we see from Eq. (69) that

$$\lim_{x \searrow 0} H^{(m)}(x) = \lim_{x \nearrow 1} H^{(m)}(x) = 0. \tag{72}$$

Now let $n \ge 4$, $L_n = n!$, $l_n = 2^n$, $I_n = (L_n + l_n, L_{n+1} - l_{n+1})$ and $J_n = [L_n - l_n, L_n + l_n]$. Note that the intervals I_n are non overlapping. For the first intervals we have $I_4 = (4! + 2^4, 5! - 2^5) = (40, 88)$, and $J_4 = [8, 40]$. Now define

$$f(x) = \begin{cases} 0 & \text{for } -\infty < x \le 8 \\ (-1)^n & \text{for } x \in I_n, n \ge 4 \\ H(0,1), (x-8)/32) & \text{for } x \in J_4 \\ (-1)^n H(-1,1), ([x - (L_n - l_n)]/(2l_n)) & \text{for } x \in J_n, n \ge 5. \end{cases} \tag{73}$$

By construction we have

$$|f'(x)| \leq \begin{cases} 0 & \text{for } -\infty \leq x \leq 8 \\ 0 & \text{for } x \in I_n,\ n \geq 4 \\ h_{\max} 2^{-n} & \text{for } x \in J_n,\ n \geq 4 \end{cases}. \tag{74}$$

Therefore, since $|J_n| = 2^{n+1}$ we have that, for $p > 1$,

$$(\|u_0\|_{L^p})^p = \int_0^\infty \left|\sqrt{4\pi} f'(x)\right|^p dx \leq \left(\sqrt{4\pi}\right)^p h_{\max} \sum_{n\geq 4} 2^{n+1} 2^{-pn} < \infty,$$

and therefore $u_0 = \sqrt{4\pi} f' \in L^p(\mathbf{R}, dx)$ for all $p > 1$, but by construction $u_0 \notin L^1(\mathbf{R}, dx)$. We now consider the value at zero of the scaled solution $\tilde{u}$ of the heat equation with initial condition $u_0 = \sqrt{4\pi} f'$. For $t > 0$ we have

$$\begin{aligned} \tilde{u}(0,t) = \sqrt{t} u(0,t) &= \int_{\mathbf{R}} e^{-\frac{y^2}{4t}} f'(y)\, dy = \int_0^\infty e^{-\frac{y^2}{4t}} f'(y)\, dy \\ &= \left[e^{-\frac{y^2}{4t}} f(y) \Big|_{y=0}^{y=\infty} + \int_0^\infty \frac{y}{2t} e^{-\frac{y^2}{4t}} f(y)\, dy \right] \\ &= \frac{1}{2} \int_0^\infty \frac{y}{t} e^{-\frac{y^2}{4t}} f(y)\, dy. \end{aligned}$$

Now let $T_n = L_n L_{n+1}$. Then, we have

$$\begin{aligned} \tilde{u}(0,T_n) &= \frac{1}{2} \int_0^{L_n + l_n} \frac{y}{T_n} e^{-\frac{y^2}{4T_n}} f(y)\, dy \\ &\quad + \frac{1}{2} \int_{I_n} \frac{y}{T_n} e^{-\frac{y^2}{4T_n}} f(y)\, dy + \frac{1}{2} \int_{L_{n+1} - l_{n+1}}^\infty \frac{y}{T_n} e^{-\frac{y^2}{4T_n}} f(y)\, dy, \end{aligned}$$

and therefore

$$\begin{aligned} \tilde{u}(0,T_n) &= \frac{1}{2} \int_0^{L_n + l_n} \frac{y}{T_n} e^{-\frac{y^2}{4T_n}} (f(y) - (-1)^n)\, dy \\ &\quad + \frac{(-1)^n}{2} \int_{\mathbf{R}} \frac{y}{T_n} e^{-\frac{y^2}{4T_n}}\, dy \\ &\quad + \frac{1}{2} \int_{L_{n+1} - l_{n+1}}^\infty \frac{y}{T_n} e^{-\frac{y^2}{4T_n}} (f(y) - (-1)^n)\, dy. \end{aligned}$$

The function xe^{-x^2} has its maximum value at $x = \sqrt{2}$. Therefore, since $L_n < T_n < L_{n+1}$, we find for the first term

$$\left| \frac{1}{2} \int_0^{L_n+l_n} \frac{y}{T_n} e^{-\frac{y^2}{4T_n}} (f(y) - (-1)^n)\, dy \right| \\ \leq \left| \int_0^{L_n+l_n} \frac{y}{T_n}\, dy \right| = \frac{(L_n + l_n)^2}{L_n L_{n+1}} \longrightarrow_{n\to\infty} 0,$$

and for the third term

$$\left| \frac{1}{2} \int_{L_{n+1}-l_{n+1}}^{\infty} \frac{y}{T_n} e^{-\frac{y^2}{4T_n}} (f(y) - (-1)^n)\, dy \right| \\ \leq \left| \int_{L_{n+1}-l_{n+1}}^{\infty} \frac{y}{2T_n} e^{-\frac{y^2}{4T_n}}\, dy \right| = e^{-\frac{(L_{n+1}-l_{n+1})^2}{4L_nL_{n+1}}} \longrightarrow_{n\to\infty} 0.$$

Therefore, we conclude that, asymptotically as $n \to \infty$,

$$\tilde{u}(0, T_n) \sim \frac{(-1)^n}{2} \int_0^{\infty} \frac{y}{T_n} e^{-\frac{y^2}{4T_n}}\, dy = (-1)^n.$$

This completes the proof of Proposition 5.

2.9. *Appendix II: Proof of the technical lemmas*

In this section we prove the Lemmas of Sec. 2.6. In what follows constants indicated by "const." may depend on α, β, γ, etc., but are independent of t and k. In addition these constants may be different from instant to instant.

2.9.1. *Proof of Lemma 10*

Let $\alpha, \beta, \gamma \geq 0$ with $\alpha + 2\beta \geq \gamma$, and let for $k \in \mathbf{R}$ and $t \geq 1$,

$$f(k, t) = e^{-k^2(t-1)} \left(\frac{t-1}{t} \right)^{\beta} |k|^{\gamma} \mu_{\alpha}(k, 1),$$

and

$$g(k, t) = \frac{1}{t^{\min\{\beta, \gamma/2\}}} \mu_{\alpha+2\beta-\gamma}(k, t).$$

We have to prove that, uniformly in $k \in \mathbf{R}$ and $t \geq 1$, $f(k, t) \leq \text{const.} g(k, t)$. Note that $f(k, t) \geq 0$. For $1 \leq t \leq 2$ and $|k| \leq 1$ we have that

$$f(k, t) \leq \text{const.} \leq \text{const.} g(k, t),$$

and for $1 \leq t \leq 2$ and $|k| > 1$ we have that

$$f(k,t) \leq \text{const.}e^{-k^2(t-1)}((t-1)k^2)^\beta |k|^{\gamma-2\beta}\mu_\alpha(k,1)$$
$$\leq \text{const.}\mu_{\alpha+2\beta-\gamma}(k,1) \leq \text{const.}g(k,t).$$

For $t > 2$ we have, for $\beta \leq \gamma/2$ that

$$f(k,t) \leq \text{const.}\frac{1}{t^\beta}e^{-k^2(t-1)}((t-1)k^2)^\beta |k|^{\gamma-2\beta}\mu_\alpha(k,1)$$
$$\leq \text{const.}\frac{1}{t^\beta}e^{-k^2(t-1)}((t-1)k^2)^\beta \leq \text{const.}g(k,t),$$

and for $\beta > \gamma/2$

$$f(k,t) \leq \text{const.}\frac{1}{t^{\gamma/2}}e^{-k^2(t-1)}((t-1)k^2)^{\gamma/2}\mu_\alpha(k,1)$$
$$\leq \text{const.}\frac{1}{t^{\gamma/2}}e^{-k^2(t-1)}((t-1)k^2)^{\gamma/2}$$
$$\leq \text{const.}\frac{1}{t^{\gamma/2}}\mu_{\alpha+2\beta-\gamma}(k,t) \leq \text{const.}g(k,t).$$

This completes the proof of Lemma 10.

2.9.2. *Proof of Lemma 11*

Let $\alpha, \alpha' > 1$ and let $D(k) = \{k' \in \mathbf{R} \,|\, |k'-k| < |k|/2\}$. For $k' \in D(k)$ we have that

$$|k'| \geq |k| - |k-k'| \geq \frac{1}{2}|k|.$$

With this notation we get for the integral in Eq. (43)

$$\int_{\mathbf{R}} \mu_\alpha(k-k',t)\mu_{\alpha'}(k',t)dk'$$
$$= \int_{\mathbf{R}\setminus D(k)} \mu_\alpha(k-k',t)\mu_{\alpha'}(k',t)dk' + \int_{D(k)} \mu_\alpha(k-k',t)\mu_{\alpha'}(k',t)dk'$$
$$\leq \sup_{k' \in \mathbf{R}\setminus D(k)} \mu_\alpha(k-k',t)\int_{\mathbf{R}\setminus D(k)} \mu_{\alpha'}(k',t)dk'$$
$$+ \sup_{k' \in D(k)} \mu_{\alpha'}(k',t)\int_{D(k)} \mu_\alpha(k-k',t)dk'$$

$$\leq \text{const.}\mu_\alpha(k/2,t)\frac{1}{t^{1/2}} + \text{const.}\mu_{\alpha'}(k/2,t)\frac{1}{t^{1/2}}$$

$$\leq \text{const.}\frac{1}{t^{1/2}}\mu_{\min\{\alpha,\alpha'\}}(k,t)$$

This completes the proof of Lemma 11.

2.9.3. *Proof of Proposition 12*

Let $\beta > 1$ and $\alpha \geq 0$. In order to prove Eq. (44) we cut the integral over $[1,t]$ into an integral over $\left[1, \frac{t+1}{2}\right]$ and an integral over $\left[\frac{t+1}{2}, t\right]$. For the first integral we have

$$\left|\int_1^{\frac{t+1}{2}} e^{-k^2(t-s)}\frac{1}{s^\beta}\mu_\alpha(k,s)\,ds\right| \leq \text{const.}e^{-k^2\frac{t-1}{2}}\mu_\alpha(k,1)\int_1^{\frac{t+1}{2}}\frac{ds}{s^\beta}$$
$$\leq \text{const.}e^{-k^2\frac{t-1}{2}}\mu_\alpha(k,1) \leq \text{const.}\mu_\alpha(k,t), \tag{75}$$

where we have used Proposition 10 in the last inequality. For the second integral we have

$$\left|\int_{\frac{t+1}{2}}^t e^{-k^2(t-s)}\frac{1}{s^\beta}\mu_\alpha(k,s)\,ds\right| \leq \text{const.}\mu_\alpha\left(k,\frac{t+1}{2}\right)\int_{\frac{t+1}{2}}^t\frac{ds}{s^\beta}$$
$$\leq \frac{\text{const.}}{t^{\beta-1}}\mu_\alpha(k,t) \leq \text{const.}\mu_\alpha(k,t), \tag{76}$$

and Eq. (44) now follows using the triangle inequality. To prove Eq. (45) we note that

$$e^{-k^2(t-s)} - e^{-k^2(t-1)} = e^{-k^2(t-s)}(1 - e^{-k^2(s-1)}) \leq e^{-k^2(t-s)}.$$

Therefore we find using the bound in Eq. (76) for the second integral

$$\left|\int_{\frac{t+1}{2}}^t \left(e^{-k^2(t-s)} - e^{-k^2(t-1)}\right)\mu_\alpha(k,s)\frac{ds}{s^\beta}\right| \leq \left|\int_{\frac{t+1}{2}}^t e^{-k^2(t-s)}\frac{1}{s^\beta}\mu_\alpha(k,s)\,ds\right|$$
$$\leq \frac{\text{const.}}{t^{\beta-1}}\mu_\alpha(k,t), \tag{77}$$

and furthermore, since

$$\frac{1-e^{-k^2(s-1)}}{k^2(s-1)} \le \text{const.},$$

we find for the first integral that

$$\begin{aligned}
&\varepsilon \int_1^{\frac{t+1}{2}} (e^{-k^2(t-s)} - e^{-k^2(t-1)})\mu_\alpha(k,s)\frac{ds}{s^\beta} \\
&\quad \le \varepsilon\, e^{-k^2\frac{t-1}{2}}\mu_\alpha(k,1)\int_1^{\frac{t+1}{2}} k^2(s-1)\frac{ds}{s^\beta} \\
&\quad \le \frac{\varepsilon}{t^{\beta-2}} e^{-k^2\frac{t-1}{2}} k^2 \left(\frac{t-1}{t}\right)^2 \mu_\alpha(k,1) \\
&\quad \le \frac{\varepsilon}{t^{\beta-1}}\mu_\alpha(k,t). \qquad (78)
\end{aligned}$$

The proof of Eqs. (46) and (47) is similar but somewhat more involved. We treat the two cases in parallel. In what follows $\delta \in \{1, 2\}$ and $\alpha \ge 1$ if $\delta = 1$ and $\alpha \ge 2$ if $\delta = 2$. The only important point is that $\alpha - \delta \ge 0$. We again cut the integral into two pieces. We have

$$\int_1^{\frac{t+1}{2}} e^{-k^2(t-s)}\frac{1}{s^\beta}\mu_{\alpha-\delta}(k,s)\,ds \le \text{const.}\, f_1(k,t), \qquad (79)$$

where

$$f_1(k,t) = e^{-k^2\frac{t-1}{2}}\left(\frac{t-1}{t}\right)\mu_{\alpha-\delta}(k,1),$$

and

$$\int_{\frac{t+1}{2}}^{t} e^{-k^2(t-s)}\frac{1}{s^\beta}\mu_{\alpha-\delta}(k,s)\,ds \le \text{const.}\, f_2(k,t), \qquad (80)$$

where

$$f_2(k,t) = \frac{1}{t^\beta}\mu_{\alpha-\delta}(k,t)\frac{1}{k^2}\left(1 - e^{-k^2\frac{t-1}{2}}\right).$$

We now estimate the functions f_1 and f_2 for various regimes of t and k. We again first treat the case of small times t and then the case of large times t.

For $1 \leq t \leq 2$ we have for $|k| \leq 1$ and for $i = 1, 2$, that

$$f_i(k,t) \leq \text{const.} \leq \text{const.}\mu_\alpha(k,t).$$

Again for $1 \leq t \leq 2$, but for $|k| > 1$, we have that

$$\begin{aligned} f_1(k,t) &\leq \text{const.}e^{-k^2\frac{t-1}{2}}(t-1)k^2\frac{1}{k^2}\mu_{\alpha-2}(k,1) \\ &\leq \text{const.}\frac{1}{k^2}\mu_{\alpha-\delta}(k,1) \leq \text{const.}\mu_\alpha(k,1) \\ &\leq \text{const.}\mu_\alpha(k,t), \end{aligned}$$

and that

$$\begin{aligned} f_2(k,t) &\leq \text{const.}\frac{1}{t^\beta}\mu_{\alpha-\delta}(k,t)\frac{1}{k^2} \\ &\leq \text{const.}\frac{1}{k^2}\mu_{\alpha-\delta}(k,1) \leq \text{const.}\mu_\alpha(k,1) \\ &\leq \text{const.}\mu_\alpha(k,t). \end{aligned}$$

For $t > 2$ we have that

$$f_1(k,t) \leq \text{const.}e^{-k^2\frac{t-1}{2}}\mu_{\alpha-\delta}(k,1) \leq \text{const.}\mu_\alpha(k,t),$$

and that

$$\begin{aligned} f_2(k,t) &\leq \text{const.}\frac{t-1}{t^\beta}\mu_{\alpha-\delta}(k,t)\frac{1-e^{-k^2\frac{t-1}{2}}}{k^2(t-1)} \\ &\leq \text{const.}\frac{1}{t^{\beta-1}}\mu_{\alpha-\delta}(k,t)\frac{1}{1+k^2t} \leq \text{const.}\mu_\alpha(k,t). \end{aligned}$$

This completes the proof of Lemma 12.

2.10. *Bibliographic notes*

The following bibliographic notes provide a chronological access to the literature on the long time asymptotics for partial differential equations in unbounded domains containing proofs that are based on the renormalization group concept outlined in the previous section. The origin of the techniques can be traced back to the article by Collet and Eckmann (1992b) and the work on the long time asymptotics of front solutions of the Ginzburg–Landau equation by Collet and Eckmann (1992a), Collet *et al.* (1992),

Bricmont and Kupiainen (1992) and Eckmann and Gallay (1993). See also Pao (1993). Parallel to further work by Eckmann and Wayne (1994), Gallay (1994) and Bricmont and Kupiainen (1994a), the connection of the ideas underlying the analysis in these papers with the renormalization group as used in statistical mechanics, field theory and dynamical systems theory, were put forward in Bricmont *et al.* (1994), Bricmont and Kupiainen (1994b) and Chen *et al.* (1994) for the case of parabolic equations and then by Bona *et al.* (1994) for nonlinear dissipative wave equations. In Bona *et al.* (1995) higher order asymptotics of solutions are studied. The article by Bricmont *et al.* (1996) introduced in particular the community of mathematical physicists to this range of problems and has triggered an intense activity in this field. Important results for the Swift–Hohenberg equation were proved around the same time by Schneider (1996), and a nontrivial application of the analysis of the critical case was studied in Bricmont and Kupiainen (1996a). The case of relevant perturbations was also studied around this time in Bricmont and Kupiainen (1996b). A first formulation of a connection with invariant manifold theory can be found in Eckmann and Wayne (1997). Soon after that Schneider (1998a) used the ideas to set up a proof of the stability of Taylor vortices in an infinite cylinder, which showed the power and robustness of the method to deal with "real world" problems. For a generalization to higher dimensions see Schneider (1998b). The method has been generalized to higher order linear operators in Eckmann and Wayne (1998a), Eckmann and Wayne (1998b) and adapted to more complicated situations in Gallay and Mielke (1998), who also considered asymptotic expansions for damped wave equations. See Gallay and Raugel (1998). A review was then written by Bricmont and Kupiainen (1998). The next wave of work was on hyperbolic fronts (Gallay and Raugel (2000)), multicomponent systems involving several length scales (van Baalen *et al.* (2000)), bifurcation problems (Eckmann and Schneider (2000)), and a review was written by Mielke *et al.* (2001). The successful introduction of the method for studying the long time asymptotics of solutions of the Navier–Stokes equations based on the analysis of the vorticity goes back to Gallay and Wayne (2002a) and Gallay and Wayne (2002b). Around the same time the non-linear stability analysis of modulated fronts for the Swift–Hohenberg equations was proved in Eckmann and Schneider (2002). The renormalization group method was also successfully applied to analyze the modulational stability of quasi-steady patterns in dispersive systems in Promislow (2002). Almost global existence and transient self similar decay for Poiseuille flow at criticality for exponentially long times was proved by Schneider and Uecker (2003), and an interface model was analyzed in Gallay and Mielke (2003). Non-vanishing profiles for the Kuramoto–Sivashinsky equation were considered by van Baalen and Eckmann (2004). A review of Fourier methods and a discussion of various types of function spaces goes back to about the same time, see Guidotti (2004), as well as another review of the method by Merdan and Caginalp (2004). The question of bifurcations was reconsidered in Gallay *et al.* (2004). Recent work includes the discussion of the anomalous scaling for three-dimensional Cahn-Hilliard fronts by Korvola *et al.* (2005), of the pulse dynamics in thermally loaded optical parametric oscillators (Moore and Promislow (2005)), the discussion of the down-stream asymptotics of stationary Navier–Stokes flows, and the global stability of vortex solutions of the two-dimensional Navier–Stokes equations by Gallay and Wayne (2005).

3. Down-stream Asymptotics of Stationary Navier–Stokes Flows

In what follows we explain how the ideas of the preceding section can be used to analyze stationary solutions of the Navier–Stokes equations in the down-stream region of an exterior domain. The reason why this region is of particular interest is that, due to the slow decay of the vorticity in the direction of the flow and its fast decay in other directions, the dominant large distance asymptotics of the exterior flow can be entirely reconstructed from the knowledge of the asymptotic behavior of the vorticity in the down-stream region. This being said it is nevertheless interesting to study directions different from downstream with the methods introduced in Sec. 2. The present techniques can in particular also be used to analyze the upstream region in more detail, or — as we will discuss in the open problem section below — to analyze the asymptotic behavior of flows parallel to a wall. The original publications on which the following discussions are based are for the two dimensional case Haldi and Wittwer (2005) and for the three-dimensional case Wittwer (2006). Earlier references are van Baalen (2002), Wittwer (2002) and Hillairet and Wittwer (2003).

Consider, in $d = 2$ or $d = 3$ dimensions, the time independent incompressible Navier–Stokes equations

$$-(\mathbf{U}\cdot\nabla)\mathbf{U} + \Delta\mathbf{U} - \nabla p = \mathbf{0}, \tag{81}$$

$$\nabla\cdot\mathbf{U} = 0, \tag{82}$$

in a half-space $\Omega_+ = \{(x, \mathbf{y}) \in \mathbf{R}\times\mathbf{R}^{d-1} | x \geq 1\}$. We are interested in modeling the situation where fluid enters the half-space Ω_+ through the surface $\Sigma = \{(x, \mathbf{y}) \in \mathbf{R}\times\mathbf{R}^{d-1} | x = 1\}$ and where the fluid flows at infinity parallel to the x-axis at a nonzero constant speed $\mathbf{u}_\infty \equiv (1, 0)$. We therefore impose at infinity the boundary condition

$$\lim_{\substack{x^2+|\mathbf{y}|^2\to\infty \\ x\geq 1}} \mathbf{U}(x, \mathbf{y}) = \mathbf{u}_\infty. \tag{83}$$

On Σ we impose the boundary condition

$$\mathbf{U}|_\Sigma = \mathbf{U}_0, \tag{84}$$

with $\mathbf{U}_0$ in a class $\mathcal{U}$ of vector fields for which

$$\lim_{|\mathbf{y}|\to\infty} \mathbf{U}_0(\mathbf{y}) = \mathbf{u}_\infty. \tag{85}$$

We are then interested in proving the existence of solutions of Eqs. (81) and (82) for this setting and in particular we are interested in studying the behavior of the solutions when $x \to \infty$. Naturally, one would like to settle this question for a large class of vector fields $\mathcal{U}$. Here, we only present the results for the case of vector fields that are perturbations of the constant vector field $\mathbf{u}_\infty$, i.e.,

$$\mathbf{U}_0 = \mathbf{u}_\infty + \mathbf{u}_0, \tag{86}$$

where $\mathbf{u}_0 = (u_0, \mathbf{v}_0)$ is small in an appropriate sense. By appropriate we mean such as to contain the cases of interest, i.e., the vector fields of exterior flows at low Reynolds numbers evaluated on vertical cross sections in the down-stream region. In particular, as we will see, we cannot require the functions u_0 to be integrable, and this is one of the reasons why we have discussed such cases in Sec. 2. We now also set

$$\mathbf{U} = \mathbf{u}_\infty + \mathbf{u}, \tag{87}$$

with $\mathbf{u} = (u, \mathbf{v})$. After substitution of Eq. (87) into Eqs. (81) and (82) we get for $\mathbf{u}$ the equations

$$-(\mathbf{u} \cdot \nabla)\mathbf{u} - \partial_x \mathbf{u} + \Delta \mathbf{u} - \nabla p = \mathbf{0}, \tag{88}$$

$$\nabla \cdot \mathbf{u} = 0, \tag{89}$$

and the boundary conditions (84) and (85) become

$$\mathbf{u}(1, \mathbf{y}) = \mathbf{u}_0(\mathbf{y}), \tag{90}$$

with $\mathbf{u}_0$ satisfying

$$\lim_{|\mathbf{y}| \to \infty} \mathbf{u}_0(\mathbf{y}) = \mathbf{0}. \tag{91}$$

We now explain the connection of the problems (88)–(91) with the theory of large time asymptotics presented in Sec. 2. Basically, the idea is to analyze the equations by rewriting them as evolution equations, where the coordinate x plays the role of time. In this interpretation the vector field $\mathbf{u}_0$ is the initial condition at the "time" $x = 1$, and one is interested in studying the large time behavior (asymptotic behavior as $x \to \infty$) of the solution of this Cauchy problem.

More precisely one applies these ideas to the vorticity formulation of Eq. (88), i.e., one considers the equation which one gets by taking the curl of Eq. (88), namely the equation

$$\nabla \times (\mathbf{u} \times \omega) + \Delta \omega = \mathbf{0}, \tag{92}$$

with the vorticity

$$\omega = \nabla \times \mathbf{u}. \tag{93}$$

The pressure in Eq. (88) can be determined in a second step by solving the equation which one gets by taking the divergence of Eq. (88), i.e., the equation

$$\Delta p = \nabla \cdot ((\mathbf{u} \cdot \nabla)\mathbf{u} - \Delta \mathbf{u}), \tag{94}$$

which one solves in Ω_+, with the boundary condition on Σ,

$$\partial_x p = -(\mathbf{u} \cdot \nabla)u + \Delta u. \tag{95}$$

See Haldi and Wittwer (2005) and Wittwer (2006) for details. As in the preceding section the main tool for the analysis of the dynamical system that one obtains this way is the Fourier transform, i.e., one studies the Eqs. (92), (93) and (89) after taking the Fourier transform with respect to the variable $\mathbf{y}$. This leads as in Sec. 2 to integral equations with nonlinearities involving convolutions, and the "initial condition" $\mathbf{u}_0$ is given as an inverse Fourier transform, i.e., $\mathbf{u}_0 = \mathcal{F}^{-1}(\hat{\mathbf{u}}_0)$, where

$$\mathcal{F}^{-1}(\hat{\mathbf{u}}_0)(\mathbf{y}) = \frac{1}{(2\pi)^{d-1}} \int_{\mathbf{R}^{d-1}} e^{-i\mathbf{k}\cdot\mathbf{y}} \hat{\mathbf{u}}_0(\mathbf{k})\, d^{d-1}\mathbf{k},$$

and where $\hat{\mathbf{u}}_0$ is chosen in a certain class of vector fields such that Eq. (91) is satisfied. Note that at this point it is important that the techniques of Sec. 2 allow classes of initial conditions which need not be in $L^1(\mathbf{R}^{d-1}, d^{d-1}\mathbf{y})$. Indeed, as has already been mentioned above, even though it might seem *a priori* natural to look at initial conditions that are integrable — the quantity

$$\int_{\mathbf{R}^{d-1}} u_0(\mathbf{y})\, d^{d-1}\mathbf{y} = \int_{\mathbf{R}^{d-1}} (\mathbf{U}_0(\mathbf{y}) - \mathbf{u}_\infty) \cdot \mathbf{e}_1\, d^{d-1}\mathbf{y}$$

with $\mathbf{e}_1$ the unit vector in the x-direction measures after all the difference of flux through the surface Σ when compared to the free flow — this choice turns out to be too restrictive and does not include the case of vector fields in the downstream region of exterior flows. In particular, the vector fields that describe the dominant asymptotic structure at large values of x turn out nonintegrable in $\mathbf{y}$.

In contrast to the example cases studied in Sec. 2 the evolution equations associated with Eqs. (92), (93) and (89) consist of four equations in $d = 2$ and seven equations in $d = 3$. On a linear level these equations can be decoupled by diagonalizing the system. On the nonlinear level the equations remain coupled. A further distinction from the case that has been discussed in Sec. 2 is that the linear system involves in Fourier space two types of behaviors at $k = 0$. There are the equations related to the vorticity that lead to a scaling of k proportional to $x^{-1/2}$ as discussed in Sec. 2 (spectral branch proportional to k^2 at small values of k), but there are also the equations related to the harmonic part in the velocity field that lead to a scaling of k proportional to x^{-1} (spectral branch proportional to $|k|$ at small values of k). For these equations the appropriate scaling is x^{-1}, and therefore the power counting is different, but otherwise the theory can be developed exactly as has been discussed in Sec. 2. As a result of the presence of two different scaling behaviors, the velocity components of the solution of the stationary Navier–Stokes equations show a somewhat more complicated asymptotic structure than what we have seen in Sec. 2. This does however not affect the overall structure of the proofs in any significant way. A final distinction to the case discussed in Sec. 2 is the fact that, because the original system is elliptic, the evolution equation associated with Eqs. (92), (93) and (89) also involves unstable branches. This again does not change the structure of the proofs in any significant way since this problem can be easily solved when passing to the integral equations. Such unstable directions are simply integrated backwards in time starting with zero initial conditions at infinity. In conclusion, the interested reader will find that Sec. 2 provides all the necessary tools for an easy reading of the papers of Haldi and Wittwer (2005) and Wittwer (2006).

We recall again that one of the main motivations for studying the Navier–Stokes equations in the way presented here is the result that the solutions of the stationary Navier–Stokes equations admit invariant quantities with respect to the "time" x. This is related to the fact that the vector fields are divergence free, but the result is nontrivial, exactly because the invariant quantities are associated with functions that are not integrable. The invariant quantities can still be calculated by an application of appropriate averaging procedures. Part of the result is therefore a prescription that allows the computation of the invariant quantities.

3.1. *Leading order term in two dimensions*

The main result in two dimensions is the following theorem (see Haldi and Wittwer (2005)).

Theorem 20. *Let Σ and Ω_+ be as defined above. Then, there exists a class of vector fields $\mathcal{C}$ (containing in particular the physically interesting case of stationary flows at low Reynolds numbers) such that for all initial conditions $\mathbf{u}_0 = \mathcal{F}^{-1}(\hat{\mathbf{u}}_0)$ with $\hat{\mathbf{u}}_0 = (\hat{u}_0, \hat{v}_0) \in \mathcal{C}$, there exist* *a vector field* $\mathbf{u} = (u, v)$ *and a function* p *satisfying the Navier–Stokes Eqs.* (88) *and* (89) *in* Ω_+ *subject to the boundary conditions* (90) *and* (91). *Furthermore,*

$$\lim_{x\to\infty} x^{1/2}\left(\sup_{y\in\mathbf{R}} |(u - u_{as})(x, y)|\right) = 0, \tag{96}$$

$$\lim_{x\to\infty} x\left(\sup_{y\in\mathbf{R}} |(v - v_{as})(x, y)|\right) = 0, \tag{97}$$

where

$$u_{as}(x, y) = \frac{c}{2\sqrt{\pi}}\frac{1}{\sqrt{x}}e^{-\frac{y^2}{4x}} + \frac{d}{\pi}\frac{x}{x^2 + y^2} + \frac{b}{\pi}\frac{y}{x^2 + y^2}, \tag{98}$$

$$v_{as}(x, y) = \frac{c}{4\sqrt{\pi}}\frac{y}{x^{3/2}}e^{-\frac{y^2}{4x}} + \frac{d}{\pi}\frac{y}{x^2 + y^2} - \frac{b}{\pi}\frac{x}{x^2 + y^2}, \tag{99}$$

and where the (*real*) *amplitudes* b, c *and* d *are invariant quantities which are given in terms of the initial condition* $\hat{\mathbf{u}}_0$ *by,*

$$b = \lim_{k\to+0} \frac{\hat{u}_0(k, t) - \hat{u}_0(-k, t)}{2i}, \tag{100}$$

$$d = \lim_{k\to+0} \frac{\hat{v}_0(k, t) - \hat{v}_0(-k, t)}{2i}, \tag{101}$$

$$c = \lim_{k\to+0} \frac{\hat{u}_0(k, t) + \hat{u}_0(-k, t)}{2} - d. \tag{102}$$

Remark 21. In the next chapter we will see that the constants b, c and d have a physical interpretation and are typically different from zero. Therefore, it follows from Eqs. (100)–(102) that the functions $\hat{u}_0$ and $\hat{v}_0$ are discontinuous at $k = 0$. However, since b, c and d are invariant quantities, $\hat{u}_0$ (and similarly $\hat{v}_0$) can be parametrized in terms of continuous functions $u_{0,E}$ and $u_{0,O}$, i.e., $u_0(k) = u_{0,E}(k) + \text{sign}(k)u_{0,O}(k)$, and this decomposition is preserved by the time evolution, so that the function spaces of the type introduced in Sec. 2 are sufficiently general.

Remark 22. As explained above, the asymptotic behavior in Theorem 20 involves the two scalings $y \sim \sqrt{x}$ and $y \sim x$. To make this explicit we can for example rewrite u_{as} as

$$u_{as}(x, y) = \frac{1}{\sqrt{x}}\tilde{u}_{as,1}(y/\sqrt{x}) + \frac{1}{x}\tilde{u}_{as,2}(y/x),$$

with

$$\tilde{u}_{as,1}(y) = \frac{c}{2\sqrt{\pi}}e^{-\frac{y^2}{4}},$$

and

$$\tilde{u}_{as,2}(y) = \frac{d}{\pi}\frac{1}{1+y^2} + \frac{b}{\pi}\frac{y}{1+y^2}.$$

Remark 23. The terms proportional to b, c and d in Eqs. (98) and (99) are divergence free vector fields. This will be essential below when we use these results for the definition of artificial boundary conditions.

Remark 24. The functions proportional to b in Eq. (98) and proportional to d in Eq. (99) are not integrable. These and similar terms are the reason why the existence of invariant quantities is not *a priori* obvious. As has been explained in Sec. 2 working in Fourier space allows for an easy mathematical treatment of such cases. It turns out that the invariant quantities b, c and d can still be obtained by certain limiting procedures from the functions in direct space. Namely, using classical results for the Fourier transforms of functions in the spaces $\mathcal{A}_\alpha$, $\alpha > 1$ (see, for example, Titchmarsh (1937)), one can, for example, show that for all $x \geq 1$,

$$c + b = \lim_{R\to\infty} \int_{-R}^{R} \left(1 - \frac{|y|}{R}\right) u(x, y)\, dy, \tag{103}$$

$$b = -\lim_{R\to\infty} \int_{-R}^{R} \left(1 - \frac{|y|}{R}\right) v(x, y)\, dy. \tag{104}$$

Remark 25. The invariance of the constants c and d can be used together with Eqs. (103) and (104) to show that for all fixed $x \geq 1$ the functions $y \mapsto u(x, y) - u_{as}(x, y)$ and $y \mapsto v(x, y) - v_{as}(x, y)$ are integrable, and that

$$\int_{\mathbf{R}} (u(x, y) - u_{as}(x, y))dy = 0,$$

$$\int_{\mathbf{R}} (v(x, y) - v_{as}(x, y))dy = 0.$$

The estimates in Haldi and Wittwer (2005) also imply the following result for the vorticity.

Theorem 26. *Let the vector field* $\mathbf{u} = (u, v)$ *and the constant* c *be as in Theorem* 20. *Let* ω *be the vorticity of the fluid, i.e.,* $\omega(x, y) = -\partial_y u(x, y) + \partial_x v(x, y)$. *Let* $\tilde{\omega}_x(y) = x\omega(x, y\sqrt{x})$, *and let*

$$\tilde{\omega}_{\mathrm{as}}(y) = -\frac{c}{4\sqrt{\pi}} y e^{-\frac{y^2}{4}}.$$

Then

$$\lim_{x\to\infty} \|\tilde{\omega}_x - \tilde{\omega}_{\mathrm{as}}\|_{L^\infty} = 0. \tag{105}$$

Remark 27. Theorem 26 is the exact analog of Proposition 16.

Remark 28. Since, as we will see, $c \neq 0$ for physically interesting cases, the result in Theorem 26 implies in particular that the vorticity decays slowly along the flow, namely only algebraically like $1/x$.

3.2. *Leading order term in three dimensions*

The main result in three dimensions is the following theorem (see Wittwer (2006) for details):

Theorem 29. *Let* Σ *and* Ω_+ *be as defined above. Then, there exists a class of vector fields* $\mathcal{C}$ *(containing in particular the physically interesting case of stationary flows at low Reynolds numbers), such that for all initial conditions* $\mathbf{u}_0 = \mathcal{F}^{-1}(\hat{\mathbf{u}}_0)$ *with* $\hat{\mathbf{u}}_0 = (\hat{u}_0, \hat{\mathbf{v}}_0) \in \mathcal{C}$, *there exist a vector field* $\mathbf{u} = (u, \mathbf{v})$ *and a function* p *satisfying the Navier–Stokes Eqs.* (88) *and* (89) *in* Ω_+ *subject to the boundary conditions* (90) *and* (91). *Furthermore,*

$$\lim_{x\to\infty} x \left(\sup_{\mathbf{y}\in\mathbf{R}^2} |(u - u_{\mathrm{as}})(x, y)| \right) = 0, \tag{106}$$

$$\lim_{x\to\infty} x^{3/2} \left(\sup_{\mathbf{y}\in\mathbf{R}^2} |(\mathbf{v}_1 - \mathbf{v}_{1,\mathrm{as}})(x, y)| \right) = 0, \tag{107}$$

$$\lim_{x\to\infty} x \left(\sup_{\mathbf{y}\in\mathbf{R}^2} |(\mathbf{v}_2 - \mathbf{v}_{2,\mathrm{as}})(x, y)| \right) = 0, \tag{108}$$

where $\mathbf{v} = \mathbf{v}_1 + v_2$, *with* $\mathbf{v}_1$ *and* $\mathbf{v}_2$ *the irrotational and divergence free parts of* $\mathbf{v}$, *respectively, and where*

$$u_{\mathrm{as}}(x, \mathbf{y}) = \frac{1}{4\pi x} e^{-\frac{y^2}{4x}} c + \frac{1}{2\pi} \frac{x}{r^3} d + \frac{1}{2\pi} \frac{\mathbf{y} \cdot \mathbf{b}}{r^3}, \tag{109}$$

$$\mathbf{v}_{1,\mathrm{as}}(x, \mathbf{y}) = \frac{\mathbf{y}}{8\pi x^2} e^{-\frac{y^2}{4x}} c + \frac{1}{2\pi} \frac{\mathbf{y}}{r^3} d - \frac{1}{2\pi} \frac{1}{r} \frac{1}{r+x} \left(\mathbf{1} - \frac{1}{r} \left(\frac{1}{r} + \frac{1}{r+x} \right) \mathbf{y}\mathbf{y}^T \right) \mathbf{b}, \tag{110}$$

$$\mathbf{v}_{2,\mathrm{as}}(x, \mathbf{y}) = \frac{1}{4\pi x} e^{-\frac{y^2}{4x}} \mathbf{a} + \frac{1}{2\pi} \left(\frac{1}{y^2} \left(e^{-\frac{y^2}{4x}} - 1 \right) \mathbf{1} - 2 \frac{1}{y^4} \left(e^{-\frac{y^2}{4x}} - 1 + \frac{y^2}{4x} e^{-\frac{y^2}{4x}} \right) \mathbf{y}\mathbf{y}^T \right) \mathbf{a}, \tag{111}$$

with $y = \sqrt{y_1^2 + y_2^2}$, *and* $(y_1, y_2) = \mathbf{y}$, *with* $r = \sqrt{x^2 + y^2}$, *with* $\mathbf{1}$ *the unit* 2×2 *matrix, with* $\mathbf{y}\mathbf{y}^T$ *the* 2×2 *matrix with entries* $(\mathbf{y}\mathbf{y}^T)_{ij} = y_i y_j$, *and where the numbers c and d and the vectors* $\mathbf{a}$ *and* $\mathbf{b}$ *are invariant quantities which are given in terms of the initial condition* $\hat{\mathbf{u}}_0 = (\hat{u}_0, \hat{\mathbf{v}}_0)$ *by*

$$d = \left\langle -i\mathbf{e}^T \lim_{k \to 0} \hat{\mathbf{v}}_{0,1}(k\mathbf{e}) \right\rangle, \tag{112}$$

$$c = -d + \left\langle \lim_{k \to 0} \hat{u}_0(k\mathbf{e}) \right\rangle, \tag{113}$$

$$\mathbf{b} = \left\langle -i\mathbf{e} \lim_{k \to 0} \hat{u}_0(k\mathbf{e}) \right\rangle, \tag{114}$$

$$\mathbf{a} = -\mathbf{b} + \left\langle 2 \lim_{k \to 0} \hat{\mathbf{v}}_{0,2}(k\mathbf{e}) \right\rangle - \left\langle 2 \lim_{k \to 0} \hat{\mathbf{v}}_{0,1}(k\mathbf{e}) \right\rangle, \tag{115}$$

where $k = |\mathbf{k}|$, *where* $\hat{\mathbf{v}}_{0,1}$ *and* $\hat{\mathbf{v}}_{0,2}$, *are the irrotational and divergence free parts of* $\hat{\mathbf{v}}_0$, *respectively, i.e.,*

$$\hat{\mathbf{v}}_{0,1}(\mathbf{k}) = \frac{\mathbf{k}\mathbf{k}^T}{k^2} \hat{\mathbf{v}}_0,$$

$$\hat{\mathbf{v}}_{0,2}(\mathbf{k}) = \left(1 - \frac{\mathbf{k}\mathbf{k}^T}{k^2} \right) \hat{\mathbf{v}}_0,$$

where $\mathbf{e} \equiv \mathbf{e}(\vartheta) = (\cos(\vartheta), \sin(\vartheta))$ *and where the average* $\langle\cdot\rangle$ *is defined by*

$$\langle\cdot\rangle = \frac{1}{2\pi}\int_0^{2\pi} \cdot d\vartheta. \tag{116}$$

The estimates in Wittwer (2006) also imply the following result for the vorticity.

Theorem 30. *Let the vector field* $\mathbf{u} = (u, v)$, *the constant* c, *and the vector* $\mathbf{a} = (a_1, a_2)$ *be as in Theorem 29. Let* ω *be the vorticity of the fluid, i.e.,* $\omega = \nabla \times \mathbf{u}$. *Let* $\tilde{\omega}_x(\mathbf{y}) = x^{3/2}\omega(x, \mathbf{y}\sqrt{x})$, *let* $\mathbf{c} = (c, \mathbf{a})$, *and*

$$\begin{aligned}\tilde{\omega}_{\text{as}}(y) &= \mathbf{c} \times \left(0, \frac{1}{8\pi}\mathbf{y}e^{-\frac{y^2}{4}}\right)\\ &= -\frac{1}{8\pi}(a_2 y_1 - a_1 y_2, c y_2, -c y_1)e^{-\frac{y^2}{4}}.\end{aligned}$$

Then,

$$\lim_{x\to\infty} \|\tilde{\omega}_x - \tilde{\omega}_{\text{as}}\|_{L^\infty} = 0.$$

3.3. *Connection with existing results*

There are many results on the large distance behavior of solutions of the stationary Navier–Stokes equations. See in particular Galdi (1998e) and references therein, where it is shown that at large distances the solution of the Navier–Stokes equations converge in certain norms to a multiple of the solution of the Oseen problem. To illustrate the connection with the above asymptotic results let $(\mathbf{E}, \mathbf{e})$ be the fundamental solution of the Oseen equation in $d = 2$, i.e., of the equation

$$\begin{aligned}-\partial_x\mathbf{u} + \Delta\mathbf{u} - \nabla p &= \mathbf{f},\\ \nabla\cdot\mathbf{u} &= 0.\end{aligned}$$

Namely, see Kress and Meyer (2000) or Galdi (1998d) and references therein,

$$\mathbf{E} = \begin{pmatrix} \psi - \psi_1 & -\psi_2 \\ -\psi_2 & \psi + \psi_1 \end{pmatrix}, \qquad \mathbf{e} = \nabla\phi,$$

where, with $r = \sqrt{x^2 + y^2}$,

$$\psi(x, y) = \frac{1}{4\pi} e^{x/2} K_0\left(\frac{r}{2}\right),$$

$$\psi_i(x, y) = \frac{1}{2\pi} \frac{x_i}{r} \left(\frac{1}{r} - \frac{1}{2} e^{x/2} K_1\left(\frac{r}{2}\right)\right), \quad i = 1, 2,$$

where $x_1 = x$, $x_2 = y_1$ K_0 and K_1 are modified Bessel functions of order zero and one, and where

$$\phi(x, y) = \frac{1}{2\pi} \log(r).$$

Now since

$$K_0(z) = \sqrt{\frac{\pi}{2}} \frac{1}{\sqrt{z}} \exp(-z/2) + o(1/\sqrt{z}),$$

as $z \to \infty$, we find for example that

$$\lim_{x\to\infty} \sqrt{x}\psi(x, y\sqrt{x}) = \frac{1}{4\sqrt{\pi}} e^{-\frac{y^2}{4}}, \tag{117}$$

$$\lim_{x\to\infty} \sqrt{x}\psi_1(x, y\sqrt{x}) = \frac{1}{4\sqrt{\pi}} e^{-\frac{y^2}{4}}, \tag{118}$$

and that

$$\lim_{x\to\infty} x\psi_1(x, yx) = \frac{1}{2\pi} \frac{1}{1 + y^2}, \tag{119}$$

and similarly for the other functions, so that we recover the asymptotic terms that we have found in Eq. (98). Therefore, in the sense of the norms used in Galdi (1998e) and in view of the scaling limits (117), (118), and (119), the results of Theorem 20 are not surprising, except for the existence of invariant quantities which, as explained above is nontrivial since it is linked to non-integrable functions. As we will see below it is precisely this result which allows to use the asymptotic results as artificial boundary conditions for the numerical solution of the Navier–Stokes equations in the regimes of interest.

3.4. *Higher order terms in two dimensions*

The same way as Proposition 6 generalizes Proposition 1, we expect generalizations of Theorems 20 and 29 that provide higher order corrections at large x. Such results will be important for the construction of higher order artificial boundary conditions, but the corresponding results are at this point still conjectural. As has been explained in Sec. 2.3.4, the main input for such higher order theorems is the construction of an asymptotic expansion on a formal level. For the two-dimensional case this work has been done up to second and in part to third-order in Bönisch *et al.* (2006). The complete formal construction up to third-order as well as the three-dimensional case are also open problems. The reason why the construction of the corresponding asymptotic expansions is nontrivial is, first, that the asymptotic behavior involves two length scales which tend to complicate things considerably, and second, that in order for such expansions to be useful for the construction of artificial boundary conditions, it is mandatory to construct vector fields that are order-by-order divergence free, sufficiently regular, and which satisfy all the boundary conditions. The following proposition is one of the main results of Bönisch *et al.* (2006).

Proposition 31. *Let* $\mathbf{u} = (u, v)$ *be the vector field of Theorem* 20 *for an initial condition for which* $c = -2d$, *and assume that Conjecture* 34 *below is valid. Then, we have for* $N = 1, 2, 3$,

$$\lim_{x\to\infty} x^{N/2}\left(\sup_{y\in\mathbf{R}} |(u - u_{\mathrm{as}}^N)(x, y)|\right) = 0, \tag{120}$$

$$\lim_{x\to\infty} x^{(N+1)/2}\left(\sup_{y\in\mathbf{R}} |(v - v_{\mathrm{as}}^N)(x, y)|\right) = 0, \tag{121}$$

where

$$u_{\mathrm{as}}^N(x, y) = \sum_{n=1}^{N}\sum_{m=1}^{n} u_{n,m}(x, y), \tag{122}$$

$$v_{\mathrm{as}}^N(x, y) = \sum_{n=1}^{N}\sum_{m=1}^{n} v_{n,m}(x, y), \tag{123}$$

with

$$u_{1,1}(x, y) = u_{1,1,E}(x, y) - \theta(x)\frac{d}{\sqrt{\pi}}\frac{1}{\sqrt{x}}e^{-\frac{y^2}{4x}},$$

$$v_{1,1}(x, y) = v_{1,1,E}(x, y) - \theta(x)\frac{d}{2\sqrt{\pi}}\frac{y}{x^{3/2}}e^{-\frac{y^2}{4x}}, \tag{124}$$

with θ the Heaviside function (i.e., $\theta(x) = 1$ for $x > 0$ and $\theta(x) = 0$ for $x \leq 0$), with

$$
\begin{aligned}
u_{1,1,E}(x, y) &= \frac{d}{\pi} \frac{x}{x^2 + y^2} + \frac{b}{\pi} \frac{y}{x^2 + y^2}, \\
v_{1,1,E}(x, y) &= \frac{d}{\pi} \frac{y}{x^2 + y^2} - \frac{b}{\pi} \frac{x}{x^2 + y^2},
\end{aligned}
\tag{125}
$$

with

$$
\begin{aligned}
u_{2,1}(x, y) &= \theta(x) \frac{bd}{2} \frac{1}{(\sqrt{\pi})^3} \frac{\log(x)}{x} \frac{y}{\sqrt{x}} e^{-\frac{y^2}{4x}}, \\
v_{2,1}(x, y) &= \theta(x) \frac{bd}{2} \frac{1}{(\sqrt{\pi})^3} \frac{1}{x^{3/2}} \left(\log(x) \left(-1 + \frac{1}{2} \frac{y^2}{x} \right) + 2 \right) e^{-\frac{y^2}{4x}},
\end{aligned}
\tag{126}
$$

and with

$$
\begin{aligned}
u_{2,2}(x, y) &= u_{2,2,E}(x, y) + \theta(x) d^2 \frac{1}{x} f'\left(\frac{y}{\sqrt{x}}\right) \\
&\quad + \lambda \theta(x) f_\infty d^2 \frac{3}{8} \frac{1}{x^2} \left(\left(1 + \frac{|y|}{\sqrt{x}}\right) \left(1 - \frac{1}{2} \frac{y^2}{x}\right) + \frac{|y|}{\sqrt{x}} \right) e^{-\frac{y^2}{4x}}, \\
v_{2,2}(x, y) &= v_{2,2,E}(x, y) + \theta(x) \frac{d^2}{2} \frac{1}{x^{3/2}} \left(\left(f\left(\frac{y}{\sqrt{x}}\right) - f_\infty \mathrm{sign}(y) \right) \right. \\
&\quad \left. + \frac{y}{\sqrt{x}} f'\left(\frac{y}{\sqrt{x}}\right) \right) + \lambda \theta(x) f_\infty d^2 \frac{3}{4} \frac{1}{x^{5/2}} \\
&\quad \times \left(\left(1 + \frac{|y|}{\sqrt{x}}\right) \frac{y}{\sqrt{x}} \left(1 - \frac{1}{8} \frac{y^2}{x}\right) + \frac{1}{4} \frac{y^2}{x} \mathrm{sign}(y) \right) e^{-\frac{y^2}{4x}},
\end{aligned}
\tag{127}
$$

and

$$
\begin{aligned}
u_{2,2,E}(x, y) &= f_\infty \frac{d^2}{2} \frac{|y|}{r^2} \left(\frac{1}{r_2} - \frac{r_2}{r} \right), \\
v_{2,2,E}(x, y) &= f_\infty \frac{d^2}{2} \frac{\mathrm{sign}(y)}{r} \left(-\frac{1}{r_2} - \frac{x}{r_2 r} + \frac{x\, r_2}{r^2} \right),
\end{aligned}
\tag{128}
$$

and with

$$u_{3,1}(x,y) = \theta(x)\frac{b^2 d}{4}\frac{1}{(\sqrt{\pi})^5}\frac{\log(x)^2}{x^{3/2}}\left(1-\frac{1}{2}\frac{y^2}{x}\right)e^{-\frac{y^2}{4x}},$$

$$v_{3,1}(x,y) = \theta(x)\frac{b^2 d}{2}\frac{1}{(\sqrt{\pi})^5}\frac{\log(x)}{x^2}\frac{y}{\sqrt{x}}\left(\frac{\log(x)}{4}\left(3-\frac{1}{2}\frac{y^2}{x}\right)-1\right)e^{-\frac{y^2}{4x}}. \tag{129}$$

Here, $r = \sqrt{x^2+y^2}$, $r_2 = \sqrt{2r+2x}$, $\lambda = 1$, *and* $f\colon \mathbf{R} \to \mathbf{R}$ *is the unique solution of the third-order linear inhomogeneous ordinary differential equation*

$$f'''(z) + \frac{1}{2}zf''(z) + f'(z) + \frac{1}{2\pi}e^{-\frac{1}{2}z^2} = 0, \tag{130}$$

satisfying $f(0) = 0,\ f'(0) = -\frac{1}{2\pi},\ f''(0) = 0$. *Explicitly,*

$$f(z) = -\frac{1}{\sqrt{2\pi}}\operatorname{erf}\left(\frac{z}{\sqrt{2}}\right) + \frac{1}{2\sqrt{\pi}}\operatorname{erf}\left(\frac{z}{2}\right)e^{-\frac{z^2}{4}}. \tag{131}$$

Note that the function f *is odd, and that* f' *and* f'' *decay faster than exponential at infinity, and that*

$$f_\infty = \lim_{z\to\infty} f(z) = -\frac{1}{\sqrt{2\pi}}. \tag{132}$$

Remark 32. Since $x \geq 1$ the functions θ could be suppressed from the above. However, when we specify artificial boundary conditions in Sec. 4, we will consider the same expressions in the domain $\mathbf{R}^2\backslash\mathbf{B}$, with $\mathbf{B}$ a compact region with smooth boundary containing the origin. With the functions θ the above expressions will also be the correct asymptotic description in $\mathbf{R}^2\backslash\mathbf{B}$. This avoids having to rewrite these lengthy expressions a second time in Sec. 5.

Remark 33. The terms proportional to λ are higher order and one might be tempted to neglect them, i.e., to set $\lambda = 0$. This is not possible, however, without giving up the regularity of the second-order derivatives $\partial_y^2 u$ and $\partial_y^2 v$ across the positive x-axis.

Proposition 31 is based on the following conjecture on the vorticity. See Bönisch *et al.* (2006).

Conjecture 34. *Let the vector field* $\mathbf{u} = (u, v)$, *the constants* b *and* d, *and the function* f *be as in Proposition* 31. *Let the vorticity* ω *be given by* $\omega(x, y) = -\partial_y u(x, y) + \partial_x v(x, y)$. *Then, for* $1 \leq N \leq 3$, *we have that*

$$\lim_{x\to\infty} x^{\frac{1+N}{2}} \sup_{y\in\mathbf{R}} \left| \omega(x, y) - \sum_{n=1}^{N} \omega_n(x, y) \right| = 0, \tag{133}$$

where the functions ω_n *are given by*

$$\omega_n(x, y) = \sum_{m=1}^{n} \rho_{n,m}(x)\varphi''_{n,m}\left(\frac{y}{\sqrt{x}}\right), \tag{134}$$

with

$$\rho_{n,m}(x) = \frac{\log(x)^{n-m}}{x^{(1+n)/2}}, \tag{135}$$

with

$$\begin{aligned} \varphi_{1,1}(z) &= d \ \mathrm{erf}\left(\frac{z}{2}\right), \\ \varphi_{2,1}(z) &= bd \ \frac{1}{\pi^{3/2}} e^{-\frac{z^2}{4}}, \\ \varphi_{2,2}(z) &= -d^2 \ f(z) + b \ c_{2,2} \ e^{-\frac{z^2}{4}}, \\ \varphi_{3,1}(z) &= -b^2 d \ \frac{z}{4\pi^{5/2}} e^{-\frac{z^2}{4}}, \end{aligned} \tag{136}$$

and where $\varphi_{3,2}$ *and* $\varphi_{3,3}$ *are smooth functions with derivatives decaying at infinity faster that exponential.*

It is interesting to compare the above results with similar results in the literature. For the particular case where the wake has an axial symmetry, results for the so-called center-line velocity (the velocity along the axis of symmetry in the wake region) are given on the basis of boundary layer theory up to third-order in Stewartson (1957). These results have been reviewed recently in Sobey (2000), see Fig. 1. Our results show that the expansions computed from the Navier–Stokes equations differ from the ones computed from boundary layer theory already at second-order, which shows that higher order results based on boundary layer theory are

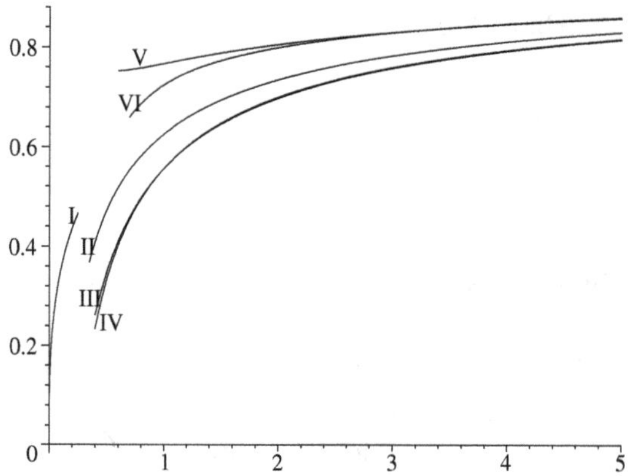

Fig. 1. Center-line velocity for the far wake as given in Sobey (2000): two-term expansion, (II), three-term expansion (III), expansion with logarithmic corrections (IV). Near wake center-line velocity (I) as reviewed in Sobey (2000). Centerline velocity for the far wake based on the Navier–Stokes equations as given in Conjecture 31: expansion to first-order (V), expansion to second-order (VI).

inadequate for modeling Navier–Stokes flows. See also Cowley (2006) for a discussion of certain insufficiencies of boundary layer theory.

3.5. *Open problems*

The following three problems are all in the reach of the techniques introduced above. Their solution will be directly useful for the definition of artificial boundary conditions in the corresponding cases.

3.5.1. *The time periodic case*

The asymptotic downstream behavior of time periodic data has been studied by van Baalen (2006). This work contains also a partial proof of some of the higher order terms presented above. The theory of periodic solutions for the Navier–Stokes equations in exterior domains is still rather incomplete so that it is not quite clear that the results of van Baalen (2006) are general enough to describe the downstream region of time periodic exterior flows, but we do expect this to be the case. The results of van Baalen (2006) have unfortunately not yet been used for the construction of artificial boundary conditions, so that this question has not yet been checked numerically either. All these questions are fascinating and the answers are in reach of the techniques described here. We therefore expect that further work will clarify the open points.

3.5.2. *The case of free-falling bodies*

A very interesting question that goes back to an article of Weinberger (1978) is the asymptotic description of solutions of the Navier–Stokes equations in $d = 3$ describing the steady free-fall of a rigid body. A free falling body typically also rotates along a certain axis that is aligned with the direction of the fall. What we therefore mean by a steady free fall is a solution of the Navier–Stokes equation which is stationary in a frame that is attached to the "body" and which is rotating with constant frequency around an axis that is aligned with the flow at infinity. For recent results see Galdi and Silvestre (2005). We expect the present techniques to be well adapted for a detailed study of these questions. Again this will allow the construction of artificial boundary conditions for this case which will make precise numerical solutions possible.

3.5.3. *Motion in the presence of a nearby wall*

Another very interesting open problem is the detailed description of the motion of a body parallel to a nearby wall. An example of recent experimental results of bubbles rising close to a wall can be found in Takemura and Magnaudet (2003). For very slow movements this problem is discussed in the literature (a classical reference is Clift *et al.*, 2005) and is mathematically modeled by the Stokes equations. For higher Reynolds numbers one again needs to consider the asymptotics of the Navier–Stokes equations. We give some details of forthcoming results by Hillairet and Wittwer (2007) for the two-dimensional case. The situation can be modeled by solving Eqs. (88) and (89) in the complement of a compact region of the upper half plane. The associated downstream problem leads again essentially to the heat equation, however not on $\mathbf{R}$ but on the half-line $\mathbf{R}_+$, with Dirichlet boundary conditions at $x = 0$. This leads, instead of Eq. (98), to a dominant term for the horizontal velocity proportional to $1/x$ instead of $1/\sqrt{x}$, and therefore like $1/x^{3/2}$ for the transverse velocity and for the vorticity. This leading term decays for fixed x exponentially fast as a function of y, as y goes to infinity. The second-order term for the vorticity is proportional to $1/x^2$ but the corresponding term in the expansion is a function that decays only algebraically in y, as y goes to infinity. As a result it can be shown that for fixed x the vorticity decays only like $1/y^4$, as y goes to infinity, and not exponentially fast as in the case of exterior flows without a nearby wall. This is due to the presence of a boundary layer ahead of the body, extending all the way to minus infinity in the sense that along the wall ahead of the obstacle the vorticity again only decays algebraically like $1/x^2$. Detailed results for this case as well as the artificial boundary conditions that can be derived from the results are in preparation.

4. Exterior Flows at Low Reynolds Numbers

After the preparatory work in Secs. 2 and 3 we now consider exterior flows at low Reynolds numbers. The main goal is to explain on a heuristic level the mechanism that allows the reconstruction of the large distance asymptotics from the knowledge of the asymptotic behavior in the downstream region. We also present some details concerning the link between the invariant quantities of the preceding section and the forces that act on the "body". This is the key reason for the efficiency of the boundary conditions in Sec. 5.

4.1. *The mathematical problem*

Consider a rigid body $\tilde{\mathbf{B}}$ (a compact set with smooth boundary) of diameter R that is placed into a uniform stream of a homogeneous incompressible fluid filling up all of $\mathbf{R}^d$, $d = 2, 3$. This situation is modeled by the stationary Navier–Stokes equations

$$-\rho(\tilde{\mathbf{U}} \cdot \nabla)\tilde{\mathbf{U}} + \mu\Delta\tilde{\mathbf{U}} - \nabla\tilde{p} = 0, \tag{137}$$

$$\nabla \cdot \tilde{\mathbf{U}} = 0, \tag{138}$$

in $\tilde{\Omega} = \mathbf{R}^d \setminus \tilde{\mathbf{B}}$, subject to the boundary conditions

$$\tilde{\mathbf{U}}|_{\partial\tilde{\mathbf{B}}} = 0, \tag{139}$$

$$\lim_{|\tilde{\mathbf{x}}|\to\infty} \tilde{\mathbf{U}}(\tilde{\mathbf{x}}) = \tilde{\mathbf{u}}_\infty. \tag{140}$$

Here, $\tilde{\mathbf{U}}$ is the velocity field, $\tilde{p}$ is the pressure and $\tilde{\mathbf{u}}_\infty$ is some constant non-zero vector field which we choose without restriction of generality to be parallel to the $\tilde{x}$-axis, i.e., $\tilde{\mathbf{u}}_\infty = u_\infty \mathbf{u}_\infty$, where $\mathbf{u}_\infty = (1, 0)$ and $u_\infty > 0$.

The density ρ and the viscosity μ are arbitrary positive constants. From μ, ρ and u_∞ we can form ℓ,

$$\ell = \frac{\mu}{\rho u_\infty}, \tag{141}$$

the so-called viscous length of the problem. The viscous forces and the inertial forces are quantities of comparable size if the diameter R of $\tilde{\mathbf{B}}$ is comparable with ℓ, i.e., if the Reynolds number

$$\mathrm{Re} = \frac{R}{\ell}, \tag{142}$$

is neither very small nor very large, i.e., depending on the geometry of the body, in the range from one to several thousand. Note that for bodies with a smooth boundary $\partial\tilde{\mathbf{B}}$ and for small enough Reynolds numbers (142), Eqs. (137) and (138) subject to the boundary conditions (139) and (140) are known to have a unique classical solution. See Galdi (1998a) and (1999b) for an interesting open problem for the two-dimensional case of symmetric stationary flows at arbitrary Reynolds numbers. Before proceeding any further we now rewrite the Navier–Stokes equations in dimensionless form. Let $\tilde{\mathbf{U}}$ be the velocity field and $\tilde{p}$ the pressure introduced in Eqs. (137)–(140), and let ℓ be as defined in Eq. (141). Then, we define dimensionless coordinates $\mathbf{x} = \tilde{\mathbf{x}}/\ell$, and introduce a dimensionless vector field $\mathbf{U}$ and a dimensionless pressure p through the definitions

$$\tilde{\mathbf{U}}(\tilde{\mathbf{x}}) = u_\infty \mathbf{U}(\mathbf{x}), \tag{143}$$

$$\tilde{p}(\tilde{\mathbf{x}}) = (\rho u_\infty^2) p(\mathbf{x}). \tag{144}$$

In the new coordinates we get instead of Eqs. (137)–(140) the equations

$$-(\mathbf{U}\cdot\nabla)\mathbf{U} + \Delta\mathbf{U} - \nabla p = 0, \tag{145}$$

$$\nabla\cdot\mathbf{U} = 0, \tag{146}$$

in $\Omega = \mathbf{R}^2 \setminus \mathbf{B}$, where $\mathbf{B} = \{\mathbf{x} \in \mathbf{R}^2 \mid \ell\mathbf{x} = \tilde{\mathbf{x}} \text{ for some } \tilde{\mathbf{x}} \in \tilde{\mathbf{B}}\}$, and the boundary conditions

$$\mathbf{U}|_{\partial\mathbf{B}} = 0, \tag{147}$$

$$\lim_{|\mathbf{x}|\to\infty} \mathbf{U}(\mathbf{x}) = (1, 0). \tag{148}$$

In Eqs. (145) and (146) all derivatives are with respect to the new coordinates. For convenience below we now introduce some additional notation and conventions. In practice, and in particular when solving the equations numerically, it will always be more convenient to work with zero boundary conditions at infinity. We therefore set $\tilde{\mathbf{U}} = \tilde{\mathbf{u}}_\infty + \tilde{\mathbf{u}}$ and $\mathbf{U} = \mathbf{u}_\infty + \mathbf{u}$ and consider either the dimensionfull equations

$$-\rho(\tilde{\mathbf{u}} \cdot \nabla)\tilde{\mathbf{u}} - \rho u_\infty \partial_x \tilde{\mathbf{u}} + \mu \Delta \tilde{\mathbf{u}} - \nabla \tilde{p} = 0, \tag{149}$$

$$\nabla \cdot \tilde{\mathbf{u}} = 0, \tag{150}$$

with the boundary conditions

$$\tilde{\mathbf{u}}|_{\partial \tilde{\mathbf{B}}} = -\tilde{\mathbf{u}}_\infty, \tag{151}$$

$$\lim_{|\tilde{\mathbf{x}}| \to \infty} \tilde{\mathbf{u}}(\tilde{\mathbf{x}}) = 0, \tag{152}$$

or the dimensionless equations

$$-(\mathbf{u} \cdot \nabla)\mathbf{u} - \partial_x \mathbf{u} + \Delta \mathbf{u} - \nabla p = 0, \tag{153}$$

$$\nabla \cdot \mathbf{u} = 0, \tag{154}$$

with the boundary conditions

$$\mathbf{u}|_{\partial \mathbf{B}} = -(1, 0), \tag{155}$$

$$\lim_{|\mathbf{x}| \to \infty} \mathbf{u}(\mathbf{x}) = 0. \tag{156}$$

Finally, for Ω as defined above, we will always chose a coordinate system with origin in $\mathbf{B}$, but such that the surface $\Sigma = \{(x, \mathbf{y}) \in \mathbf{R} \times \mathbf{R}^{d-1} | x = 1\}$ defined in the previous section is "to the right of" $\mathbf{B}$, i.e., such that $\sup\{x \in \mathbf{R} | (x, \mathbf{y}) \in \mathbf{B} \text{ for some } \mathbf{y} \in \mathbf{R}^{d-1}\} < 1$.

4.2. *Consequences of incompressibility*

We limit the discussion to the two-dimensional case. The three-dimensional case is similar. Consider a solution $\mathbf{u} = (u, v)$ of the Navier–Stokes Eqs. (153) and (154) and consider the region $\Omega_{x_0,R} = \{(x, y) \in \Omega | \, |x| \leq x_0, |y| \leq R\}$. We assume that x_0 and R are large enough such that

$\mathbf{B} \subset \Omega_{x_0,R}$. Since $\nabla \cdot \mathbf{u} = 0$, and in view of the boundary condition (155), it follows using Gauss' theorem that

$$\int_{-R}^{R} (u(-x_0, y) - u(x_0, y))\, dy + \int_{-x_0}^{x_0} (v(x, -R) - v(x_0, R))\, dy = 0.$$

Taking the limit $R \to \infty$ and using that for classical solutions the boundary condition (156) implies that $\lim_{R\to\pm\infty} v(x, R) = 0$, we conclude that

$$\lim_{R\to\infty} \int_{-R}^{R} (u(-x_0, y) - u(x_0, y))\, dy = 0.$$

Therefore, since $\lim_{R\to\infty} \int_{-R}^{R} \left(1 - \frac{|y|}{R}\right) u(x_0, y)\, dy = c + b$ by Eq. (103), it follows now that also

$$\lim_{R\to\infty} \int_{-R}^{R} \left(1 - \frac{|y|}{R}\right) u(-x_0, y)\, dy = c + b, \tag{157}$$

i.e., the invariant quantity $c+b$ can also be extracted from the velocity field upstream of the body. Similarly, using Stokes' theorem instead of Gauss' theorem, it can be shown that

$$-\lim_{R\to\infty} \int_{-R}^{R} \left(1 - \frac{|y|}{R}\right) v(-x_0, y)\, dy = b, \tag{158}$$

i.e., the invariant quantity b can also be extracted from the velocity field upstream of the body.

4.3. *Connection between global and downstream asymptotics*

The discussion in this section is on a heuristic level, and we consider the two-dimensional case only. See Bönisch *et al.* (2006) and Bichsel and Wittwer (2007) for more details. See also Amick (1991) for one of the first references where the structure of the vorticity is rigorously exploited in order to obtain an improved description of the velocity field. So assume that the vorticity ω of an exterior flow is known. Then, the vector field $\mathbf{u} = (u, v)$ is given by

$$u(x, y) = \partial_y \psi(x, y), \tag{159}$$

$$v(x, y) = -\partial_x \psi(x, y), \tag{160}$$

with ψ solution of the Poisson equation

$$\Delta\psi(x, y) = -\omega(x, y), \tag{161}$$

with the boundary conditions

$$\bar{\psi}|_{\partial\mathbf{B}} = \partial_{\mathbf{n}}\bar{\psi}|_{\partial\mathbf{B}} = 0, \tag{162}$$

$$\lim_{x,y\to\infty} \partial_y\psi(x, y) = \lim_{x,y\to\infty} -\partial_x\psi(x, y) = 0, \tag{163}$$

where $\bar{\psi} = \psi + y$. By Theorem 26 the vorticity decays slowly in the downstream direction. Assuming that it decays rapidly in the other directions, more precisely, if ω satisfies a bound of the type

$$\sup_{(x,y)\in\Omega} \left(\left|\omega(x, y|x|^{1/2})\right| \, e^{\delta|y|}(1 + e^{-\delta x}) \right), \tag{164}$$

for some $0 < \delta \ll 1$, then it follows using general results from potential theory that

$$\psi(x, y) = \psi_\omega(x, y) + h(x, y), \tag{165}$$

with ψ_ω a particular solution of Eq. (161) that is independent of the geometry of **B**, and h a harmonic function in Ω satisfying the bound

$$|h(x, y)| \leq \frac{\text{const.}}{r}, \tag{166}$$

with $r = \sqrt{x^2 + y^2}$. The partial derivatives of h with respect to x and y obey analogous bounds. The function ψ_ω can be further decomposed into a part that dominates at large distances and a rest, i.e.,

$$\psi_\omega = \psi_0 + \psi_{\omega,1}. \tag{167}$$

where

$$\psi_0(x, y) = 2b\, G(x, y) + 2d\, H(x, y) + \frac{c}{2}\theta(x) \left(\varphi_{1,1}\left(\frac{y}{\sqrt{x}}\right) - \text{sign}(y) \right), \tag{168}$$

with

$$\varphi_{1,1}(z) = \text{erf}\left(\frac{z}{2}\right),$$

with

$$G(x, y) = \frac{1}{4\pi} \log(x^2 + y^2),$$

and with

$$H(x, y) = \frac{1}{2\pi} \arctan\left(\frac{y}{x}\right) - \frac{1}{2}\theta(x)\mathrm{sign}(y).$$

From ψ_0 we get the vector field $(\partial_y \psi_0, -\partial_x \psi_0)$ which describes the solution at large distances. Furthermore, using Eqs. (157) and (158) it can be shown that $c = -2d$ in two- and three-dimensions and that $\mathbf{a} = -2\mathbf{b}$ in three-dimensions. In the next subsection we now establish an additional link between the invariant quantities and drag and lift.

4.4. *Drag, lift, and torque*

We again limit the discussion to the two-dimensional case. Let $\mathbf{U}$, p be a solution of the Navier–Stokes Eqs. (145) and (146) subject to the boundary conditions (147) and (148), and let $\mathbf{e}$ be some arbitrary unit vector in $\mathbf{R}^2$. Multiplying Eq. (145) with $\mathbf{e}$ leads to

$$-(\mathbf{U} \cdot \nabla)(\mathbf{U} \cdot \mathbf{e}) + \Delta(\mathbf{U} \cdot \mathbf{e}) - \nabla \cdot (p\mathbf{e}) = 0. \tag{169}$$

Equation (169) can be written as $\nabla \cdot \mathbf{P}(\mathbf{e}) = 0$, where

$$\mathbf{P}(\mathbf{e}) = -(\mathbf{U} \cdot \mathbf{e})\mathbf{U} + [\nabla\mathbf{U} + (\nabla\mathbf{U})^T] \cdot \mathbf{e} - p\mathbf{e}, \tag{170}$$

i.e., the vector field $\mathbf{P}(\mathbf{e})$ is divergence free. Therefore, applying Gauss' theorem to the region $\Omega_{x_0,R}$ of the preceding subsection we find that

$$\int_{\partial\mathbf{B}} \mathbf{P}(\mathbf{e}) \cdot \mathbf{n}\, d\sigma = \int_S \mathbf{P}(\mathbf{e}) \cdot \mathbf{n}\, d\sigma, \tag{171}$$

with $S = \partial\Omega_{x_0,R} \backslash \partial\mathbf{B}$, and with $\mathbf{n}$ outside unit normal vectors. We have that $\mathbf{P}(\mathbf{e}_1) \cdot \mathbf{e}_2 = \mathbf{P}(\mathbf{e}_2) \cdot \mathbf{e}_1$ for any two unit vectors $\mathbf{e}_1$ and $\mathbf{e}_2$, and therefore, since the vector $\mathbf{e}$ in Eq. (171) is arbitrary, it follows that

$$\int_{\partial\mathbf{B}} \mathbf{P}(\mathbf{n})\, d\sigma = \int_S \mathbf{P}(\mathbf{n})\, d\sigma. \tag{172}$$

Since $\mathbf{u} = 0$ on $\partial\mathbf{B}$, one finally get from Eqs. (170) and (172) that the total force which the fluid exerts on the body is

$$\mathbf{F} = \int_{\partial\mathbf{B}} \Sigma(\mathbf{U}, p)\mathbf{n}\, d\sigma = \int_S (-(\mathbf{U}\cdot\mathbf{n})\mathbf{U} + [\nabla\mathbf{U}+(\nabla\mathbf{U})^T]\mathbf{n} - p\mathbf{n})\, d\sigma,$$

with $\Sigma(\mathbf{U}, p) = \nabla\mathbf{U}+(\nabla\mathbf{U})^T - p$ the Stress tensor. The force $\mathbf{F}$ is traditionally decomposed into a component F parallel to the flow at infinity called drag and a component L perpendicular to the flow at infinity called lift. Note that $\mathbf{F}$ is independent of the choice of the surface S. This has the important consequence that F and L can be computed from the dominant terms of the velocity field and the pressure (the dominant large distance behavior of the pressure can also be recovered together with the velocity field starting from the information in the downstream region (see again Bönisch *et al.* (2006) for further details). In particular, since $\lim_{|y|\to\infty} \mathbf{u}(x, y) = (1, 0)$ and with the normalization $\lim_{|y|\to\infty} p(x, y) = 0$, one can again first take the limit $R \to \infty$, and then use the invariance properties of the equations. This leads to the following theorem.

Theorem 35. *If a stationary solution of the stationary Navier–Stokes equations satisfies the conditions of Theorem* 20 *or Theorem* 29 *then we have that*

$$d = \frac{1}{2\rho\ell^{d-1}u_\infty^2}\tilde{F}, \tag{173}$$

$$\mathbf{b} = \frac{1}{2\rho\ell^{d-1}u_\infty^2}\tilde{\mathbf{L}}, \tag{174}$$

with $\tilde{F}$ *the drag and* $\tilde{L}$ *the lift (dimension-full quantities).*

Remark 36. We are not aware of any work that links the additional constants that appear at higher order in the asymptotic expansion of the wake to other physical quantities like the torque acting on the body. This is another problem worth pursuing. Its solution will allow to obtain a complete prescription of third order artificial boundary conditions (see Sec. 5).

Expressions similar to Eqs. (173) and (174) can be found in many textbooks but are typically based on boundary layer theory (see for example Batchelor (1967), Landau and Lipschitz (1989) or Berger (1971)). In the context of the boundary layer approach the quantities d and $\mathbf{b}$ are however not invariant so that the relations therefore appear to be valid only asymptotically, in the sense that d and $\mathbf{b}$ are supposed to be computed from the flow far downstream of the body. Our results show that in the context of the Navier–Stokes equations the quantities can be computed on any transversal section.

4.5. *Open problems*

Using the results in van Baalen (2006) one can get expressions similar to Eqs. (173) and (174) also in the time periodic case, but the right hand side contains an average over the period of the motion. Therefore, either the period of this motion needs to be known in order to link the forces to the invariant quantities, or a procedure like averaging over sufficiently large intervals of time needs to be used in order to extract the constant part of the periodic signal. The theory is however still rather incomplete and the corresponding artificial boundary conditions have not been numerically tested yet. This is an other interesting open problem, in particular also in view of the study of the transition from stationary to time periodic flows. See for example Pipe and Monkewitz (2005) for recent experimental work. Finally, for the case of a falling body that rotates (see Weinberger (1978)), even the downstream asymptotic behavior of the solution is not yet known. We hope that the present work will stimulate further activity in this direction.

5. Artificial Boundary Conditions

In what follows we define artificial boundary conditions for stationary exterior flows in two- and three-dimensions. The section is basically self-contained which allows the interested reader to use the boundary conditions without having to work through all of the material of the preceding sections. We expect that beyond the stationary case discussed here similar boundary conditions will be developed in the near future for the case of time periodic flows, for the case of free-falling bodies, for the case of bodies that move close to a wall and for the case of bodies undergoing collisions. One of the main benefits of the artificial boundary conditions defined here is their simplicity. No additional differential equations need to be solved. The boundary conditions are simply explicit Dirichlet boundary conditions on the artificial boundary, depending on parameters which are updated as part of the solution process.

In order to solve the exterior problem described in Sec. 4.1 numerically, one typically uses the formulation with zero boundary conditions at infinity, i.e., one considers the equations

$$\begin{aligned} -\rho u_\infty \partial_x \tilde{\mathbf{u}} - \rho(\tilde{\mathbf{u}} \cdot \nabla)\tilde{\mathbf{u}} + \mu\Delta\tilde{\mathbf{u}} - \nabla\tilde{p} &= 0, \\ \nabla \cdot \tilde{\mathbf{u}} &= 0, \end{aligned} \tag{175}$$

in $\tilde{\Omega} = \mathbf{R}^d \setminus \tilde{\mathbf{B}}$, subject to the boundary conditions

$$\tilde{\mathbf{u}}|_{\partial\tilde{\mathbf{B}}} = -\tilde{\mathbf{u}}_\infty, \tag{176}$$

$$\lim_{|\tilde{\mathbf{x}}|\to\infty} \tilde{\mathbf{u}}(\tilde{\mathbf{x}}) = 0. \tag{177}$$

Here $\tilde{\mathbf{B}}$, is "the body", i.e., a compact set of diameter R, and $\tilde{\mathbf{u}}_\infty$ is a constant vector field. When restricting the equations for numerical purposes from the exterior infinite domain $\tilde{\Omega}$ to (a sequence of) bounded domains $\tilde{\mathbf{D}} \subset \tilde{\Omega}$, one is confronted with the necessity of finding appropriate boundary conditions on the surface $\tilde{\Gamma} = \partial\tilde{\mathbf{D}} \backslash \partial\tilde{\mathbf{B}}$ of the truncated domain. We now use the results of the proceeding sections to define appropriate boundary conditions. Namely, we set

$$\tilde{\mathbf{u}}|_{\tilde{\Gamma}} = \tilde{\mathbf{u}}_{ABC}, \tag{178}$$

with $\tilde{\mathbf{u}}_{ABC}$ the vector fields that are explicitly given below. The vector fields $\tilde{\mathbf{u}}_{ABC}$ depend on the invariant quantities discussed in the previous section and these quantities are computed as part of the solution process.

Note that, choosing instead of Eq. (178), for example, simply $\tilde{\mathbf{u}}|_{\tilde{\Gamma}} = 0$, forces the mass flux through a vertical line in $\tilde{\mathbf{D}}$ to be zero. This corresponds to invariant quantities equal to zero. However, because of the link that exists between the invariant quantities and the forces acting on the body, any choice of boundary conditions that does not respect the correct mass flux through vertical lines produces significant changes to the forces, unless extremely large computational domains are used. The adaptive boundary conditions (178) eliminate this problem. More details can be found in the sections below and in Bönisch *et al.* (2005) and Bönisch *et al.* (2006).

5.1. *Stationary flows in two dimensions*

For the two-dimensional case the artificial boundary conditions on $\tilde{\Gamma} = \partial\tilde{\mathbf{D}} \backslash \partial\tilde{\mathbf{B}}$ are

$$\tilde{\mathbf{u}}_{ABC}(\tilde{\mathbf{x}}) = u_\infty \, \mathbf{u}_{ABC}\left(\frac{\tilde{\mathbf{x}}}{\ell}\right), \tag{179}$$

where $\ell = \rho u_\infty / \mu$, and where

$$\mathbf{u}_{ABC} = (u_{ABC}, v_{ABC}), \tag{180}$$

with

$$u_{ABC}(x, y) = -\theta(x)\frac{d}{\sqrt{\pi}}\frac{1}{\sqrt{x}}e^{-\frac{y^2}{4x}} + \frac{d}{\pi}\frac{x}{x^2+y^2} + \frac{b}{\pi}\frac{y}{x^2+y^2}, \tag{181}$$

$$v_{ABC}(x, y) = -\theta(x)\frac{d}{2\sqrt{\pi}}\frac{y}{x^{3/2}}e^{-\frac{y^2}{4x}} + \frac{d}{\pi}\frac{y}{x^2+y^2} - \frac{b}{\pi}\frac{x}{x^2+y^2}, \tag{182}$$

with θ the Heaviside function (i.e., $\theta(x) = 1$ for $x > 0$ and $\theta(x) = 0$ for $x \leq 0$), and with

$$d = \frac{1}{2}\frac{1}{\rho\ell u_\infty^2}\tilde{F}, \tag{183}$$

$$b = \frac{1}{2}\frac{1}{\rho\ell u_\infty^2}\tilde{L}, \tag{184}$$

where $\tilde{F}$ and $\tilde{L}$ are, respectively, the drag and the lift acting on the body (dimensionfull quantities). The drag $\tilde{F}$ and the lift $\tilde{L}$ are computed (by an evaluation of the stress tensor for example) as part of the solution process (see Sec. 6), which in turn allows to update the boundary conditions on $\tilde{\Gamma} = \partial\tilde{\mathbf{D}} \backslash \partial\tilde{\mathbf{B}}$ using Eqs. (183) and (184).

Using the results in Sec. 3 the boundary conditions can be improved without the introduction of constants other than b and d. Namely one sets

$$\tilde{\mathbf{u}}_{ABC}(\tilde{\mathbf{x}}) = u_\infty \, \mathbf{u}_N\left(\frac{\tilde{\mathbf{x}}}{\ell}\right),$$

where

$$\mathbf{u}_N(x, y) = \sum_{n=1}^{N}\sum_{m=1}^{n} \mathbf{u}_{n,m}(x, y),$$

with $\mathbf{u}_{n,m}$ the vector fields that have been defined in Conjecture 31. The case $N = 0$ corresponds to choosing homogeneous Dirichlet data on $\tilde{\Gamma} = \partial\tilde{\mathbf{D}}\backslash\partial\tilde{\mathbf{B}}$, the case $N = 1$ is identical to Eqs. (181) and (182) (first-order adaptive boundary conditions (see Bönisch *et al.* (2005)), and $N = 2$ and $N = 3$ are second, respectively, third-order adaptive boundary conditions (see Bönisch *et al.* (2006)).

5.2. *Stationary flows in three dimensions*

For the three-dimensional case the artificial boundary conditions on $\tilde{\Gamma} = \partial\tilde{\mathbf{D}}\backslash\partial\tilde{\mathbf{B}}$ are

$$\tilde{\mathbf{u}}_{ABC}(\tilde{\mathbf{x}}) = u_\infty \mathbf{u}_{ABC}\left(\frac{\tilde{\mathbf{x}}}{\ell}\right), \tag{185}$$

where $\ell = \rho u_\infty/\mu$, and where

$$\mathbf{u}_{ABC} = (u_{ABC}, \mathbf{v}_{ABC}), \tag{186}$$

$$\mathbf{v}_{ABC} = \mathbf{v}_{1,ABC} + \mathbf{v}_{2,ABC}, \tag{187}$$

with

$$u_{ABC}(x, \mathbf{y}) = -\frac{\theta(x)}{2\pi x} e^{-\frac{y^2}{4x}}\, d + \frac{1}{2\pi}\frac{x}{r^3}\, d + \frac{1}{2\pi}\frac{\mathbf{y}\cdot\mathbf{b}}{r^3}, \tag{188}$$

$$\mathbf{v}_{1,ABC}(x, \mathbf{y}) = -\frac{\mathbf{y}}{4\pi x^2}\theta(x) e^{-\frac{y^2}{4x}}\, d + \frac{1}{2\pi}\frac{\mathbf{y}}{r^3}\, d - \frac{1}{2\pi}\frac{1}{r}\frac{\operatorname{sign}(x)}{r+|x|}\left(\mathbf{b} - \frac{1}{r}\left(\frac{1}{r} + \frac{1}{r+|x|}\right)(\mathbf{y}\cdot\mathbf{b})\,\mathbf{y}\right), \tag{189}$$

$$\mathbf{v}_{2,ABC}(x, \mathbf{y}) = -\frac{\theta(x)}{2\pi x}\left(e^{-\frac{y^2}{4x}} + \frac{1}{2}\frac{e^{-\frac{y^2}{4x}} - 1}{\frac{y^2}{4x}}\right)\mathbf{b} + \frac{\theta(x)}{2\pi x}\left(\frac{e^{-\frac{y^2}{4x}} - 1}{\frac{y^2}{4x}} + e^{-\frac{y^2}{4x}}\right)\frac{(\mathbf{y}\cdot\mathbf{b})}{y^2}\mathbf{y}, \tag{190}$$

with θ the Heaviside function (i.e., $\theta(x) = 1$ for $x > 0$ and $\theta(x) = 0$ for $x \leq 0$), with $\operatorname{sign}(x) = -1 + 2\theta(x)$, and with $\mathbf{y}\cdot\mathbf{b} = y_1 b_1 + y_2 b_2$, and with

$$d = \frac{1}{2\rho\ell^2 u_\infty^2}\tilde{F}, \tag{191}$$

$$\mathbf{b} = \frac{1}{2\rho\ell^2 u_\infty^2}\tilde{\mathbf{L}}, \tag{192}$$

where $\tilde{F}$ and $\tilde{\mathbf{L}}$ are, respectively, the drag and the lift acting on the body (dimensionfull quantities). The drag $\tilde{F}$ and the lift $\tilde{\mathbf{L}}$ are computed (by an evaluation of the stress tensor for example) as part of the solution process (see Sec. 6), which in turn allows to update the boundary conditions on $\tilde{\Gamma} = \partial\tilde{\mathbf{D}}\backslash\partial\tilde{\mathbf{B}}$ using Eqs. (191) and (192).

5.3. *Bibliographic notes*

The following notes give access to the literature on artificial boundary conditions in a chronological order. Indeed, even so different problems require different solutions, the basic ideas and techniques are pretty much independent of the specific problem. The literature on the topic is very extensive and we have therefore made no attempt to be exhaustive. The main goal has rather been to compile a list of references from a broad set of applications. Additional lists of references can be found in the various reviews that we mention.

The problem of artificial boundary conditions can be traced back all the way to the beginnings of scientific computing. Originally, a popular way of handling exterior problems was to map the exterior domain to some finite domain and to discretize the resulting equations there. We do not discuss this approach here, mainly, because in its essence this is not different from choosing artificial boundary conditions, since it corresponds simply to a particular way of specifying such conditions.

The problem with artificial boundaries is most obvious for wave-like equations, since scattering on the artificial boundary obviously produces unphysical reflections of waves back into the computational domain. This has lead to the development of so-called non-reflecting or absorbing boundary conditions. Early work which addresses the subject is due to Engquist and Majda (1979), Bayliss and Turkel (1980) and Sochacki *et al.* (1986). In Peterson (1988) absorbing boundary conditions for the vector wave equation are discussed. See also Luebbers *et al.* (1991) for another early reference with a somewhat different discussion of the subject.

Artificial boundary conditions for incompressible viscous flows can be found in Halpern and Schatzman (1989). Another classic reference is Heywood *et al.* (1992). This work discusses in particular questions related to the mass flux.

The work of Grote (1995) and Grote and Keller (1995) on nonreflective boundary conditions has stimulated the work on artificial boundary conditions for the hyperbolic case. Higher order radiation boundary conditions have been proposed in Hagstrom (1995) and absorbing boundary conditions for the Schrödinger equation can be found in Fevens and Jiang (1995). Absorbing boundary conditions for the linearized Euler equations have been proposed in Hu (1996) on the basis of a so-called perfectly matched layers. See also Hesthaven (1997). Griffiths (1997) contains a proof (in $d = 1$) of the effectiveness of well chosen boundary conditions when compared to traditional boundary conditions. A first article containing artificial boundary conditions for the computation of oscillating external flows is due to Tsynkov (1997). Ryaben'kii and Tsynkov (1997) contains a first review of a method that allows the construction of artificial boundary conditions for exterior problems in computational fluid dynamics and Hagstrom and Hariharan (1998) contains a formulation of asymptotic and exact boundary conditions using local operators. A two-dimensional treatment of transonic flow around an airfoil is discussed in Coclici *et al.* (1998), and transparent boundary conditions for the two-dimensional Helmholtz equation are discussed in Schmidt (1998). In Tsynkov and Vatsa (1998) an improved treatment of external boundary conditions for three-dimensional flow computations is discussed, and Tsynkov (1998) contains a major review of the numerical solution of problems in unbounded domains.

The work of Rols *et al.* (1998) introduces the idea of fractal absorbing boundary conditions in electromagnetic simulations by an application of the spectral moments method. In Rowley and Colonius (2000) so-called discrete non-reflecting boundary conditions are discussed for linear hyperbolic systems and in Schmidt (2000), such boundary conditions are discussed for the Helmholtz equation.

The review of Bruneau (2000) discusses boundary conditions not only for the incompressible but also for compressible Navier–Stokes equations, and Huan and Thompson (2000) discuss boundary conditions for the time-dependent wave equation. The question of adequate artificial boundary conditions for the computation of external flows with jets is reviewed in, Tsynkov *et al.* (2000) and a further review of external boundary conditions for three-dimensional problems of aerodynamics is given in Tsynkov (2000). Thompson and Huan (2000) contains exact nonreflecting boundary conditions and Ryaben'kii *et al.* (2001) discuss discrete artificial boundary conditions for the time-dependent wave equation. Higher order artificial boundary conditions for nonlinear wave propagation with backscattering are introduced in Fibich and Tsynkov (2001) and Lions *et al.* (2002) discusses the questions of the well posedness of an absorbing layer for hyperbolic problems. Nazarov and Specovius-Neugebauer (2003) introduced nonlinear artificial boundary conditions for the exterior three-dimensional Navier–Stokes problem, together with pointwise error estimates. An adaptive finite element method with perfectly matched absorbing layers for the computation of wave scattering by periodic structures is introduced in Chen and Wu (2003), and in Novak and Bonazzola (2004) absorbing boundary conditions for the simulation of gravitational waves with spectral methods are discussed. In Nataf (2005), a new construction of perfectly matched layers for the linearize Euler equations is presented.

6. Summary of Numerical Results

In what follows we give some details concerning the discretization procedure and the algorithms that we have used to solve Eqs. (175), (176) and (178) numerically. See Bönisch *et al.* (2005), Bönisch *et al.* (2006) for details. To unburden the notation we suppress throughout this section the "tildes".

6.1. *Galerkin finite element discretization*

In order to solve Eq. (175), we have considered a discretization based on conforming mixed finite elements with continuous pressure. This discretization starts from a variational formulation of the system of Eq. (175).

For a bounded domain $\mathbf{D} \subset \mathbf{R}^2$, we denote by $L^2(\mathbf{D})$ the Lebesgue space of square-integrable functions on $\mathbf{D}$ equipped with the inner product and the associated norm

$$(f, g)_{\mathbf{D}} = \int_{\mathbf{D}} fg\, d\mathbf{x}, \quad \|f\|_{\mathbf{D}} = (f, f)_{\mathbf{D}}^{1/2}.$$

The pressure is assumed to be an element of the space $L_0^2(\mathbf{D}) := \{q \in L^2(\mathbf{D}) | \int_{\mathbf{D}} q\, d\mathbf{x} = 0\}$, which defines it uniquely. The $L^2(\mathbf{D})$ functions with generalized (in the sense of distributions) first-order derivatives in $L^2(\mathbf{D})$ form the Sobolev space $H^1(\mathbf{D})$, and we define $H_0^1(\mathbf{D}) = \{v \in H^1(\mathbf{D}) | v|_{\partial\mathbf{D}} = 0\}$. Now let $W = [H_0^1(\mathbf{D})]^2 \times L_0^2(\mathbf{D})$. For $\mathbf{w} = \{\mathbf{v}, p\} \in W$ and $\phi = \{\varphi, q\} \in W$, we define the semi-linear form

$$\begin{aligned} \mathcal{A}(\mathbf{w}; \phi) = {} & \rho\, (((\mathbf{v} + \mathbf{u}_\infty) \cdot \nabla)\mathbf{v}, \varphi)_{\mathbf{D}} - (p, \nabla \cdot \varphi)_{\mathbf{D}} \\ & + 2\mu \int_{\mathbf{D}} \mathcal{D}(\mathbf{v}) : \mathcal{D}(\varphi)\, d\mathbf{x} - (\nabla \cdot \mathbf{v}, q)_{\mathbf{D}}, \end{aligned} \tag{193}$$

which is obtained by testing Eq. (175) with $\phi \in W$ and by integration by parts of the diffusive term and the pressure gradient (see e.g., Rannacher (2000), Galdi (1998f), Turek (1999) and Heywood *et al.* (1992) for more details). $\mathcal{D}(\mathbf{v})$ denotes the deformation tensor, i.e., $\mathcal{D}(\mathbf{v}) = \frac{1}{2}\left(\nabla v + (\nabla v)^T\right)$. Then, a weak form of Eq. (175) can be formulated as: find $\mathbf{w} = \{\mathbf{v}, p\} \in W$, such that

$$\mathcal{A}(\mathbf{w}; \phi) = 0, \quad \forall \phi \in W. \tag{194}$$

The discretization of problem (194) uses a conforming finite element space $W_h \subset W$ defined on quasi-uniform triangulations $\mathcal{T}_h = \{K\}$ consisting of quadrilateral cells K covering the domain $\mathbf{D}$. We have used the standard Hood-Taylor finite elements (Hood and Taylor (1973)) for the trial and test spaces, i.e., we used

$$W_h = \{(\mathbf{v}, p) \in [C(\overline{\mathbf{D}})]^3 |\, \mathbf{v}|_K \in [Q_2]^2,\ p|_K \in Q_1\},$$

where Q_r describes the space of iso-parametric tensor-product polynomials of degree r (for a detailed description of this standard construction process, see for example, Brenner and Scott (1994)). This choice for the trial and test functions guarantees a stable approximation of the pressure since the Babuska–Brezzi inf-sup stability condition is satisfied uniformly in $\mathbf{D}$ (see Brezzi and Falk (1991) and references therein). The advantage, when compared to equal order function spaces for the pressure and the velocity, is that no additional stabilization terms are needed. The discrete counterpart of the problem (194) then reads: find $\mathbf{w}_h = \{\mathbf{v}_h, p_h\} \in \mathbf{w}_{b,h} + W_h$, such that

$$\mathcal{A}(\mathbf{w}_h; \phi_h) = 0, \quad \forall \phi_h \in W_h. \tag{195}$$

Here, $\mathbf{w}_{b,h}$ describes the prescribed Dirichlet data on the boundary $\mathbf{\Gamma}$ of the domain $\mathbf{D}$.

The artificial boundary conditions of Sec. 5 are independent of the details of the geometry of the body, but they depend explicitly on drag and lift. The accurate determination of these forces is therefore a key issue in this context. We have used the approach proposed in Giles *et al.* (1997) which is based on a reformulation of the expressions for drag and lift in terms of volume integrals. This formulation allows to attain the full order of convergence for the values of drag and lift.

6.2. *The solver*

We have solved the nonlinear algebraic system (195) in a fully coupled manner by means of a damped Newton method. Denoting the derivative of $\mathcal{A}(\cdot, \cdot)$ taken at a discrete function $\mathbf{w}_h \in W_h$ by $\mathcal{A}'(\mathbf{w}_h, \cdot)(\cdot)$, the linear system arising at the Newton step number k has the following form,

$$\mathcal{A}'(\mathbf{w}_h^k, \boldsymbol{\phi}_h)(\hat{\mathbf{w}}_h^k) = (\mathbf{r}_h^k, \boldsymbol{\phi}_h), \quad \forall \phi_h \in W_h, \tag{196}$$

where $\mathbf{r}_h^k$ is the equation residual of the current approximation $\mathbf{w}_h^k$, and where $\hat{\mathbf{w}}_h^k$ corresponds to the needed correction. The updates $\mathbf{w}_h^{k+1} = \mathbf{w}_h^k + \alpha^k \hat{\mathbf{w}}_h^k$ with a relaxation parameter α^k chosen by means of Armijo's rule are carried out until convergence. In practice, the Jacobian involved in Eq. (196) is directly derived from the analytical expression for the derivative of the variational system (195).

It is well known that the rapid convergence of Newton iterations greatly depends on the quality of the initial approximation (see e.g., Kelley (1995)). In order to find such an initial approximation, we consider a mesh hierarchy $\mathcal{T}_{h_l}$ with $\mathcal{T}_{h_l} \subset \mathcal{T}_{h_{l+1}}$, and the corresponding system of Eq. (195) is successively solved by taking advantage of the previously computed solution, i.e., the nonlinear Newton steps are embedded in a nested iteration process.

More precisely, the linear subproblems (196) are solved by the generalized minimal residual method (GMRES), see Saad (1996), preconditioned by means of multigrid iterations. See Wesseling and Oosterlee (2001) and Wesseling (1992) and references therein for a description of the different multigrid techniques for flow simulations. Our preconditioner is based on a new multigrid scheme which is optimized for conformal higher order finite element methods. It is a key ingredient of the overall solution process. Two specific features characterizing the scheme are: varying order of the finite element Ansatz on the mesh hierarchy and a Vanka type smoother (Vanka (1986)) adapted to higher order discretization. This somewhat technical part of the solver is described in full detail in Heuveline (2003). Its implementation is part of the HiFlow project (see Heuveline (2000)).

6.3. *Numerical results*

Figure 2 summarizes some of the results in two dimensions. See Bönisch *et al.* (2005) and Bönisch *et al.* (2006) for details. Similar work in three dimensions is in preparation and will be published elsewhere.

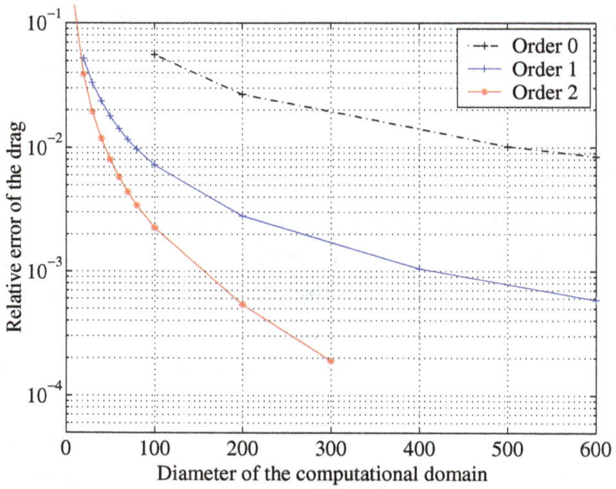

Fig. 2. The figure shows the size of the relative error for the drag as a function of the diameter of the computational domain for a test configuration consisting of a flat plate of diameter one at Reynolds number $\mathrm{Re} = 1$. A reference value for the drag has been computed with a very large scale computation on a domain of size 5,000. To compute the drag with an error of about one percent, a domain with 500 times the body size is needed with naive boundary conditions, with about 100 times the body size with first order and with about 50 times the body size with second-order artificial boundary conditions.

7. Bibliography

The following notes provide entry points to the literature on low Reynolds number flows in as much as they have not yet been provided in the more specific bibliographic sections above or within the text. The main goal is again to provide a list of references from a broad range of applications. The big number of recent references shows that the subject of low Reynolds numbers is in spite of its long history still, or maybe, again, a very active topic of research.

7.1. *Books*

A very interesting early reference are the lectures on fluid mechanics by Goldstein (1957) and a classic reference to boundary layer theory is the book by Schlichting and Gersten (1999). A classic reference for perturbation theory in fluid mechanics is van Dyke (1975). A nicely written and easy to read introduction to the problem of exterior flows is the booklet by Ockendon and Ockendon (1995). A recent book that contains an important section on low Reynolds number flows is Guyon *et al.* (2001). A book on viscous incompressible flows at (very) low Reynolds numbers is Kohr and Pop (2004).

7.2. *Boundary layer theory, wakes*

The computation of the forces on bodies has only made significant progress after the introduction of boundary layer theory, see Blasius (1908), which has allowed to explain and resolve the d'Alambert paradox, i.e., the fact that the Euler equations lead to a no drag theorem in two dimensions and a no drag and no lift theorem in three dimensions. See also Goldshtik (1990) for a review of viscous-flow paradoxes. Extensive computations of wakes based on boundary layer theory can be found in Stewartson (1957). Questions concerning the uniqueness of solutions are discussed for example in Smith (1984). The computation of axisymmetric flows for slender bodies goes back to the paper of Bodonyi *et al.* (1985). More recently boundary layer computations have been reviewed in Cole (1994), Anderson Jr (2005), Tulapurkara (2005) and Cowley (2006). In this general context it is also useful to consult the publications by Lamb *et al.* (2003) which discusses bifurcation theory from periodic solutions with spatiotemporal symmetry.

7.3. *Expansion techniques*

Expansion techniques have played a very important role for the computation of the forces that act on bodies that move through liquids. See in particular also Keller and Ward (1996) for a reference concerning low Reynolds number flows, and Boyd (1999) for a discussion of expansion techniques. See also Schwartz (2002) for a discussion of the work of van Dyke. Recent references are Vorobieff *et al.* (2002) and the publication by Kohr (2004) where the method of matched asymptotic expansions for low Reynolds number flow past a cylinder of arbitrary cross section is discussed.

7.4. *Flows around plates, cylinders and spheres*

With the introduction of boundary layer theory many authors have computed the drag on simple geometric obstacles like plates, cylinders and spheres. For (semi-infinite) flat plates interesting references are Alden (1948), Imai (1957), Olmstead and Hector (1966), Olmstead (1975) and Lagerstrom (1975). See also Bichsel and Wittwer (2007) for a recent review of the semi-infinite flat plate problem.

Interesting references discussing flows around cylinders are Dennis and Shimshoni (1965), Kropinski *et al.* (1995) and Titcombe *et al.* (1999). For a recent reference to experimental techniques, including the case of time periodic flows see Fujisawa *et al.* (2005).

Flows around spheres, including the question of the stability of such flows are discussed in Shirayama (1992) and Cliffe *et al.* (2000). An analytical solution of low Reynolds number slip flows past a sphere can be found in Barber and Emerson (2000), and the drag on a sphere moving horizontally in a stratified fluid is discussed in Greenslade (2000). Jayaraman and Belmonte (2003) contains the observation of the oscillations of a solid sphere falling through a wormlike micellar fluid.

7.5. *Numerical studies*

An early reference containing a numerical study of the drag on a sphere at low and intermediate Reynolds number is Le Clair *et al.* (1970). A recent reference for the flow around a cylinder is Padrino and Joseph (2006). Oscillatory flows are discussed in Testik *et al.* (2005). Another reference to the time dependent case is Bönisch and Heuveline (2006). Interesting work on the bases of the lattice Boltzmann method is Verberg and Ladd (2000) and Latt *et al.* (2006). The application of the lattice Boltzmann method to the simulation of particle-fluid suspensions can be found in Ladd and Verberg (2001).

7.6. *Linearized problems (Stokes, Oseen)*

Linearized problems play an important role. On one hand they are directly used as approximations to the full equations, on the other hand their study is the basis of most work on the nonlinear problems. The Stokes equations are quantitatively useful at very low Reynolds numbers (less than one). Quantitatively, the Oseen equations are less successful and have played a less important role for direct computations. See for example Weisenborn and Bosch (1995). In spite of these shortcomings the Oseen equation captures the asymptotic behavior of the flows in the regime of Reynolds numbers above one to several hundred much better than the Stokes equations. Interesting aspects of the Oseen equation are discussed in Kress and Meyer (2000). The hydrodynamic forces on submerged rigid bodies and its relation to the far field behavior are discussed in Guenther *et al.* (2002). A well-posedness analysis for the so-called Oseen coupling method for exterior flows is discussed in He *et al.* (2004). A recent discussion of the Oseen problem in the whole space is due to Boulmezaoud and Razafison (2005). Guenther and Thomann (2005) contains a new discussion of the fundamental solutions of the Stokes and Oseen problem in two spatial dimensions, and Thomann and Guenther (2006) contains the fundamental solutions, including the time dependent case, for the linearized Navier–Stokes equations for spinning bodies in three spatial dimensions. Girault *et al.* (1992) contains a stream-function-vorticity variational formulation for the exterior Stokes problem in weighted Sobolev spaces, and Amrouche and Razafison (2006) provide weighted estimates for the Oseen problem in three dimensions.

7.7. *Other references on exterior flows*

The standard reference concerning the existence of solutions for exterior flows is Leray (1934). The question of the behavior of the solutions at infinity was an open problem for a long time. See Finn (1960) and Finn (1965). Another early reference concerning solutions of the stationary and non stationary Navier–Stokes equations in exterior domains is Chen (1993). The problem is reviewed in Galdi (1998c). See also Galdi (1999b) for the description of an important open problem. More recent references are Farwig (1998), Cerejeiras and Kähler (2000), Galdi and Rabier (2000) and Giga *et al.* (2001). Stability questions are discussed in Biler *et al.* (2004). The approximation of three dimensional stationary flows by flows in bounded domains are discussed in Deuring and Kračmar (2004). Nazarov (1999) contains the discussion of the Navier–Stokes problem in a two-dimensional domain with angular outlets to infinity, and Shibata and Yamazaki (2005) provides uniform estimates for the velocity at infinity for stationary solutions. Geissert *et al.* (2004) reviews the theory of the Navier–Stokes flow in the exterior of a moving or rotating obstacle, and Galdi (2006) discusses modes, nodes and volume elements for stationary solutions of the Navier–Stokes problem past a three-dimensional body.

Acknowledgment

We would like to thank Christoph Boeckle for a careful reading of the manuscript.

References

1. M. Abdulrahim and J. Cocquyt, Development of mission-capable flexible-wing micro air vehicles, Technical report, University of Florida (2000), http://pcc2341f.unige.ch/pdf/Abdulrahim_cocquyt.pdf.
2. Y. D. Afanasyev and Yakov, Flight in a viscous fluid: Asymptotic theory of the vortex wake, *Phys. Fluids* **17** (2005): 038104, http://pcc2341f.unige.ch/pdf/paper_flight.pdf.
3. H. L. Alden, Second approximation to the laminar boundary layer flow over a flat plate, *J. Math. Phys.*, **XVII** (1948): 91–104.
4. C. J. Amick, On the asymptotic form of Navier–Stokes flow past a body in the plane, *J. Differ. Equat.* **91** (1991): 149–167.
5. C. Amrouche and U. Razafison, Weighted estimates for the Oseen problem in $\mathbf{R}^3$, *Appl. Math. Lett.* **19** (2006): 56–62, http://pcc2341f.unige.ch/pdf/art33.pdf.
6. J. D. Anderson, Jr. Ludwig Prandtl's boundäry layer, *Bernoulli* **1700** (2005): 42–48, http://pcc2341f.unige.ch/pdf/prandtl_vol58no12p42_48.pdf.
7. J. E. Avron, O. Gat and O. Kenneth, low Reynolds numbers, *Phys. Rev. Lett.*, Vol. 93 (2004), http://pcc2341f.unige.ch/pdf/optimalswimmingprl.pdf.
8. M. B. Baker, Cloud microphysics and climate, *Sci.* **276** (1997): 279–306.
9. R. W. Barber and D. R. Emerson, Analytical solutions of low Reynolds number slip flow past a sphere, Technical report, Daresbury Laboratory (2000), http://pcc2341f.unige.ch/pdf/dltr-2000001.pdf.

10. G. K. Batchelor, *An Introduction to Fluid Dynamics.* (Cambridge University Press, 1967).
11. A. Bayliss and E. Turkel, Radiation boundary conditions for wave-like equations, *Commun. Pure Appl. Math.* **33** (1980): 707–725.
12. K. V. Beard and H. R. Pruppacher, A determination of the terminal velocity and drag of small water drops by means of a wind tunnel, *J. Atmos. Sci.* **26** (1969): 1066–1072, http://pcc2341f.unige.ch/pdf/i1520-0469-26-5-1066.pdf.
13. S. Berger. *Laminar Wakes* (American Elsevier Publishing Company, Inc., 1971).
14. T. R. Bewley, Flow control: New challenges for a new Renaissance, *Prog. Aero. Sci.* **37** (2001): 21–58, http://pcc2341f.unige.ch/pdf/bewley.pdf.
15. D. Bichsel and P. Wittwer, Stationary flow past a semi-infinite flat plate: Analytical and numerical evidence for a symmetry-breaking solution, *J. Stat. Physics* (2007), http://pcc2341f.unige.ch/pdf/plate_preprint.pdf.
16. P. Biler, M. Cannone and G. Karch, Asymptotic stability of the Navier–Stokes flow past an obstacle (Banach Center Publications, 2004).
17. H. Blasius, Grenzschichten in Flüssigkeiten mit kleiner Reibung, *Zeitschrift für Mathematik und Physik* **56** (1908): 1–37, http://pcc2341f.unige.ch/pdf/Kapitel1.pdf.
18. R. J. Bodonyi, F. T. Smith and A. Kluwick, Axisymmetric flow past a slender body of finite length, *Proc. Roy. Soc. Lond., Math. Phys. Sci.* **400** (1985): 37–54, http://pcc2341f.unige.ch/pdf/bodonyi.pdf.
19. J. L. Bona, K. S. Promislow and C. E. Wayne, Higher order asymptotics of decaying solutions of some nonlinear, dispersive, dissipative wave equations, *Nonlinearity* **8** (1995): 1179–1206, http://pcc2341f.unige.ch/pdf/Bona1995.pdf.
20. J. Bona, K. Promislow and G. Wayne, On the asymptotic behavior of solutions to nonlinear, dispersive, dissipative wave equations, *Math. Comput. Simulat.* **37** (1994): 265–277, http://pcc2341f.unige.ch/pdf/93-335.ps[1].pdf.
21. T. Z. Boulmezaoud and U. Razafison, On the steady Oseen problem in the whole space, *Hiroshima Math. J.* **35** (2005): 371–401, http://pcc2341f.unige.ch/pdf/{O}seenInWholeSpace.pdf.
22. J. P. Boyd. The devil's invention: Asymptotic, superasymptotic and hyperasymptotic series, *Acta Appl. Math.* **56** (1999): 1–98, http://pcc2341f.unige.ch/pdf/boydacta-applicreview.pdf.
23. S. C. Brenner and R. L. Scott, *The Mathematical Theory of Finite Element Methods* (Springer, Berlin-Heidelberg-New York, 1994).
24. F. Brezzi and R. Falk, Stability of higher-order Hood-Taylor methods, *SIAM J. Numer. Anal.* **28** (1991): 581–590.
25. J. Bricmont and A. Kupiainen. Global large time self-similarity of a thermal-diffusive combustion system with critical nonlinearity, *J. Differ. Equat.* **130** (1996a): 9–35, http://pcc2341f.unige.ch/pdf/Bricmont1996.pdf.
26. J. Bricmont and A. Kupiainen, Stability of moving fronts in the Ginzburg–Landau equation, *Commun. Math. Phys.* **159** (1994a): 287–318, http://pcc2341f.unige.ch/pdf/1104254600.pdf.
27. J. Bricmont and A. Kupiainen, Universality in blow-up for nonlinear heat equations, *Nonlinearity*, **7** (1994b): 539–575, http://pcc2341f.unige.ch/pdf/0951-7715-7-2-011.pdf.
28. J. Bricmont and A. Kupiainen, Stable non-Gaussian diffusive profiles, *Nonlinear Anal.* **26** (1996b): 583–593, http://pcc2341f.unige.ch/pdf/bricmont-1993-.pdf.

29. J. Bricmont and A. Kupiainen, Renormalization group for fronts and patterns, In *European Congress of Mathematics, Vol. I (Budapest, 1996)*, volume 168 of *Progr. Math.* (Birkhäuser, Basel, 1998), pp. 121–130. http://pcc2341f.unige.ch/pdf/UCL-IPT-96-18.pdf.
30. J. Bricmont, A. Kupiainen and J. Xin, Global large time self-similarity of a thermal-diffusive combustion system with critical nonlinearity, *J. Differ. Equat.*, **130** (1996): 9–35, http://pcc2341f.unige.ch/pdf/Bricmont1996.pdf.
31. J. Bricmont and A. Kupiainen, Renormalization group and the Ginzburg–Landau equation, *Commun. Pure Appl. Math.* **150** (1992): 193–208, http://pcc2341f.unige.ch/pdf/1104251790.pdf.
32. J. Bricmont and A. Kupiainen, Renormalizing partial differential equations, In *Constructive Physics (Palaiseau, 1994)*, volume 446 of *Lecture Notes in Physics* (Springer, Berlin, 1995), pp. 83–115. http://pcc2341f.unige.ch/pdf/bricmont-1994.pdf.
33. J. Bricmont, A. Kupiainen and G. Lin, Renormalization group and asymptotics of solutions of nonlinear parabolic equations, *Commun. Pure Appl. Math.* **47** (1994): 893–922, http://pcc2341f.unige.ch/pdf/9306008.pdf.
34. P. P. Brown and D. F. Lawler, Sphere drag and settling velocity revisited, *J. Environ. Eng.* **129** (2003): 222–231.
35. C.-H. Bruneau, Boundary conditions on artificial frontiers for incompressible and compressible Navier–Stokes equations, *M2AN Math. Model. Numer. Anal.* **34** (2000): 303–314, http://pcc2341f.unige.ch/pdf/Boundary.pdf.
36. S. Bönisch and V. Heuveline, On the numerical simulation of the instationary free fall of a solid in a fluid. I. The Newtonian case (2006), preprint, http://pcc2341f.unige.ch/pdf/Preprint2004-24.pdf.
37. S. Bönisch, V. Heuveline and P. Wittwer, Adabtive boundary conditions for exterior flow problems, *J. Math. Fluid Mech.* **7** (2005): 85–107, http://pcc2341f.unige.ch/pdf/first_order_ABC.pdf.
38. S. Bönisch, V. Heuveline and P. Wittwer, Second-order adaptive boundary conditions for exterior flow problems: Non-symmetric stationary flows in two dimensions, *J. Math. Fluid Mech.* **8** (2006): 1–26, http://pcc2341f.unige.ch/pdf/second_order_ABC.pdf.
39. B. H. Carmichael, Low Reynolds number airfoil survey, Technical report, NASA, (1981), http://arjournals.annualreviews.org/doi/pdf/10.1146/annurev.fl.15.010183.001255.
40. P. Cerejeiras and U. Kähler, Elliptic boundary value problems of fluid dynamics over unbounded domains, *Math. Meth. Appl. Sci.* **23** (2000): 81–101, http://pcc2341f.unige.ch/pdf/Cerejeiras2000.pdf.
41. L.- Y. Chen, N. Goldenfeld and Y. Oono, Renormalization group theory for global asymptotic analysis, *Phys. Rev. lett.* **73** (1994): 1311–1315, http://pcc2341f.unige.ch/pdf/p1311_1.pdf.
42. Z. Chen and H. Wu, An adaptive finite element method with perfectly matched absorbing layers for the wave scattering by periodic structures, *SIAM J. Numer. Anal.* **41** (2003): 799–826, http://pcc2341f.unige.ch/pdf/chen-adaptive.pdf. preprint.
43. Z. M. Chen, Solutions of the stationary and nonstationary Navier–Stokes equations in exterior domains, *Pac. J. Math.* **159** (1993): 227–240, http://pcc2341f.unige.ch/pdf/1102634262[1].pdf.

44. S. Childress and R. Dudley, Transition from ciliary to flapping mode in a swimming mollusc: Flapping flight as a bifurcation in Re_ω, *J. Fluid Mech.* **498** (2004): 257–288, http://pcc2341f.unige.ch/pdf/flapfinal.pdf.
45. K. A. Cliffe, A. Spence and S. J. Tavener, O(2)-symmetry breaking bifurcation: With application to the flow past a sphere in a pipe, *Int. J. Numer. Meth. Fluids* **32** (2000): 175–200, http://pcc2341f.unige.ch/pdf/02_web.pdf.
46. R. Clift, J. R. Grace, and M. E. Weber, *Bubbles, Drops, and Particles* (Dover Publications, 2005).
47. C. Coclici, I. Sofronov and W. L. Wendland, A domain decomposition method and far-field boundary conditions for 2D transonic flow around an airfoil, In *Analysis, Numerics and Applications of Differential and Integral Equations (Stuttgart, 1996)*, Vol. 379 of *Pitman Res. Notes Math. Ser.* (Longman, Harlow, 1998), pp. 58–63, http://pcc2341f.unige.ch/pdf/coclici98domain.pdf.
48. J. D. Cole, The development of perturbation theory at GALCIT, *SIAM Rev.* **36** (1994): 425–430, http://pcc2341f.unige.ch/pdf/cole.pdf.
49. P. Collet and J.-P. Eckmann, Solutions without phase-slip for the Ginsburg–Landau equation, *Commun. Math. Phys.* **145** (1992a): 345–356, http://pcc2341f.unige.ch/pdf/1104249645[1].pdf.
50. P. Collet, J.-P. Eckmann and H. Epstein, Diffusive repair for the Ginzburg–Landau equation, *Helv. Phys. Acta.* **65** (1992): 56–92, http://pcc2341f.unige.ch/pdf/collet92diffusive.pdf.
51. P. Collet and J.-P. Eckmann, Space-time behavior in problems of hydrodynamic type: A case study, *Nonlinearity* **5** (1992b): 1265–1302 http://pcc2341f.unige.ch/pdf/no920604.pdf.
52. S. J. Cowley, Laminar boundary-layer theory: A 20th century paradox? (2006), preprint, http://pcc2341f.unige.ch/pdf/short.pdf.
53. S. C. R. Dennis and M. Shimshoni, The steady flow of a viscous fluid past a circular cylinder, Technical report, Ministry of Aviation, Aeronautical Research Council (1965), http://pcc2341f.unige.ch/pdf/good0797.pdf.
54. P. Deuring and S. Kračmar, Exterior stationary Navier–Stokes flows in 3D with non-zero velocity at infinity: Approximation by flows in bounded domains, *Math. Nachr.* **269/270** (2004): 86–115, http://pcc2341f.unige.ch/pdf/deuring001.pdf.
55. M. H. Dickinson, F. Lehmann and S. P. Sane, Wing rotation and the aerodynamic basis of insect flight, *Sci.* **284** (1999): 1954, http://pcc2341f.unige.ch/pdf/1954.pdf.
56. J.-P. Eckmann and Th. Gallay, Front solutions for the Ginzburg–Landau equation. *Commun. Math. Phys.* **152** (1993): 221–248, http://pcc2341f.unige.ch/pdf/1104252408.pdf.
57. J.-P. Eckmann and G. Schneider, Nonlinear stability of bifurcating front solutions for the Taylor–Couette problem, *ZAMM Z. Angew. Math. Mech.* **80** (2000): 745–753, http://pcc2341f.unige.ch/pdf/rXiv.pdf.
58. J.-P. Eckmann and G. Schneider, Non-linear stability of modulated fronts for the Swift–Hohenberg equation, *Commun. Math. Phys.* **225** (2002): 361–397, http://pcc2341f.unige.ch/pdf/00-116.ps[1].pdf.
59. J.-P. Eckmann and C. E. Wayne, The nonlinear stability of front solutions for parabolic partial differential equations, *Commun. Math. Phys.* **161** (1994): 323–334, http://pcc2341f.unige.ch/pdf/1104269905.pdf.

60. J.-P. Eckmann and C. E. Wayne, Non-linear stability analysis of higher-order dissipative partial differential equations, *Math. Phys. Electron. J.* **4** (1998a), http://pcc2341f.unige.ch/pdf/Eckmann1998a.pdf.
61. J.-P. Eckmann and C. E. Wayne, Higher order dissipative partial differential equations, *Math. Phys. Eelectron. J.* **4** (1998b): 1–20, http://pcc2341f.unige.ch/pdf/Eckmann1998a.pdf.
62. J.-P. Eckmann and C. E. Wayne and P. Wittwer, Geometric stability analysis for the periodic solutions of the Swift–Hogenberg equation, *Commun. Math. Phys.* **190** (1997): 173–211, http://pcc2341f.unige.ch/pdf/PDEwithJPWayne.pdf.
63. B. Engquist and A. Majda, Radiation boundary conditions for acoustic and elastic calculations, *Comm. Pure Appl. Math.* **32** (1979): 313–357.
64. R. Eppler and D. M. Somers, *A Computer Program for the Design and Analysis of Low-Speed Airfoils* (NASA, USA, 1981).
65. R. Eppler and D. M. Somers, Low speed airfoil design and analysis, *Adv. Technol. Airfoil Res. Vol. 1*, **1** (1979): 73–100.
66. R. Eppler and D. M. Somers, Airfoil design for Reynolds numbers between 50,000 and 500,000. *Conf. Low Reynolds Number Airfoil Aerodynamics*, (1985), pp. 1–14.
67. A. Esmaeeli and G. Tryggvason, Direct numerical simulations of bubbly flows, part 2. Moderate Reynolds number arrays, *J. Fluid Mech.* **385** (1999): 325–358.
68. A. Esmaeeli and G. Tryggvason, Direct numerical simulations of bubbly flows, part 1. Low Reynolds number arrays, *J. Fluid Mech.* **377** (2000): 313–345.
69. R. Farwig, The stationary Navier–Stokes equations in a 3D-exterior domain. *Lect. Notes Numer. Appl. Anal.* **16** (1998): 53–115.
70. T. Fevens and H. Jiang, Absorbing boundary conditions for the Schrödinger equation, Technical Report 95–376, Dept. of Computing and Information Science, Queen's University at Kingston, Kingston, Ontario, Canada (1995). http://pcc2341f.unige.ch/pdf/fevens95absorbing.pdf.
71. G. Fibich and S. Tsynkov, High-order two-way artificial boundary conditions for nonlinear wave propagation with backscattering, *J. Comput. Phys.* **171** (2001): 632–677, http://pcc2341f.unige.ch/pdf/fibich.pdf.
72. R. Finn, An energy theorem for viscous fluid motions, *Arch. Ration. Mech. Anal.* **6** (1960): 371–381, http://www.springerlink.com/content/r8u3041g7653/.
73. R. Finn. On the exterior stationary problem for the Navier–Stokes equations, and associated perturbation problems. *Arch. Ration. Mech. Anal.* **19** (1965): 363–406, http://www.springerlink.com/content/g3267010k313/.
74. G. B. Foote and P. S. du Toit, Terminal velocity of raindrops aloft, *J. Appl. Meteorol.* **8** (1969): 249–253.
75. N Fujisawa, S. Tanahashi, and K. Srinivas, Evaluation of pressure field and fluid forces on a circular cylinder with and without rotational oscillation using velocity data from PIV measurement, *Meas. Sci. Technol.* **16** (2005): 989–996, http://pcc2341f.unige.ch/pdf/mst5_4_011.pdf.
76. G. P. Galdi, *An Introduction to the Mathematical Theory of the Navier–Stokes Equations: Linearized Steady Problems*, Springer Tracts in Natural Philosophy, Vol. 38 (Springer-Verlag, 1998a).

77. G. P. Galdi, *An Introduction to the Mathematical Theory of the Navier–Stokes Equations: Nonlinear Steady Problems*, Springer Tracts in Natural Philosophy, Vol. 39 (Springer-Verlag, 1998b).
78. G. P. Galdi, Mathematical questions relating to the plane steady motion of a Navier–Stokes fluid past a body, *Lect. Notes Num. Appl. Anal.* **16** (1998c): 117–160.
79. G. P. Galdi and P. J. Rabier, Sharp existence results for the stationary Navier–Stokes problem in three-dimensional exterior domains, *Arch. Ration. Mech. Anal.* **154** (2000): 343–368, http://pcc2341f.unige.ch/pdf/Galdi2000.pdf.
80. G. P. Galdi and A. L. Silvestre, Strong solutions of the Navier–Stokes equations around a roating obstacle, *Arch. Ration. Mech. Anal.* **176** (2005): 331–350.
81. G. P. Galdi, Determining modes, nodes and volume elements for stationary solutions of the Navier–Stokes problem pat a three-dimensional body, *Arch. Ration. Mech. Anal.* **180** (2006): 97–126, http://pcc2341f.unige.ch/pdf/galdi009.pdf.
82. G. P. Galdi, *An Introduction to the Mathematical Theory of the Navier–Stokes Equations: Linearized Steady Problems*, Springer Tracts in Natural Philosophy, Vol. 38 (Springer-Verlag, 1998d).
83. G. P. Galdi, *An Introduction to the Mathematical Theory of the Navier–Stokes Equations: Nonlinear Steady Problems*, Springer Tracts in Natural Philosophy, Vol. 39 (Springer-Verlag 1998e).
84. G. P. Galdi, *An Introduction to the Mathematical Theory of the Navier–Stokes Equations: Nonlinear Steady Problems* Springer Tracts in Natural Philosophy, Vol. 39, (Springer-Verlag 1998f).
85. G. P. Galdi, On the existence of symmetric steady-state solutions to the plane exterior Navier–Stokes problem for a large Reynolds number, *Adv. Fluid Dynam. Quaderni Di Matematica della II Universita' di Napoli* **4** (1999a): 1–25.
86. G. P. Galdi, On the existence of symmetric steady-state solutions to the plane exterior Navier–Stokes problem for a large Reynolds number, *Adv. Fluid Dynam. Quaderni Di Matematica della II Universita' di Napoli* **4** (1999b): 1–25.
87. T. Gallay and A. Mielke, Convergence results for a coarsening model using global linearization, *J. Nonlinear Sci.* **13** (2003): 311–346, http://pcc2341f.unige.ch/pdf/Gallay2003.pdf.
88. T. Gallay, Local stability of critical fronts in nonlinear parabolic partial differential equations, *Nonlinearity*, **7** (1994): 741–764, http://pcc2341f.unige.ch/pdf/Gallay1994.pdf.
89. T. Gallay, A center-stable manifold theorem for differential equations in banach spaces, *Commun. Math. Phys.* **152** (1993): 249–268, http://pcc2341f.unige.ch/pdf/thcm.pdf.
90. T. Gallay and G. Raugel, Scaling variables and asymptotic expansions in damped wave equations, *J. Differ. Equat.*, **150** (1998): 42–97, http://pcc2341f.unige.ch/pdf/97-522.ps[1].pdf.
91. T. Gallay and C. E. Wayne, Long-time asymptotics of the Navier–Stokes and vorticity equations on R^3, *Phil. Trans. Roy. Soc. Lond.* **360** (2002a): 2155–2188, http://pcc2341f.unige.ch/pdf/70hghxlm02a6qwgp.pdf.
92. T. Gallay and C. E. Wayne, Invariant manifolds and the long-time asymptotics of the Navier–Stokes and vorticity equations on R^2, *Arch. Rat. Mech. Anal.* **163** (2002b): 209–258, http://pcc2341f.unige.ch/pdf/42rggl3gjcph75g4.pdf.

93. T. Gallay and A. Mielke, Diffusive mixing of stable states in the Ginzburg–Landau equation, *Commun. Math. Phys.*, **199** (1998): 71–97, http://pcc2341f.unige.ch/pdf/6ql2lee3gbx18k46.pdf.
94. T. Gallay and G. Raugel, Scaling variables and stability of hyperbolic fronts, *SIAM J. Math. Anal.* **32** (2000): 1–29, http://pcc2341f.unige.ch/pdf/9812007.pdf.
95. T. Gallay and C. E. Wayne, Global stability of vortex solutions of the two-dimensional Navier–Stokes equation, *Commun. Math. Phys.* **255** (2005): 97–129, http://pcc2341f.unige.ch/pdf/jm8amjjc2g9554mx.pdf.
96. T. Gallay, G. Schneider and H. Uecker, Stable transport of information near essentially unstable localized structures, *Discrete Contin. Dyna. Syst. Ser. B* **4** (2004): 349–390, http://pcc2341f.unige.ch/pdf/0306335.pdf.
97. M. Geissert, H. Heck and M. Hieber, Lp-theory of the Navier–Stokes flow in the exterior of a moving or rotating obstacle (2004), preprint, http://pcc2341f.unige.ch/pdf/geissert2004ptn.pdf.
98. Y. Giga, S. Matsui and O. Sawada, Global existence of two-dimensional Navier–Stokes flow with nondecaying initial velocity, *J. Math. Fluid Mech.* **3** (2001): 203–315, http://pcc2341f.unige.ch/pdf/4rnbb3avqc665mwf.pdf.
99. M. Giles, M. Larson, M. Levenstam and E. Süli, Adaptive error control for finite element approximations of the lift and drag coefficients in viscous flow, Technical Report NA-97/06, Oxford University Computing Laboratory (1997).
100. V. Girault, J. Giroire and A. Sequeira, A stream-function-vorticity variational formulation for the exterior Stokes problem in weighted Sobolev spaces, *Math. Meth. Appl. Sci.* **15** (1992): 345–363, http://dx.doi.org/10.1002/mma.1670150506.
101. M. A. Goldshtik, Viscous-flow paradoxes. *Ann. Rev. Fluid Mech.* **22** (1990): 441–472, http://arjournals.annualreviews.org/doi/abs/10.1146/annurev.fl.22.010190.002301.
102. S. Goldstein, *Lectures on Fluid Mechanics* (Interscience Publishers, Ltd., London, 1957).
103. M. D. Greenslade, Drag on a sphere moving horizontally in a stratified fluid, *J. Fluid Mech.* **418** (2000): 339–350, http://pcc2341f.unige.ch/pdf/jfm_00_draft.pdf.
104. D. Greer, P. Hamory, K. Hrake and M. Drela, Design and predicitions for a high-altitude (low-Reynolds-number) aerodynamic flight experiment, Technical report, NASA (1999), http://pcc2341f.unige.ch/pdf/H-2340.pdf.
105. D. F. Griffiths, The "no boundary condition" outflow boundary condition, *Int. J. Numer. Meth. Fluids* **24** (1997): 393–411, http://pcc2341f.unige.ch/pdf/Griffiths1996.pdf.
106. M. J. Grote and J. B. Keller, On nonreflecting boundary conditions, *J. Comput. Phys.* **122** (1995): 231–243, http://pcc2341f.unige.ch/pdf/grote95nonreflecting.pdf.
107. M. J. Grote, *Nonreflecting Boundary Conditions*, Ph.D. thesis, Stanford University (1995). http://pcc2341f.unig.ch/pdf/Aron1999.pdf.
108. R. B. Guenther, R. T. Hudspeth and E. A. Thomann, Hydrodynamic forces on submerged rigid bodies — steady flow, *J. Math. Fluid Mech.* **4** (2002): 187–202, http://pcc2341f.unige.ch/pdf/bluff001.pdf.
109. R. B. Guenther and E. A. Thomann, Fundamental solutions of Stokes and Oseen problem in two spatial dimensions, *J. Math. Fluid. Mech.* **99** (2005): 1–16, http://pcc2341f.unige.ch/pdf/FundSol_2d.pdf.

110. P. Guidotti, Elliptic and parabolic problems in unbounded domains. *Mathematische Nachrichten* **272** (2004): 32–45, http://pcc2341f.unige.ch/pdf/G041.pdf.
111. R. Gunn and G. D. Kinzer, The terminal velocity of fall for water droplets in stagnant air, *J. Atmos. Sci.* **6** (1949): 243–248.
112. E. Guyon, J. P. Hulin, L. Petit and C. Mitescu, *Physical Hydrodynamics* (Oxford University Press, 2001).
113. T. Hagstrom, On high-order radiation boundary conditions, Technical report, Institute for Computational Mechanics in Propulsion (1995). http://gltrs.grc. nasa.gov/cgi-bin/GLTRS/browse.pl?1995/CR-198404.html.
114. T. Hagstrom and S. I. Hariharan, A formulation of asymptotic and exact boundary conditions using local operators, *Appl. Numer. Math. Trans. IMACS* **27** (1998): 403–416, http://pcc2341f.unige.ch/pdf/hagstrom98formulation.pdf.
115. F. Haldi and P. Wittwer, Leading order down-stream asymptotics of non-symmetric stationary Navier–Stokes flows in two dimensions. *J. Math. Fluid Mech.* **7** (2005): 611–648, http://pcc2341f.unige.ch/pdf/2dTheoryRG.pdf.
116. L. Halpern and M. Schatzman, Artificial boundary conditions for incompressible viscous flows, *SIAM J. Math. Anal.* **20** (1989): 308–353.
117. E. S. Hanf, Piv application in advanced low Reynolds number facility, *IEE Trans. Aerospace Electr. Systs.* **40** (2004): 310–319, http://pcc2341f.unige.ch/pdf/4010310.pdf.
118. Y. He, K. Li and K.-M. Liu, Oseen coupling method for the exterior flow. Part II: Well-posedness analysis, *Math. Meth. Appl. Sci.* **27** (2004): 2027–2044, http://pcc2341f.unige.ch/pdf/fulltext0001.pdf.
119. J. S. Hesthaven, The analysis and construction of perfectly matched layers for the linearized Euler equations, Technical Report TR-97-49, NASA (1997), http://pcc2341f.unige.ch/pdf/hesthaven97analysis.pdf.
120. V. Heuveline, **HiFlow** a genereal finite element C–+ package for 2D/3D flow simulation (2000), www.hiflow.de
121. V. Heuveline, On higher-order mixed FEM for low Mach number flows: Application to a natural convection benchmark problem, *Int. J. Numer. Math. Fluids* **41** (2003): 1339–1356.
122. A. J. Heymsfield, Ice crystal terminal velocities, *J. Atmos. Sci.* **29** (1972): 1348–1357, http://pcc2341f.unige.ch/pdf/i1520-0469-29-7-1348.pdf.
123. A. J. Heymsfield and J. Iaquinta, Cirrus crystal terminal velocities, *J Atmos. Sci.* **57** (1999): 916–938, http://pcc2341f.unige.ch/pdf/i1520-0469-57-7-916.pdf.
124. J. G. Heywood, R. Rannacher and S. Turek, Artificial boundaries and flux and pressure conditions for the incompressible Navier–Stokes equations, *Int. J. Numer. Math. Fluids* **22** (1992): 325–352, http://pcc2341f.unige.ch/pdf/heywood92artificial.pdf.
125. M. Hillairet and P. Wittwer, Asymptotic behavior of solutions to the Oseen problem in a half space (2007), preprint.
126. P. Hood and C. Taylor, A numerical solution of the Navier–Stokes equations using the finite element techniques, *Comput. Fluids* **1** (1973): 73–100.
127. F. Q. Hu, On absorbing boundary conditions for linearized Euler equations by a perfectly matched layer, *J. Comput. Phys.* **129** (1996): 201–219, http://pcc2341f.unige.ch/pdf/hu95absorbing.pdf.

128. R. Huan and L. L. Thompson, Accurate radiation boundary conditions for the time-dependent wave equation on unbounded domains, *Int. J. Numer. Meth. Eng.* **47** (2000): 1569–1603, http://pcc2341f.unige.ch/pdf/huan99accurate.pdf.
129. I. Imai, Second approximation to the laminar boundary-layer flow over a flat plate, *J. Aeronaut. Sci.* **24** (1957): 155–156.
130. L. Jaxquim, D. Fabre, D. Sipp, V. Theofilis and H. Vollmers, Instability and unsteadiness of aricraft wake vortices, *Aerospace Sci. Tech.* **7** (2003): 577–593, http://pcc2341f.unige.ch/pdf/JacquinFabreSippTheofilisVollmers_AerospaceScience-Technology_2003.pdf.
131. A. Jayaraman and A. Belmonte, Oscillations of a solid sphere falling through a wormlike micellar fluid, *Phys. Rev. E*, **67** (2003): 65301, http://pcc2341f.unige.ch/pdf/micel.pdf.
132. R. E. Johnson and T. Y. Wu, Hydromechanics of low-Reynolds-number flows. Part 5. Motion of a slender torus, *J. Fluid Mech.* **95** (1979): 263–277, http://pcc2341f.unige.ch/pdf/JOHjfm79.pdf.
133. J. B. Keller and M. J. Ward, Asymptotics beyond all orders for a low Reynolds number flow, *J. Eng. Math.* **30** (1996): 253–265, http://pcc2341f.unige.ch/pdf/asylowre.pdf.
134. C. T. Kelley. *Iterative Methods for Linear and Nonlinear Equations*, Vol. 16, *Frontiers in Applied Mathematics* (SIAM, Philadelphia, 1995).
135. M. Kohr and I. Pop, *Viscous Incompressible Flow for Low Reynolds Numbers*, vol. 16, *Advances in Boundary Elements* (WIT Press, Southampton, 2004), http://www.witpress.com/c9917.html.
136. M. Kohr, An application of the method of matched asymptotic expansions for low Reynolds number flow past a cylinder of arbitrary cross section, *Int. J. Math. Math. Sci.* **47** (2004): 2525–2535, http://pcc2341f.unige.ch/pdf/Sciences.pdf.
137. T. Korvola, A. Kupiainen and J. Taskinen, Anomalous scaling for three-dimensional Cahn-Hilliard fronts, *Commun. Pure Appl. Math.* **58** (2005): 1077–1115, http:// pcc2341f. unige. ch/ pdf/ Korvola2005.pdf.
138. R. Kress and S. Meyer, An inverse boundary value problem for the Oseen equation, *Math. Meth. Appl. Sci.* **23** (2000): 103–120, http://pcc2341f.unige.ch/pdf/fulltext.pdf.
139. M. C. A. Kropinski, On the design of lifting airfoils with high critical mach number using full potential theory, *Theor. Comput. Fluid Dynam.* **9** (1997): 17–32, http://pcc2341f.unige.ch/pdf/full.pdf.
140. M. C. A. Kropinski, M. J. Ward and J. B. Keller, A hybrid asymptotic-numerical method for low Reynolds number flows past a cylindrical body, *SIAM, J. Appl. Math.* **6** (1995): 1484–1510, http://pcc2341f.unige.ch/pdf/E000114281_1.pdf.
141. A. J. C Ladd and R. Verberg, Lattice–Boltzmann simulations of particle-fluid suspensions, *J. Stat. Phys.* **104** (2001): 1191–1251, http://pcc2341f.unige.ch/pdf/ladd2001lbs.pdf.
142. C. L. Ladson, C. W. Brooks, Jr., A. S. Hill and D. W. Sproles, Computer program to obtain ordinates for NACA airfoils, Technical report, NASA (1996).
143. P. A. Lagerstrom, Solutions of the Navier–Stokes equations at large Reynolds number, *SIAM J. Appl. Math.* **28** (1975): 202–214, http://pcc2341f.unige.ch/pdf/Lagerstrom001.pdf.

144. J. S. W. Lamb, I. Melbourne and C. Wulff, Bifurcation from periodic solutions with spatiotemporal symmetry, including resonances and mode interactions, *J. Differ. Equat.* **191** (2003): 377–407, http://pcc2341f.unige.ch/pdf/lamb98bifurcation.pdf.
145. A. Lamura, G. Gompper, T. Ihle and D. M. Kroll, Multi-particle collision dynamics: Flow around a circular and a square cylinder, *Europhys. Lett.* **56** (2001): 319–325, http://pcc2341f.unige.ch/pdf/lamb98bifurcation.pdf.
146. L. Landau and E. Lipschitz, *Physique Théorique, Tome 6, Mécanique des Fluides* (Editions MIR, 1989).
147. J. Latt, Y. Grillet, B. Chopard and P. Wittwer, Simulating an exterior domain for drag force computations in the lattice Boltzmann method, *Math. Comput. Sim.* **72** (2006): 169–172, http://pcc2341f.unige.ch/pdf/jonas002.pdf.
148. B. P. Le Clair, A. E. Hamielec and H. R. Pruppacher, A numerical study of the drag on a sphere at low and intermediate Reynolds numbers, *J. Atmos. Sci.* **27** (1970): 308–315, http://pcc2341f.unige.ch/pdf/10.1175%2F1520-0469(1970)027%3C0308_ANSOTD%3E2.0.pdf.
149. J. Leray, Sur le mouvement d'un liquide visqueux enlissant l'espace, *Acta Math.* **63** (1934): 1–25.
150. J.-L. Lions, J. Métral and O. Vacus, Well-posed absorbing layer for hyperbolic problems, *Numerische Mathematik*, **92** (2002): 535–562, http://pcc2341f.unige.ch/pdf/wellposedabsorbinglayerPML.pdf.
151. P. B. S. Lissaman, Low-Reynolds-number airfoils (1983), preprint, http://arjournals.annualreviews.org/doi/abs/10.1146/annurev.fl.15.010183.001255?journalCode=fluid.
152. R. J. Luebbers, K. S. Kunz, F. Hunsberger and M. Schneider, A finite-difference time-domain near zone to far zone transformation. *IEEE Trans. Anten. Propag.* **39** (1991): 429–433, http://ieeexplore.ieee.org/search/ wrapper.jsp?arnumber=81453.
153. A. Lunardi, *Analytic Semigroups and Optimal Regularity in Parabolic Problems* (Birkhauser, 2003).
154. H. Merdan and G. Caginalp, Renormalization group methods for nonlinear parabolic equations, *Appl. Math. Lett.* **17** (2004): 217–223, http://pcc2341f.unige.ch/pdf/AppMathLetFinalVersionSubmitted.pdf.
155. R. C. Michelson and M. A. Naqvi, Beyond biologically-inspired insect flight. Technical report, Karman Institute (2003a), http://pcc2341f.unige.ch/pdf/michelson2003bbi.pdf.
156. R. C. Michelson and M. A. Naqvi, Extraterrestrial flight (entomopter-based mars surveyor), Technical report, von Karman Institute (2003b), http://pcc2341f.unige.ch/pdf/MICHELSON-NAQVI-2.pdf.
157. A. Mielke, G. Schneider and H. Uecker, Stability and diffusive dynamics on extended domains, *Ergod. Theor. Anal. Efficient Simulat. Dynam. Syst.* (2001), pp. 563–583. http://pcc2341f.unige.ch/pdf/Mielke2001.pdf.
158. R. O. Moore and K. Promislow, Renormalization group reduction of pulse dynamics in thermally loaded optical parametric oscillators, *Phys. D* **206** (2005): 62–81, http://pcc2341f.unige.ch/pdf/MR2166526.pdf.
159. T. J. Mueller, *Fixed and Flapping Wing Aerodynamics for Micro Air Vehicle Applications*, Vol. 195 (AIAA, 2001), http://pcc2341f.unige.ch/pdf/paa-v195-61-81.pdf.

160. T. J. Mueller, Aerodynamic measurements at low Reynolds numbers for fixed wing micro-air vehicles, Technical report, Hessert center for Aerospace research, University of Notre Dame (1999), http://pcc2341f.unige.ch/pdf/mueller.pdf.
161. T. J. Mueller and J. D. DeLaurier, Aerodynamics of small vehicles, *Ann. Rev. Fluid Mech.* **35** (2003): 89–111.
162. F. Nataf, A new construction of perfectly matched layers for the linearize Euler equations, *J. Comput. Phys.* **214** (2006): 757–772, http://pcc2341f.unige.ch/pdf/nataf.pdf.
163. S. A. Nazarov, The Navier–Stokes problem in a two-dimensional domain with angular outlets to infinity, *Zap. Nauchn. Sem. S.-Peterburg. Otdel. Mat. Inst. Steklov. (POMI)*, **257** (1999): 207–227, 351 https://www.springerlink.com/content/w8p47028h0271g70/resource-secured/?target= fulltext.pdf.
164. S. A, Nazarov and M. Specovius-Neugebauer, Nonlinear artificial boundary conditions with pointwise error estimates for the exterior three dimensional Navier–Stokes problem, *Math. Nachr.* **252** (2003): 86–105, http://pcc2341f.unige.ch/pdf/MR1903042.pdf.
165. S. Necasova, Asymptotic properties of the steady fall of a body in viscous fluids, *Math. Meth. Appl. Sci.* **27** (2004): 1969–1995, http://pcc2341f.unige.ch/pdf/necasova.pdf.
166. J. Novak and S. Bonazzola, Absorbing boundary conditions for simulation of gravitational waves with spectral methods in spherical coordinates, *J. Comput. Phys.* **197** (2004): 186–196, http://pcc2341f.unige.ch/pdf/2004JCoPh.197..186N.pdf.
167. H. Ockendon and J. Ockendon, *Viscous Flow* (Cambridge University Press, 1995).
168. W. E. Olmstead, A homogeneous solution for viscous flow around a half-plane, *Quart. Appl. Math.* **33** (1975): 165–169.
169. W. E. Olmstead and D. L. Hector, On the nonuniqueness of Oseen flow past a half plane, *J. Math. Phys.* **45** (1966): 408–417.
170. J. C. Padrino and D. D. Joseph, Numerical study of the steady-state uniform flow past a rotating cylinder, *J. Fluid Mech.* **557** (2006): 191–223, http://pcc2341f.unige.ch/pdf/921_Juan_numerical.pdf.
171. C. V. Pao. *Nonlinear Parabolic and Elliptic Equations* (Plenum Publishing Corporation, 1993).
172. A. Pelletier and T. J. Mueller. Low Reynolds number aerodynamics of low-aspect-ratio, thin/flat/cambered-plate wings, *J. Aircraft* **37** (2000): 825–832, http://pcc2341f.unige.ch/pdf/pelletier001.pdf.
173. A. F. Peterson, Absorbing boundary conditions for the vector wave equation, *Micro. Opt. Tech. Lett.* **1** (1988): 62–64.
174. C. Pipe and P. A. Monkewitz, An experimental study of the effect of viscoelasticity on the cylinder wake at low Reynolds numbers, Ph.D. thesis, EPFL (2005), http://pcc2341f.unige.ch/pdf/Pipe.pdf.
175. K. Promislow, A renormalization method for modulational stability of quasi-steady patterns in dispersive systems, *SIAM J. Math. Anal.* **33** (2002): 1455–1482, http://pcc2341f.unige.ch/pdf/promislow001.pdf.
176. R. Rannacher, *Finite Element Methods for the Incompressible Navier–Stokes* Equations, in *Advances in Mathematical Fluid Mechanics* (Birkhäuser, 2000), http://pcc2341f.unige.ch/pdf/CFD-Course.pdf.

177. M. Ripoll, K. Mussawisade, R. G. Winkler and G. Gompper, Low-Reynolds-number hydrodynamics of complex fluids by multi-particle-collision dynamics, *Europhys. Lett.* **68** (2004): 106–112, http://pcc2341f.unige.ch/pdf/epl8385.pdf.
178. S. Rols, C. Benoit, G. Poussigue, E. Bachelier and P. Borderies, Introduction of fractal absorbing boundary conditions in electromagnetic simulation by SMM, *Model. Simulat. Mater. Sci. Eng.* **6** (1998): 111–121, http://pcc2341f.unige.ch/pdf/1998MSMSE...6..111R.pdf.
179. C. W. Rowley and T. Colonius, Discretely nonreflecting boundary conditions for linear hyperbolic systems, *J. Comput. Phys.* **157** (2000): 500–538, http://pcc2341f.unige.ch/pdf/bc2d.pdf.
180. V. S. Ryaben'kii, S. V. Tsynkov and V. I. Turchaninav, Global discrete artificial boundary conditions for time-dependent wave propagation, Technical report, NASA (2001), http://pcc2341f.unige.ch/pdf/ryaben'kii-global.pdf.
181. V. S. Ryaben'kii and S. V. Tsynkov, *An Application of the Difference Potentials Method to Solving External Problems in CFD* (National Aeronautics and Space Administration, Langley Research Center; National Technical Information Service, 1997), http://pcc2341f.unige.ch/pdf/ryaben'kii-application.pdf.
182. Y. Saad, *Iterative Methods for Sparse Linear Systems*, Computer Science/Numerical Methods (PWS Publishing Company, 1996).
183. H. Schlichting and K. Gersten, *Boundary Layer Theory* (Springer, 1999).
184. F. Schmidt, Discrete nonreflecting boundary conditions for the Helmholtz equation, Technical report, Konrad-Zuse-Zentrum, Philadelphia, PA (2000), http://pcc2341f.unige.ch/pdf/ZR-00-06.pdf.
185. F. Schmidt, Computation of discrete transparent boundary conditions for the 2d Helmholtz equation, *Opt. Quant. Electron.* 30 (1998): 427–441, http://pcc2341f.unige.ch/pdf/schmidt98computation.pdf.
186. G. Schneider, Diffusive stability of spatial periodic solutions of the Swift–Hohenberg equation, *Commun. Math. Phys.* **178** (1996): 679–702, http://pcc2341f.unige.ch/pdf/1104286771.pdf.
187. G. Schneider, Nonlinear stability of Taylor vortices in infinite cylinders, *Arch. Ration. Mech. Anal.* **144** (1998a): 121–200, http://www.springerlink.com/content/fvb7091rtmba6jn3/.
188. G. Schneider, Nonlinear diffusive stability of spatially periodic solutions — abstract theorem and higher space dimensions, *Proc. Int. Conf. Asymptotics in Nonlinear Diffusive Systems (Sendai, 1997)*, Sendai, 1998b. Tohoku University (1996), pp. 159–167, http://pcc2341f.unige.ch/pdf/Schneider1998a.pdf.
189. G. Schneider and H. Uecker, Almost global existence and transient self similar decay for Poiseuille flow at criticality for exponentially long times, *Phys. D* **185** (2003): 209–226.
190. L. W. Schwartz, Milton van Dyke, the man and his work, *Ann. Rev. Fluid Mech.* **34** (2002): 1–18, http://pcc2341f.unige.ch/pdf/milt_anrev.pdf.
191. A. Sergej, A. Nazarov and M. Specovius-Neugebauer, Nonlinear artificial boundary conditions with pointwise error estimates for the exterior three dimensional Navier–Stokes problem, *Math. Nachr.* (2001).
192. Y. Shibata and M. Yamazaki, Uniform estimates in the velocity at infinity for stationary solutions to the Navier–Stokes exterior problem, *Japan J. Math. (N.S.)* **31** (2005): 225–279, http://pcc2341f.unige.ch/pdf/shibata-yamazaki.pdf.

193. S. Shirayama, Flow past a sphere — Topological transitions of the vorticity field, *AIAA J.* **30** (1992): 349–358, http://pcc2341f.unige.ch/pdf/PVJAPRE10925.pdf.
194. F. T. Smith, Non-uniqueness in wakes and boundary layers, *Proc. Roy. Soc. Lond. Series A* **391** (1984): 1–26, http://links.jstor.org/sici?sici=0080-4630%2819840109%29391:1800%3c1:NIWABL%3e2.0.CO%3b2-1&origin=ads.
195. I. J. Sobey, *Introduction to Interactive Boundary Layer Theory* (Oxford University press, 2000).
196. J. Sochacki, R. Kubichek, J. George, W. R. Fletcher and S. Smithson, Absorbing boundary conditions and surface waves, *Geophys.* **52** (1986): 60–71.
197. M. C. Sostarecz and A. Belmonte, Motion and shape of a viscoelastic drop falling through a viscous fluid, *J. Fluid Mech.* **497** (2003): 235–252, http://pcc2341f.unige.ch/pdf/JFMdrop.pdf.
198. E. M. Stein and G. Weiss, *Introduction to Fourier Analysis on Euclidean Spaces* (Princeton University Press, 1975).
199. K. Stewartson, On asymptotic expansions in the theory of boundary layers, *J. Math. Phys.* **36** (1957): 173–191.
200. A. Suhariyono, J. H. Kim, N. S. Goo, H. C. Park and K. J. Yoon, Design of precision balance and aerodynamic characteristic measurement system for micro aerial vehicles, *Aero. Sci. and Tech.*, **10** (2006): 92–99, http://pcc2341f.unige.ch/pdf/sdarticle.pdf.
201. F. Takemura and J. Magnaudet, The transverse force on clean and contaminated bubbles rising near a vertical wall at moderate Reynolds number. *J. Fluid Mech.* **495** (2003): 235–253.
202. F. Y. Testik, S. I. Voropayev and H. J. S. Fernando. Flow around a short horizontal bottom cylinder under steady and oscillatory flows (2005), preprint, http://pcc2341f.unige.ch/pdf/FD2.pdf.
203. E. A. Thomann and R. B. Guenther, The fundamental solution of the linearized Navier–Stokes equations for spinning bodies in three spatial dimensions — time dependent case, *J. Math. Fluid Mech.* **8** (2006): 77–98, http://pcc2341f.unige.ch/pdf/spin.pdf.
204. L. L. Thompson and R. Huan, Computation of far-field solutions based on exact nonreflecting boundary conditions for the time-dependent wave equation, *Comput. Meth. Appl. Mech. Eng.* **190** (2000): 1551–1577, http://pcc2341f.unige.ch/pdf/thompson99computation.pdf.
205. E. C. Titchmarsh, *Theory of Fourier Integrals* (Clarendon Press, 1937).
206. M. S. Titcombe, M. J. Ward and M. C. Kropisnki, A hybrid method for low Reynolds number flows past an asymmetric cylindrical body, Technical report, The University of British Columbia (1999), http://pcc2341f.unige.ch/pdf/lrpaper.pdf.
207. S. V. Tsynkov, Artificial boundary conditions for computation of oscillating external flows, *SIAM J. Sci. Comput.* **18** (1997): 1612–1656, http://pcc2341f.unige.ch/pdf/oscillate.pdf.
208. S. Tsynkov, S. Abarbanel, J. Nordström, V. Ryaben'kii and V. Vasta, Global artificial boundary conditions for computation of external flows with jets, *AIAA J.* **38** (2000): 2014, http://pcc2341f.unige.ch/pdf/jets001.pdf.
209. S. V. Tsynkov, External boundary conditions for three-dimensional problems of computational aerodynamics, *SIAM J. Sci. Comput.* **21** (2000): 166–206, http://pcc2341f.unige.ch/pdf/tsynkov00external.pdf.

210. S. V. Tsynkov, Numerical solution of problems on unbounded domains. A review, *Appl. Numer. Math. Trans. IMACS* **27** (1998): 465–532, http://pcc2341f.unige.ch/pdf/tsynkov98numerical.pdf.
211. S. V. Tsynkov and V. N. Vatsa, Improved treatment of external boundary conditions for three-dimensional flow computations, *AIAA J.* **36** (1998): 1998–2004, http://pcc2341f.unige.ch/pdf/semyon.pdf.
212. E. G. Tulapurkara, Hundred years of the boundary layer — some aspects, *Sdhana* **30** (2005): 499–512, http://pcc2341f.unige.ch/pdf/PE1287.pdf.
213. S. Turek, Efficient solvers for incompressible flow problems. An algorithmic and computational approach, Lecture notes in Computational Science and Engineering, Springer (1999).
214. H. Uecker, Self-similar decay of localized perturbations of the Nusselt solution for the inclined film problem, *Arch. Rat. Mech. Anal.* (2006).
215. A. Vaidya, Existence of steady freefall of rigid bodies in a second order fluid with applications to particle sedimentation, *Nonlinear Anal. R. World Appl.* **7** (2006): 748–768, http://pcc2341f.unige.ch/pdf/vaidya.pdf.
216. G. van Baalen. Stationary solutions of the Navier–Stokes equations in a half-plane down-stream of an object: Universality of the wake, *Nonlinearity* **15** (2002): 315–366, http://pcc2341f.unige.ch/pdf/vanbaalen001.pdf.
217. G. van Baalen, Downstream asymptotics in exterior domains: From stationary wakes to time periodic flows, *J. Math. Fluid Mech.* **9** (2007) 295–242.
218. G. van Baalen and J.-P. Eckmann, Non-vanishing profiles for the Kuramoto — Sivashinsky equation on the infinite line, *Nonlinearity* **17** (2004): 1367–1375, http://pcc2341f.unige.ch/pdf/Baalen2004.pdf.
219. G. van Baalen, A. Schenkel and P. Wittwer, Asymptotics of solutions in na+nb->c reaction-diffusion systems, *Commun. Math. Phys.* **210** (2000): 145–176, http://pcc2341f.unige.ch/pdf/nAnBtoC.pdf.
220. M. van Dyke, *Perturbation Methods in Fluid Mechanics* (The Parabolic Press, Stanford, California, 1975).
221. S. Vanka, Block-implicit multigrid calculation of two-dimensional recirculating flows, *Comput. Meth. Appl. Mech. Eng.* **59** (1986): 29–48.
222. R. Verberg and A. J. C. Ladd, Lattice–Boltzmann model with sub-grid-scale boundary conditions, *Phys. Rev. Lett.* **84** (2000): 2148–2151, http://pcc2341f.unige.ch/pdf/prl_00.pdf.
223. P. Vorobieff, D. Georgiev and M. S. Ingber, Onset of the second wake: Dependence on the Reynolds number, *Phys. Fluids* **14** (2002): 53–56.
224. P. K. Wang and W. Ji, Collision efficiencies of ice crystals at low-intermediate Reynolds numbers colliding with supercooled cloud droplets: A numerical study, *J. Atmos. Sci.* **57** (2000): 1001–1009, http://pcc2341f.unige.ch/pdf/i1520-0469-57-8-1001.pdf.
225. Z. J. Wang, Two dimensional mechanism for insect hovering, *Phys. Rev. Lett.* **83** (2000): 2216–2219, http://pcc2341f.unige.ch/pdf/2000_PRL_Wang_Hv.pdf.
226. E. Wayne, Invariant manifolds for parabolic partial differential equations on unbounded domains, *Arch. Ration. Mech. Anal.* **138** (1997): 279–306, http://pcc2341f.unige.ch/pdf/wayne001.pdf.
227. H. F. Weinberger, On the steady fall of a body in a Navier–Stokes fluid (1978).

228. A. I. Weinstein, A numerical model of cumulus dynamics and microphysics, *J. Atmos. Sci.* **27** (1969): 246–255, http://pcc2341f.unige.ch/pdf/10.1175%2F1520-0469(1970)027%3C0246_ANMOCD%3E2.0.pdf.
229. A. J. Weisenborn and B. I. M. Bosch, The Oseen drag at infinite Reynolds number, *SIAM J. Appl. Math.* **55** (1995): 1227–1232, http://pcc2341f.unige.ch/pdf/weisenborn.pdf.
230. H. Werle, *Le tunnel hydrodynamique au service de la recherche aérospatiale*, Publication No. 156, ONERA (Oficce National d'Études et de Recherches Aérospaciales), 1974.
231. P. Wesseling, *An Introduction to Multigrid Methods* (Wiley, Chichester, 1992).
232. P. Wesseling and C. W. Oosterlee, Geometric multigrid with applications to computational fluid dynamics, *J. Comput. Appl. Math.* **128** (2001): 311–334.
233. P. Wittwer, On the structure of stationary solutions of the Navier–Stokes equations, *Commun. Math. Phys.* **226** (2002): 455–474, http://pcc2341f.unige.ch/pdf/NSstructureofSolutions.pdf.
234. P. Wittwer, Supplement: On the structure of stationary solutions of the Navier–Stokes equations, *Commun. Math. Phys.* **234** (2003): 557–565, http://pcc2341f.unige.ch/pdf/supplement.pdf.
235. P. Wittwer, Leading order down-stream asymptotics of stationary Navier–Stokes flows in three dimensions, *J. Math. Fluid Mech.* **8** (2006): 147–186, http://pcc2341f.unige.ch/pdf/3dTheoryRG.pdf.
236. X. Yuan and D. Li, Particle motions in low-Reinolds number pressure-driven flows through converging-diverging microchannels, *J. Micromech. Microeng.* **16** (2006): 62–69, http://pcc2341f.unige.ch/pdf/jmm6_1_009.pdf.

CHAPTER 3

NUMERICAL SIMULATION AND BENCHMARKING OF FLUID-STRUCTURE INTERACTION WITH APPLICATION TO HEMODYNAMICS

M. Razzaq[*,**,a], S. Turek[*,b], J. Hron[*,†,c], J. F. Acker[*,d], F. Weichert[‡], I. Q. Grunwald[§], C. Roth[§], M. Wagner[¶] and B. F. Romeike[||]

[*] *Institute of Applied Mathematics, TU Dortmund, Germany*

[†] *Institute of Mathematics, Charles University Prague, Czech Republic*

[‡] *Department of Computer Graphics, TU Dortmund, Germany*

[§] *Department of Neuroradiology, Saar State University Medical School Homburg Saar, Germany*

[¶] *Department of Pathology, Saar State University Medical School Homburg Saar, Germany*

[||] *Department of Neuropathology, Friedrich-Schiller University Jena, Germany*

[a] *mrazzaq@math.uni-dortmund.de*
[b] *ture@featflow.de*
[c] *hron@karlin.mff.cuni.cz*
[d] *jens.acker@math.uni-dortmund.de*

Numerical techniques for solving the problem of fluid-structure interaction with an elastic material in a laminar incompressible viscous flow are described. An Arbitrary Lagrangian–Eulerian (ALE) formulation is employed in a fully coupled monolithic way, considering the problem as one continuum. The mathematical description and the numerical schemes are designed in such a way that more complicated constitutive relations (and more realistic for biomechanics applications) for the fluid as well as the structural part can be easily incorporated. We utilize the well-known Q_2P_1 finite element pair for discretization in space to gain

**Corresponding author.

high accuracy and perform as time-stepping the second-order Crank–Nicholson, respectively, Fractional-Step-θ-scheme for both solid and fluid parts. The resulting nonlinear discretized algebraic system is solved by a Newton method which approximates the Jacobian matrices by a divided differences approach, and the resulting linear systems are solved by iterative solvers, preferably of Krylov-multigrid type.

For validation and evaluation of the accuracy of the proposed methodology, we present corresponding results for a new set of FSI benchmarking configurations which describe the self-induced elastic deformation of a beam attached to a cylinder in laminar channel flow, allowing stationary as well as periodically oscillating deformations. Then, as an example for fluid-structure interaction (FSI) in biomedical problems, the influence of endovascular stent implantation onto cerebral aneurysm hemodynamics is numerically investigated. The aim is to study the interaction of the elastic walls of the aneurysm with the geometrical shape of the implanted stent structure for prototypical 2D configurations. This study can be seen as a basic step towards the understanding of the resulting complex flow phenomena so that in future aneurysm rupture shall be suppressed by an optimal setting for the implanted stent geometry.

Keywords: Fluid-structure interaction (FSI); monolithic FEM; ALE; multigrid; incompressible laminar flow.

1. Introduction

In this paper, we consider the general problem of viscous flow interacting with an elastic body which is being deformed by the fluid action. Such a problem is of great importance in many real life applications, and typical examples of this type of problem are the areas of biomedical fluids which include the influence of hemodynamic factors in blood vessels, cerebral aneurysm hemodynamics, joint lubrication and deformable cartilage and blood flow interaction with elastic veins.[1,8,20,21,29] The theoretical investigation of fluid-structure interaction problems is complicated by the need of a mixed description for both parts: While for the solid part the natural view is the material (Lagrangian) description, for the fluid it is usually the spatial (Eulerian) description. In the case of their combination some kind of mixed description (usually referred to as the Arbitrary Lagrangian–Eulerian description or ALE) has to be used which brings additional nonlinearity into the resulting equations (see Ref. 14).

The numerical solution of the resulting equations of the fluid-structure interaction problem poses great challenges since it includes the features of structural mechanics, fluid dynamics and their coupling. The most straightforward solution strategy, mostly used in the available software packages (see for instance Ref. 13), is to decouple the problem into the fluid part and solid part, for each of those parts using some well established method of solution; then the interaction process is introduced as external boundary conditions in each of the subproblems. This has the advantage that there are many well tested numerical methods for both separate problems of fluid flow and elastic deformation, while on the other hand the treatment of the interface and the interaction is problematic due to high stiffness and sensitivity. In contrast, the monolithic approach discussed here treats the problem as a single continuum with the coupling automatically taken care of as internal interface.

Beside a short description of the underlying numerical aspects regarding discretization and solution procedure for this monolithic approach (see Refs. 14 and 19), we present corresponding results for a new set of FSI benchmarking test cases ("channel flow around cylinder with attached elastic beam", see Ref. 25), and we concentrate on prototypical numerical studies for 2D aneurysm configurations. The corresponding parametrization is based on abstractions of biomedical data (i.e., cutplanes of 3D specimens from New Zealand white rabbits as well as computer tomographic and magnetic resonance imaging data of human neurocrania). In our studies, we allow the walls of the aneurysm to be elastic and hence deforming with the flow field in the vessel. Moreover, we examine several configurations for stent geometries which clearly influence the flow behavior inside of the aneurysm such that a very different elastic displacement of the walls is observed too. We demonstrate that either the elastic modeling of the aneurysm walls as well as the proper description of the geometrical details of the shape of the aneurysm and particularly of the stents is of great importance if the complex interaction between structure and fluid shall be quantitatively analyzed in future, especially in view of more realistic blood flow models and anisotropic constitutive laws of the elastic walls.

2. Fluid-Structure Interaction Problem Formulation

The general fluid-structure interaction problem consists of the description of the fluid and solid fields, appropriate interface conditions at the interface and conditions for the remaining boundaries, respectively. In this paper, we consider the flow of an incompressible Newtonian fluid interacting with an elastic solid. We denote the domain occupied by the fluid by Ω_t^f and the solid by Ω_t^s at the time $t \in [0, T]$. Let $\Gamma_t^0 = \bar{\Omega}_t^f \cap \bar{\Omega}_t^s$ be the part of the boundary where the elastic solid interacts with the fluid. In the following, the description for both fields and the interface conditions are introduced. Furthermore, discretization aspects and computational methods used are described in the following subsections.

2.1. *Fluid mechanics*

The fluid is assumed to be laminar and it is governed by the Navier–Stokes equations of incompressible flows derived in the ALE framework:

$$\rho^f\left(\frac{\partial \mathbf{v}^f}{\partial t} + \mathbf{v}\cdot\nabla\mathbf{v}\right) - \nabla\cdot\sigma^f = \mathbf{0}, \quad \nabla\cdot\mathbf{v} = \mathbf{0} \quad \text{in } \Omega_t^f, \tag{1}$$

where ρ^f is the constant density and $\mathbf{v}$ is the velocity of the fluid. The state of the flow is described by the *velocity* and *pressure* fields $\mathbf{v}^f$, p^f, respectively. The external forces, due to gravity or human motion, are assumed to be not significant and are neglected. Although the blood is known to be non-Newtonian in general, we assume it to be Newtonian in this study. This is because we consider large arteries with radii of the order 2.0 mm, where the velocity and shear rate are high. The kinematic viscosity ν^f is nearly a constant in arteries with relatively large diameters 5 mm,[16] and therefore the non-Newtonian effects are neglected. The constitutive relations for the stress tensors read

$$\sigma^f = -p^f\mathbf{I} + 2\mu\varepsilon(\mathbf{v}^f), \tag{2}$$

where μ is the dynamic viscosity of the fluid, p^f is the Lagrange multiplier corresponding to the incompressibility constraint in Eq. (1), and $\varepsilon(\mathbf{v}^f)$ is the strain-rate tensor:

$$\varepsilon(\mathbf{v}^f) = \frac{1}{2}(\nabla\mathbf{v}^f + (\nabla\mathbf{v}^f)^T). \tag{3}$$

The material time derivative depends on the choice of the reference system. There are basically 3 alternative reference systems: the Eulerian, the Lagrangian, and the Arbitrary Lagrangian–Eulerian formulation. The most commonly used description for the fluid-structure interaction is the ALE description. For the ALE formulation presented in this paper, corresponding discretization techniques are discussed in Sec. 3. Let us remark that also non-Newtonian flow models can be used for modeling blood flow, for instance of Power Law type or even including viscoelastic effects (see Ref. 6), which is planned for future extensions.

2.2. *Structural mechanics*

The governing equations for the structural mechanics are the balance equations:

$$\rho^s\left(\frac{\partial \mathbf{v}^s}{\partial t} + (\nabla \mathbf{v}^s)\mathbf{v}^s - \mathbf{g}\right) - \nabla\cdot\sigma^s = \mathbf{0}, \quad \text{in } \Omega_t^s, \tag{4}$$

where the subscript s denotes the structure. ρ^s is the density of the material, $\mathbf{g}^s$ represents the external body forces acting on the structure, and σ^s is the Cauchy stress tensor. The configuration of the structure is described by the displacement $\mathbf{u}^s$, with velocity field $\mathbf{v}^s = \partial \mathbf{u}^s/\partial t$. Written in the more common Lagrangian description, i.e., with respect to some fixed reference (initial) state Ω^s, we have

$$\rho^s\left(\frac{\partial^2 \mathbf{u}^s}{\partial t^2} - \mathbf{g}\right) - \nabla\cdot\Sigma^s = \mathbf{0}, \quad \text{in } \Omega^s, \tag{5}$$

where the tensor $\Sigma^s = J\sigma^s\mathbf{F}^{-T}$ is called the first Piola–Kirchhoff tensor and Eq. (5) is the momentum equation (or the equation of elastodynamics). Unlike the Cauchy stress tensor σ^s, the first Piola–Kirchhoff tensor Σ^s is non-symmetric. Since constitutive relations are often expressed in terms of symmetric stress tensor, it is natural to introduce the second Piola–Kirchhoff tensor $\mathbf{S}^s$

$$\mathbf{S}^s = \mathbf{F}^{-T}\Sigma^s = J\mathbf{F}^{-1}\sigma^s\mathbf{F}^{-T}, \tag{6}$$

which is symmetric. For an elastic material (arterial wall is known to be made of elastic material which is nonlinear, we assume it to be linear in this study) the stress is a function of the deformation (and possibly of thermodynamic variables such as the temperature) but it is independent of deformation history and thus of time. The material characteristics may still vary in space. In a homogeneous material, mechanical properties do not vary, the strain energy function depends only on the deformation. A material is mechanically isotropic if its response to deformation is the same in all directions. The constitutive equation is then a function of **F**. More precisely, it is usually written in terms of the Green–Lagrange strain tensor, as

$$\mathbf{E} = \frac{1}{2}(\mathbf{C} - \mathbf{I}), \tag{7}$$

where **I** is the identity tensor and $\mathbf{C} = \mathbf{F}^T\mathbf{F}$ is the right Cauchy–Green strain tensor. J denotes the determinant of the deformation gradient tensor **F**, defined as $\mathbf{F} = \mathbf{I} + \nabla\mathbf{u}^s$.

For the subsequent FSI benchmark we employ a St. Venant–Kirchhoff material model as an example for hyperelastic homogeneous isotropic material whose reference configuration is the natural state (i.e., where the Cauchy stress tensor is zero everywhere). The St. Venant–Kirchhoff material model is specified by the following constitutive law

$$\sigma^s = \frac{1}{J}\mathbf{F}(\lambda^s(tr\mathbf{E})\mathbf{I} + 2\mu^s\mathbf{E})\mathbf{F}^T \quad \mathbf{S}^s = \lambda^s(tr\mathbf{E})\mathbf{I} + 2\mu^s\mathbf{E}, \tag{8}$$

where λ^s denotes the Lamé coefficients, and μ^s the shear modulus. More complex constitutive relations for hyperelastic materials may be found in Ref. 11, and particular models for biological tissues and blood vessels are reported in Refs. 9 and 12. The material elasticity is characterized by a set of two parameters, the Poisson ratio ν^s and the Young modulus E. These parameters satisfy the following relations

$$\nu^s = \frac{\lambda^s}{2(\lambda^s + \mu^s)} \quad E = \frac{\mu^s(3\lambda^s + 2\mu^2)}{(\lambda^s + \mu^s)}, \tag{9}$$

$$\mu^s = \frac{E}{2(1 + \nu^s)} \quad \lambda^s = \frac{\nu^s E}{(1 + \nu^s)(1 - 2\nu^s)}, \tag{10}$$

where $\nu^s = 1/2$ for a incompressible and $\nu^s < 1/2$ for a compressible structure. In the large deformation case it is common to describe the constitutive equation using a stress–strain relation based on the Green–Lagrangian strain tensor **E** and the 2nd Piola–Kirchhoff stress tensor **S**(**E**) as a function of **E**. However, also incompressible structures can be handled in the same way (see Ref. 14).

For the hemodynamic applications, a Neo–Hooke material model ($J = \det \mathbf{F}$) is taken which can be used for compressible or incompressible (for $\nu^s = 1/2 \Rightarrow \lambda^s \to \infty$) material and which is described by the constitutive laws

$$\sigma^s = -p^s\mathbf{I} + \frac{\mu^s}{J}(\mathbf{F}\mathbf{F}^T - \mathbf{I}), \tag{11}$$

$$\mathbf{0} = -p^s + \frac{\lambda^s}{2}\left(J - \frac{1}{J}\right). \tag{12}$$

Both models, the St. Venant–Kirchhoff and the Neo–Hooke material model, share the isotropic and hyperelastic properties, and both can be used for the computation of large deformations. However, the St. Venant–Kirchhoff model does not allow for large strain computation, while the Neo–Hooke model is also valid for large strains. After linearization, both material models have to converge to the same expression, which is then valid only for small strains and small deformations. We implemented the St. Venant–Kirchhoff material model as the standard model for the compressible case, since the setup of the benchmark does not involve large strains in the oscillating beam structure. Its implementation is simpler and, therefore, the FSI benchmark will hopefully be adopted by a wider group of researchers. If someone wants or has to use the Neo–Hooke material, the results for a given set of E and ν or λ and μ are comparable, if the standard Neo–Hooke material model as in Eq. (12) is used. Similarly as in the case of more complex blood flow models, more realistic constitutive relations for the anisotropic behavior of the walls of aneurysms also can be included. However, this is beyond the scope of this paper.

2.3. *Interaction conditions*

The boundary conditions on the fluid-solid interface are assumed to be

$$\sigma^f n = \sigma^s n, \quad \mathbf{v}^f = \mathbf{v}^s, \quad \text{on } \Gamma_t^0, \tag{13}$$

where n is a unit normal vector to the interface Γ_t^0. This implies the no-slip condition for the flow and that the forces on the interface are in balance.

3. Discretization and Solution Techniques

In this study, we restrict at the moment to two dimensions which allow systematic tests of the proposed methods for biomedical applications in a very efficient way such that the qualitative behavior can be carefully analyzed. The corresponding fully implicit, monolithic treatment of the fluid-structure interaction problem suggests that an A-stable second-order time stepping scheme and that the same finite elements for both the solid part and the fluid region should be utilized. Moreover, to circumvent the fluid incompressibility constraints, we have to choose a stable finite element pair. For that reason, the conforming biquadratic, discontinuous linear Q_2P_1 pair, see Fig. 1 for the location of the degrees of freedom, is chosen which will be explained in the next section.

3.1. *The conforming Stokes element Q_2P_1*

Let us define the usual finite dimensional spaces U for displacement, V for velocity, P for pressure approximation as follows

$$
\begin{aligned}
U &= \{\mathbf{u} \in L^\infty(I, [W^{1,2}(\Omega)]^2), \mathbf{u} = \mathbf{0} \text{ on } \partial\Omega\},\\
V &= \{\mathbf{v} \in L^2(I, [W^{1,2}(\Omega_t)]^2) \cap L^\infty(I, [L^2(\Omega_t)]^2), \mathbf{v} = \mathbf{0} \text{ on } \partial\Omega\},\\
P &= \{p \in L^2(I, L^2(\Omega))\},
\end{aligned}
$$

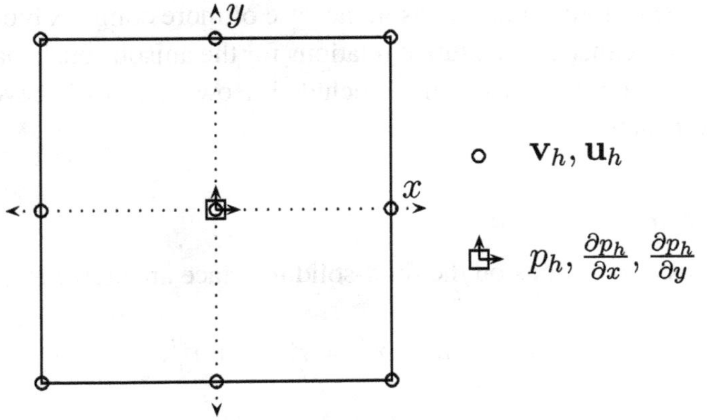

Fig. 1. Location of the degrees of freedom for the Q_2P_1 element.

then the variational formulation of the fluid-structure interaction problem is to find $(\mathbf{u}, \mathbf{v}, p) \in U \times V \times P$ such that the equations are satisfied for all $(\zeta, \xi, \gamma) \in U \times V \times P$ including appropriate initial conditions. The spaces U, V, P on an interval $[t^n, t^{n+1}]$ would be approximated in the case of the Q_2, P_1 pair as

$$
\begin{aligned}
U_h &= \{\mathbf{u}_h \in [C(\Omega_h)]^2, \mathbf{u}_h|_T \in [Q_2(T)]^2\ \forall T \in \mathcal{T}_h, \mathbf{u}_h = \mathbf{0} \text{ on } \partial\Omega_h\},\\
V_h &= \{\mathbf{v}_h \in [C(\Omega_h)]^2, \mathbf{v}_h|_T \in [Q_2(T)]^2\ \forall T \in \mathcal{T}_h, \mathbf{v}_h = \mathbf{0} \text{ on } \partial\Omega_h\},\\
P_h &= \{p_h \in L^2(\Omega_h), p_h|_T \in P_1(T)\ \forall T \in \mathcal{T}_h\}.
\end{aligned}
$$

Let us denote by $\mathbf{u}_h^n$ the approximation of $\mathbf{u}(t^n)$, $\mathbf{v}_h^n$ the approximation of $\mathbf{v}(t^n)$ and p_h^n the approximation of $p(t^n)$. Consider for each $T \in T_h$ the bilinear transformation $\psi_T : \hat{T} \to T$ to the unit square T. Then, $Q_2(T)$ is defined as

$$
\mathrm{Q}_2(T) = \left\{q \circ \psi_T^{-1} : q \in \operatorname{span}\langle 1, x, y, xy, x^2, y^2, x^2y, y^2x, x^2y^2\rangle\right\} \tag{14}
$$

with nine local degrees of freedom located at the vertices, midpoints of the edges and in the center of the quadrilateral. The space $P_1(T)$ consists of linear functions defined by

$$
\mathrm{P}_1(T) = \left\{q \circ \psi_T^{-1} : q \in \operatorname{span}\langle 1, x, y\rangle\right\} \tag{15}
$$

with the function value and both partial derivatives located in the center of the quadrilateral, as its three local degrees of freedom, which leads to a discontinuous pressure. The inf-sup condition is satisfied (see Ref. 4); however, the combination of the bilinear transformation ψ with a linear function on the reference square $P_1(\widehat{T})$ would imply that the basis on the reference square did not contain the full basis. So, the method can at most be first-order accurate on general meshes (see Refs. 2 and 4)

$$
\|p - p_h\|_0 = O(h). \tag{16}
$$

The standard remedy is to consider a local coordinate system (ξ, η) obtained by joining the midpoints of the opposing faces of T (see Refs. 2, 17 and 24).

Then, we set on each element T

$$P_1(T) := \text{span}\langle 1, \xi, \eta \rangle. \tag{17}$$

For this case, the inf-sup condition is also satisfied and the second-order approximation is recovered for the pressure as well as for the velocity gradient (see Refs. 4 and 10)

$$\|p - p_h\|_0 = O(h^2) \quad \text{and} \quad \|\nabla(u - u_h)\|_0 = O(h^2). \tag{18}$$

For a smooth solution, the approximation error for the velocity in the L_2-norm is of order $O(h^3)$ which can easily be demonstrated for prescribed polynomials or for smooth data on appropriate domains.

3.2. *Time discretization*

In view of a more compact presentation, the applied time discretization approach is described only for the fluid part (see Ref. 18 for more details). In the following, we restrict to the (standard) incompressible Navier–Stokes equations

$$\mathbf{v}_t - \nu\Delta\mathbf{v} + \mathbf{v}\cdot\nabla\mathbf{v} + \nabla p = \mathbf{f}, \quad \nabla\cdot\mathbf{v} = 0, \quad \text{in } \Omega \times (0, T], \tag{19}$$

for given force $\mathbf{f}$ and viscosity ν, with prescribed boundary values on the boundary $\partial\Omega$ and an initial condition at $t = 0$. Then, the usual θ-scheme for time discretization reads:

Basic θ-scheme: Given $\mathbf{v}^n$ and $K = t_{n+1} - t_n$, then solve for $\mathbf{v} = \mathbf{v}^{n+1}$ and $p = p^{n+1}$

$$\frac{\mathbf{v} - \mathbf{v}^n}{K} + \theta[-\nu\Delta\mathbf{v} + \mathbf{v}\cdot\nabla\mathbf{v}] + \nabla p = \mathbf{g}^{n+1}, \quad \text{div}\,\mathbf{v} = 0, \quad \text{in } \Omega \tag{20}$$

with right hand side $\mathbf{g}^{n+1} := \theta\mathbf{f}^{n+1} + (1-\theta)\mathbf{f}^n - (1-\theta)[-\nu\Delta\mathbf{v}^n + \mathbf{v}^n\cdot\nabla\mathbf{v}^n]$. The parameter θ has to be chosen depending on the time-stepping scheme, e.g., $\theta = 1$ for the Backward Euler (BE), or $\theta = 1/2$ for the Crank–Nicholson-scheme (CN) which we prefer. The pressure term $\nabla p = \nabla p^{n+1}$ may be replaced by $\theta\nabla p^{n+1} + (1-\theta)\nabla p^n$, but with appropriate postprocessing, both strategies lead to solutions of the same accuracy. In all cases, we end up with the task of solving, at each time step, a nonlinear saddle point problem of given type which has then to be discretized in space as described above. These two methods, CN and BE, belong to the group of *One-Step-θ-schemes*. The CN scheme can occasionally suffer from numerical instabilities because of its only weak damping property (not strongly A-stable), while the BE-scheme is of first-order accuracy only (however; it is a good candidate for steady-state simulations). Another method which has proven to have the potential to excel in this competition is the Fractional-Step-θ-scheme (FS). It uses three different values for θ and for the time step K at each time level. In Refs. 19 and 26 we additionally described a modified Fractional-Step-θ-scheme which seems to be advantageous particularly for fluid-structure interaction problems. A detailed description will appear in the thesis.[18]

3.3. *Solution algorithms*

After applying the standard finite element method with the Q_2P_1 element pair as described in Sec. 3.1, the system of nonlinear algebraic equations arising from the governing equations described in Secs. 2.1 and 2.2, reads

$$\begin{pmatrix} S_{\mathbf{uu}} & S_{\mathbf{uv}} & 0 \\ S_{\mathbf{vu}} & S_{\mathbf{vv}} & kB \\ c_{\mathbf{u}}B_s^T & c_{\mathbf{v}}B_f^T & 0 \end{pmatrix} \begin{pmatrix} \mathbf{u} \\ \mathbf{v} \\ p \end{pmatrix} = \begin{pmatrix} \mathbf{f_u} \\ \mathbf{f_v} \\ f_p \end{pmatrix}, \tag{21}$$

which is a typical saddle point problem, where S describes the diffusive and convective terms from the governing equations. The above system of nonlinear algebraic Eq. (21) is solved using Newton method as a basic iteration which can exhibit quadratic convergence provided that the initial guess is sufficiently close to the solution. The basic idea of the Newton iteration is to find a root of a function, $\mathbf{R}(\mathbf{X}) = \mathbf{0}$, using the available known function value and its first derivative. One step of the Newton iteration can be written as

$$\mathbf{X}^{n+1} = \mathbf{X}^n + \omega^n \left[\frac{\partial \mathbf{R}(\mathbf{X^n})}{\partial \mathbf{X}}\right]^{-1} \mathbf{R}(\mathbf{X}^n), \tag{22}$$

where $\mathbf{X} = (\mathbf{u}_h, \mathbf{v}_h, p_h)$ and $\partial \mathbf{R}(\mathbf{X^n})/\partial \mathbf{X}$ is the Jacobian matrix. To ensure the convergence globally, some improvements of this basic iteration are used. The damped Newton's method with line search improves the chance of convergence by adaptively changing the length of the correction vector (see Refs. 14 and 24 for more details). The damping parameter $\omega^n \in (-1, 0)$ is chosen such that

$$\mathbf{R}(\mathbf{X}^{n+1}) \cdot \mathbf{X}^{n+1} \leq \mathbf{R}(\mathbf{X}^n) \cdot \mathbf{X}^n. \tag{23}$$

The damping greatly improves the robustness of the Newton iteration in the case when the current approximation $\mathbf{X}^n$ is not close enough to the final solution since the Newton method without damping is not guaranteed to converge (see Refs. 14 and 24 for more details). The Jacobian matrix $\partial \mathbf{R}(\mathbf{X^n})/\partial \mathbf{X}$ can be computed by finite differences from the residual vector $\mathbf{R}(\mathbf{X})$

$$\left[\frac{\partial \mathbf{R}(\mathbf{X^n})}{\partial \mathbf{X}}\right]_{ij} \approx \frac{[\mathbf{R}]_i(\mathbf{X}^n + \alpha_j \mathbf{e}_j) - [\mathbf{R}]_i(\mathbf{X}^n - \alpha_j \mathbf{e}_j)}{2\alpha_j}, \tag{24}$$

where $\mathbf{e}_j$ are the unit basis vectors in $\mathbf{R}^n$ and the coefficients α_j are adaptively taken according to the change in the solution in the previous time step. Since we know the sparsity pattern of the Jacobian matrix in advance, which is given by the used finite element method, this computation can be done in an efficient way so that the linear solver remains the dominant part in terms of the CPU time (see Refs. 24 and 27 for more details). A good candidate, at least in 2D, seems to be a direct solver for sparse systems like UMFPACK (see Ref. 7); while this choice provides very robust linear solvers, its memory and CPU time requirements are too high for larger systems (i.e., more than 20,000 unknowns). Large linear problems can be solved by Krylov-space methods (BiCGStab, GMRes, see Ref. 3) with suitable preconditioners. One possibility is the ILU preconditioner with special treatment of the saddle point character of our system, where we allow certain fill-in for the zero diagonal blocks, see Ref. 5. As an alternative, we also utilize a standard geometric multigrid approach based on a hierarchy of grids obtained by successive regular refinement of a given coarse mesh. The complete multigrid iteration is performed in the standard defect-correction setup with the V or F-type cycle. While a direct sparse solver[7] is used for the coarse grid solution, on finer levels a fixed number (2 or 4) of iterations by local MPSC schemes (Vanka-like smoother)[14,24,30] is performed. Such iterations can be written as

$$\begin{pmatrix} \mathbf{u}^{l+1} \\ \mathbf{v}^{l+1} \\ p^{l+1} \end{pmatrix} = \begin{pmatrix} \mathbf{u}^{l} \\ \mathbf{v}^{l} \\ p^{l} \end{pmatrix} - \omega$$

$$\times \sum_{\text{element } \Omega_i} \begin{pmatrix} S_{\mathbf{uu}|\Omega_i} & S_{\mathbf{uv}|\Omega_i} & 0 \\ S_{\mathbf{vu}|\Omega_i} & S_{\mathbf{vv}|\Omega_i} & kB_{|\Omega_i} \\ c_{\mathbf{u}} B^T_{s|\Omega_i} & c_{\mathbf{v}} B^T_{f|\Omega_i} & 0 \end{pmatrix}^{-1} \begin{pmatrix} \mathbf{def}^l_{\mathbf{u}} \\ \mathbf{def}^l_{\mathbf{v}} \\ def^l_p \end{pmatrix}.$$

The inverse of the local systems (39 × 39) can be done by hardware optimized direct solvers. The full nodal interpolation is used as the prolongation operator **P** with its transposed operator used as the restriction $\mathbf{R} = \mathbf{P}^T$ (see Refs. 13 and 24 for more details).

4. FSI Benchmarking

In order to validate and to analyze different techniques to solve such FSI problems, also in a quantitative way, a set of benchmark configurations has been proposed in Ref. 25. The configurations consist of laminar incompressible channel flow around an elastic object which results in self-induced oscillations of the structure. Moreover, characteristic flow quantities and corresponding plots are provided for a quantitative comparison.

The domain is based on the 2D version of the well-known CFD benchmark in Ref. 28 and is shown in Fig. 2, an overview of the geometrical parameters is given in Table 1. By omitting the elastic bar behind the cylinder one can easily recover the setup of the "classical" *flow around cylinder* configuration which allows for validation of the flow part by comparing the results with the older flow benchmark. The setting is

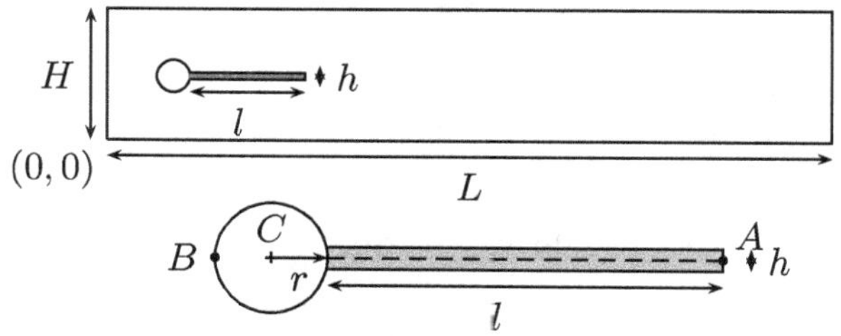

Fig. 2. Computational domain with geometrical details of the structure part.

Table 1. Overview of the geometrical parameters.

		Value [m]
Channel length	L	2.5
Channel width	H	0.41
Cylinder center position	C	(0.2, 0.2)
Cylinder radius	r	0.05
Elastic structure length	l	0.35
Elastic structure thickness	h	0.02
Reference point (at $t = 0$)	A	(0.6, 0.2)
Reference point	B	(0.2, 0.2)

Table 2. Parameter settings for the FSI benchmarks.

Parameter	FSI1	FSI2	FSI3
ϱ^s $[10^3 \frac{\text{kg}}{\text{m}^3}]$	1	10	1
ν^s	0.4	0.4	0.4
μ^s $[10^6 \frac{\text{kg}}{\text{ms}^2}]$	0.5	0.5	2.0
ϱ^f $[10^3 \frac{\text{kg}}{\text{m}^3}]$	1	1	1
ν^f $[10^{-3} \frac{\text{m}^2}{\text{s}}]$	1	1	1
$\bar{U}$ $[\frac{\text{m}}{\text{s}}]$	0.2	1	2
$\beta = \frac{\varrho^s}{\varrho^f}$	1	10	1
ν^s	0.4	0.4	0.4
$\text{Ae} = \frac{E^s}{\varrho^f \bar{U}^2}$	3.5×10^4	1.4×10^3	1.4×10^3
$\text{Re} = \frac{\bar{U}d}{\nu^f}$	20	100	200
$\bar{U}$	0.2	1	2

intentionally nonsymmetric[28] to prevent the dependence of the onset of any possible oscillation on the precision of the computation. The mesh used for the computations is shown in Fig. 3.

A parabolic velocity profile is prescribed at the left channel inflow

$$v^f(0, y) = 1.5\bar{U}\frac{y(H - y)}{\left(\frac{H}{2}\right)^2} = 1.5\bar{U}\frac{4.0}{0.1681}y(0.41 - y), \tag{25}$$

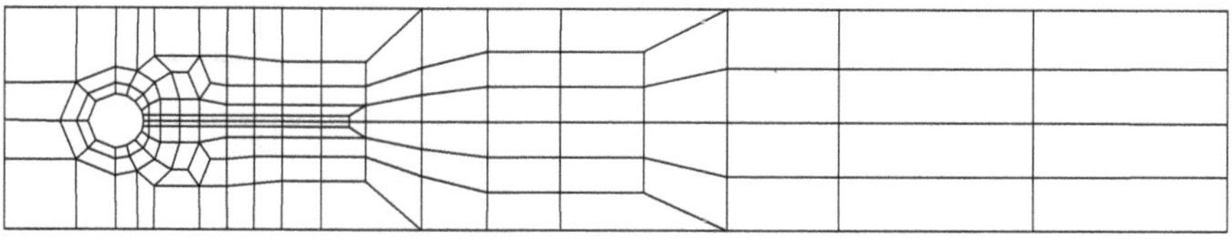

level	#refine	#el	#dof
0	0	62	1338
1	1	248	5032
2	2	992	19488
3	3	3968	76672
4	4	15872	304128

Fig. 3. Coarse mesh with number of degrees of freedom for refined levels.

such that the mean inflow velocity is $\bar{U}$ and the maximum of the inflow velocity profile is $1.5\bar{U}$. The *no-slip* condition is prescribed for the fluid on the other boundary parts. i.e., top and bottom wall, circle and fluid-structure interface Γ_t^0. The outflow condition can be chosen by the user, for example *stress free* or *do nothing* conditions. The outflow condition effectively prescribes some reference value for the pressure variable p. While this value could be arbitrarily set in the incompressible case, in the case of compressible structure this will have influence on the stress and consequently the deformation of the solid. In this description, we set the reference pressure at the outflow to have *zero mean value*. Suggested starting procedure for the non-steady tests is to use a smooth increase of the velocity profile in time as

$$v^f(t, 0, y) = \begin{cases} v^f(0, y)\dfrac{1 - \cos\left(\frac{\pi}{2}t\right)}{2} & \text{if } t < 2.0 \\ v^f(0, y) & \text{otherwise} \end{cases}, \tag{26}$$

where $v^f(0, y)$ is the velocity profile given in Eq. (25).

The following FSI tests are performed for three different inflow speeds. FSI1 is resulting in a steady state solution, while FSI2 and FSI3 result in

Table 3. Results for FSI1.

Level	nel	ndof	ux of A $[\times 10^{-3}]$	uy of A $[\times 10^{-3}]$	Drag	Lift
2	992	19,488	0.022871	0.81930	14.27360	0.76178
3	3,968	76,672	0.022775	0.82043 14.29177	0.76305	
4	15,872	304,128	0.022732	0.82071	14.29484	0.76356
5	63,488	1,211,392	0.022716	0.82081	14.29486	0.76370
6	253,952	4,835,328	0.022708	0.82086	14.29451	0.76374
Ref.			0.0227	0.8209	14.295	0.7638

periodic solutions. The parameter values for the FSI1, FSI2 and FSI3 are given in Table 2. Here, the computed values are summarized in Table 3 for the steady state test FSI1. In Figs. 4 and 5, resulting plots of x–y displacement of the trailing edge point A of the elastic bar and plots of the forces (lift, drag) acting on the cylinder attached with an elastic bar are drawn and computed values for three different mesh refinement levels and two different time steps for the nonsteady tests FSI2 and FSI3 are presented, respectively, which show the (almost) grid independent solution behavior (for more details see Ref. 25).

5. Applications to Hemodynamics

In the following, we consider the numerical simulation of special problems encountered in the area of cardiovascular hemodynamics, namely flow interaction with thick-walled deformable material, which can become a useful tool for deeper understanding of the onset of diseases of the human circulatory system, as for example, blood cell and intimal damages in stenosis, aneurysm rupture, evaluation of the new surgery techniques of heart, arteries and veins (see Refs. 1, 15 and 29 and therein cited literature). In this contribution, prototypical studies are performed for brain aneurysm. The word "aneurysm" comes from the latin word *aneurysma* which means dilatation. Aneurysm is a local dilatation in the wall of a blood vessel, usually an artery, due to a defect, disease or injury. Typically, as the aneurysm enlarges, the arterial wall becomes thinner and eventually leaks or ruptures, causing subarachnoid hemorrhage (SAH) (bleeding into brain fluid) or formation of a blood clot within the brain. In the case of a vessel

FSI2: x and y displacement of the point A

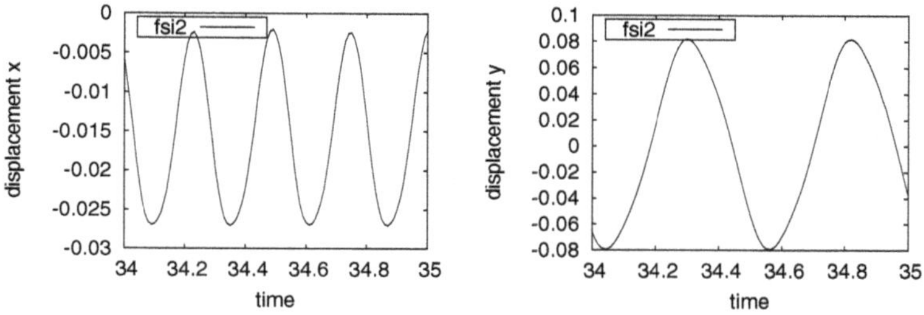

FSI2: lift and drag force on the cylinder and elastic bar

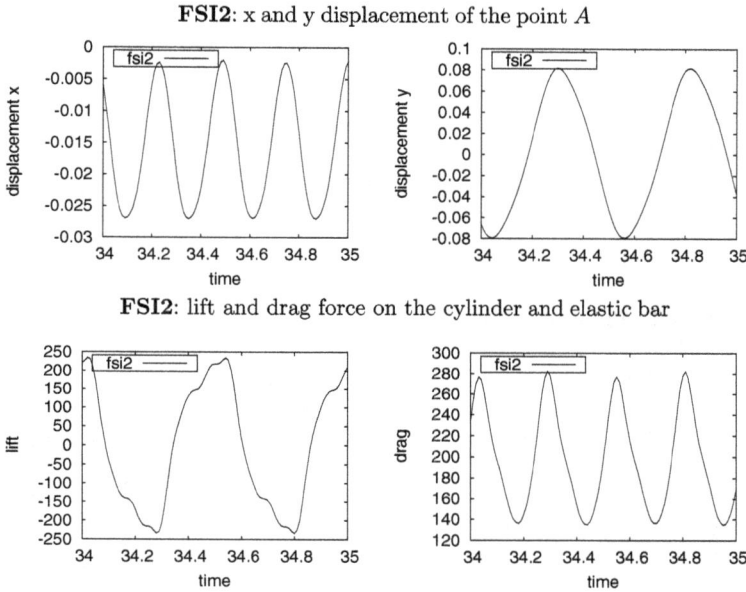

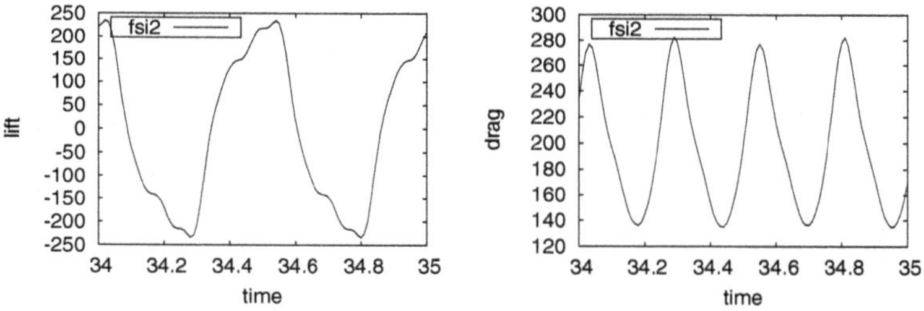

lev.	ux of A $[\times 10^{-3}]$	uy of A $[\times 10^{-3}]$	drag	lift
2	$-14.00 \pm 12.03[3.8]$	$1.18 \pm 78.7[2.0]$	$209.46 \pm 72.30[3.8]$	$-1.18 \pm 269.6[2.0]$
3	$-14.25 \pm 12.03[3.8]$	$1.20 \pm 79.2[2.0]$	$202.55 \pm 67.02[3.8]$	$0.71 \pm 227.1[2.0]$
4	$-14.58 \pm 12.37[3.8]$	$1.25 \pm 80.7[2.0]$	$201.29 \pm 67.61[3.8]$	$0.97 \pm 233.2[2.0]$
lev.	ux of A $[\times 10^{-3}]$	uy of A $[\times 10^{-3}]$	drag	lift
2	$-14.15 \pm 12.23[3.7]$	$1.18 \pm 78.8[1.9]$	$210.36 \pm 70.28[3.7]$	$0.80 \pm 286.0[1.9]$
3	$-13.97 \pm 12.01[3.8]$	$1.25 \pm 79.3[2.0]$	$203.54 \pm 68.43[3.8]$	$0.41 \pm 229.3[2.0]$
4	$-14.58 \pm 12.44[3.8]$	$1.23 \pm 80.6[2.0]$	$208.83 \pm 73.75[3.8]$	$0.88 \pm 234.2[2.0]$
ref.	$-14.58 \pm 12.44[3.8]$	$1.23 \pm 80.6[2.0]$	$208.83 \pm 73.75[3.8]$	$0.88 \pm 234.2[2.0]$

Fig. 4. Results for FSI2 with time step $\Delta t = 0.002$, $\Delta t = 0.001$.

rupture, there is a hemorrhage, and when an artery ruptures, then the hemorrhage is more rapid and more intense. In arteries the wall thickness can be up to 30% of the diameter and its local thickening can lead to the creation of an aneurysm so that the aim of numerical simulations is to relate the aneurysm state (unrupture or rupture) with wall pressure, wall

FSI3: x and y displacement of the point A

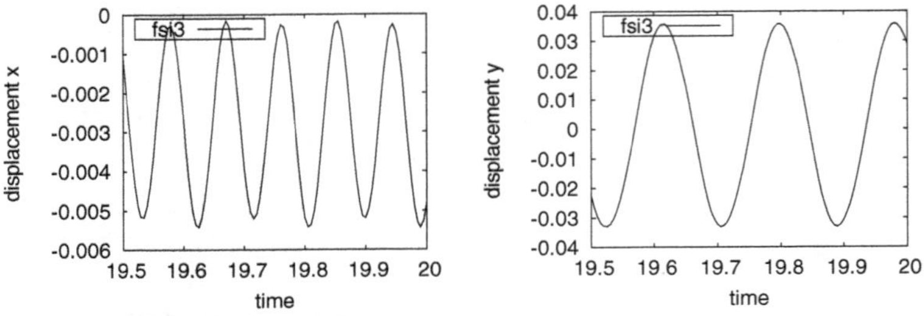

FSI3: lift and drag force on the cylinder and elastic bar

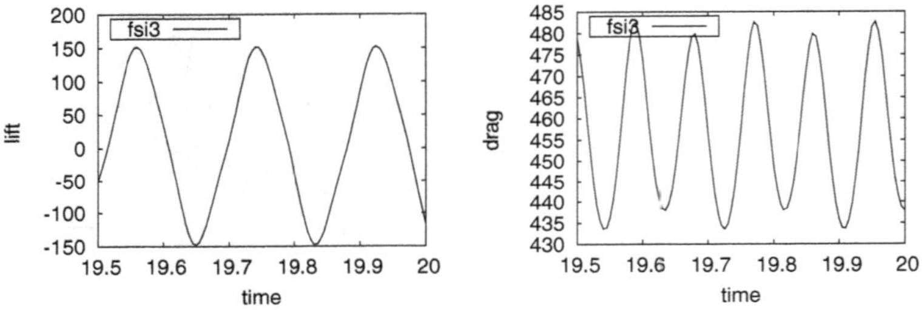

lev.	ux of A $[\times 10^{-3}]$	uy of A $[\times 10^{-3}]$	drag	lift
2	$-3.02 \pm 2.78[10.6]$	$0.99 \pm 35.70[5.3]$	$444.6 \pm 31.69[10.6]$	$9.48 \pm 151.55[5.3]$
3	$-3.02 \pm 2.83[10.6]$	$1.43 \pm 35.43[5.3]$	$457.1 \pm 20.05[10.6]$	$1.23 \pm 146.04[5.3]$
4	$-2.85 \pm 2.56[10.9]$	$1.53 \pm 34.35[5.3]$	$459.8 \pm 20.00[10.9]$	$1.51 \pm 148.76[5.3]$
lev.	ux of A $[\times 10^{-3}]$	uy of A $[\times 10^{-3}]$	drag	lift
2	$-3.00 \pm 2.79[10.7]$	$1.19 \pm 35.72[5.3]$	$445.0 \pm 35.09[10.7]$	$8.26 \pm 163.72[5.3]$
3	$-2.86 \pm 2.68[10.7]$	$1.45 \pm 35.34[5.3]$	$455.7 \pm 24.69[10.7]$	$1.42 \pm 146.43[5.3]$
4	$-2.69 \pm 2.53[10.9]$	$1.48 \pm 34.38[5.3]$	$457.3 \pm 22.66[10.9]$	$2.22 \pm 149.78[5.3]$
ref.	$-2.69 \pm 2.53[10.9]$	$1.48 \pm 34.38[5.3]$	$457.3 \pm 22.66[10.9]$	$2.22 \pm 149.78[5.3]$

Fig. 5. Results for FSI3 with time step $\Delta t = 0.001$, $\Delta t = 0.0005$.

deformation and effective wall stress. Such a relationship would provide information for the diagnosis and treatment of unrupture and rupture of an aneurysm by elucidating the risk of bleeding or rebleeding, respectively.

As a typical example for the related CFD simulations, a real view is provided in Fig. 6 which also contains the automatically extracted

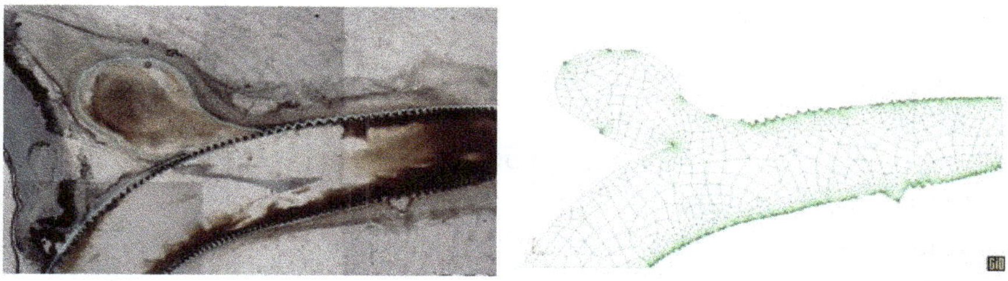

Fig. 6. Left: Real view of aneurysm. Right: Schematic drawing of the mesh.

computational domain and (coarse) mesh in 2D, however without stents. In order to use the proposed numerical methods for aneurysm hemodynamics, in a first step, only simplified two-dimensional examples, which however include the interaction of the flow with the deformable material, are considered in the following. Flow through a deformable vein with elastic walls of a brain aneurysm is simulated to analyse qualitatively the described methods; here, the flow is driven by prescribing the flow velocity at the inflow section while the elastic part of the boundary is either fixed or stress-free. Both ends of the walls are fixed, and the flow is driven by a periodical change of the inflow at the left end.

5.1. *Geometry of the problem*

For convenience, the geometry of the fluid domain under consideration is currently based on simplified 2D models (see Fig. 7) which allows us to concentrate on the detailed qualitative evaluation of our approach based on the described monolithic ALE formulation. The underlying construction of

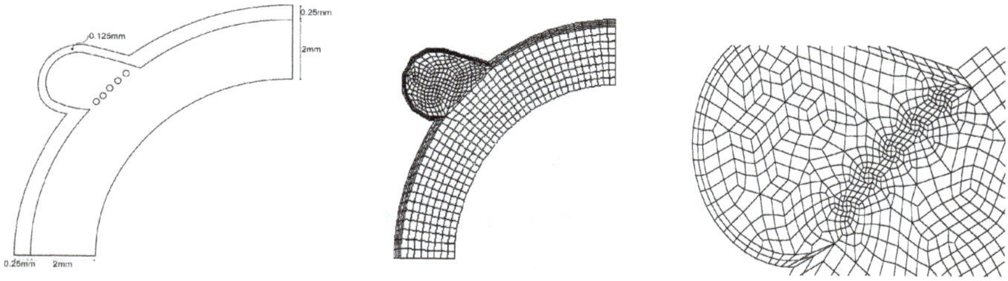

Fig. 7. Schematic drawing of the measurement section (left). Mesh without stents (776 elements) (middle). Mesh with stents (1,431 elements) which are part of the simulations (right).

the (2D) shape of the aneurysm can be explained as follows:

- The bent blood vessel is approximated by quarter circles around the origin.
- The innermost circle has the radius 6 mm, the next has 8 mm, and the last one has 8.25 mm.
- This results in one rigid inner wall and an elastic wall between 8 mm and 8.25 mm of thickness 0.25 mm.

The aneurysm shape is approximated by two arcs and lines intersecting the arcs tangentially. The midpoints of the arcs are the same $(-6.75; 6)$, they have the radius 1.125 mm and 1.25 mm. They are intersected tangentially by lines at angular value 1.3 radians. This results in a wall thickness of 0.125 mm for the elastic aneurysm walls (see Fig. 7). The examined stents are of circular shape, placed on the neck of the aneurysm, and we use three, respectively, five stents (simplified "circles" in 2D as cutplanes from 3D configurations) of different size and position. The stents also consist of a grid, immersed in the blood flow, which is located at the inlet of the aneurysm so that in future elastic deformations of the stents can be included, too, since in real life, the stent is a medical device which consists of a wire metal wire tube. Stents are typically used to keep arteries open and are located on the vessel wall while this stent is immersed in the blood flow (Fig. 7). The purpose of this device is to reduce the flux into and within the aneurysm in order to occlude it by a clot or rupture. The aneurysm is then intersected with the blood vessel and all missing angular values and intersection points can be determined.

5.2. *Boundary and initial conditions*

The (steady) velocity profile, to flow from the right to the left part of the channel, is defined as parabolic inflow, namely

$$\mathbf{v}^f(0, y) = \bar{U}(y - 6)(y - 8). \tag{27}$$

Correspondingly, the pulsatile inflow profile for the nonsteady tests for which peak systole and diastole occur for $\Delta t = 0.25$ s and $\Delta t = 0.75$ s respectively, is prescribed as

$$\mathbf{v}^f(t, 0, y) = \mathbf{v}^f(0, y)(1 + 0.75 \sin(2\pi t)). \tag{28}$$

The natural outflow condition at the lower left part effectively prescribes some reference value for the pressure variable p, here $p = 0$. While this value could be arbitrarily set in the incompressible case, in the case of a compressible structure this might have influence onto the stress and consequently the deformation of the solid. The *no-slip* condition is prescribed for the fluid on the other boundary parts, i.e., top and bottom walls, stents and fluid-structure interface.

5.3. *Numerical results*

The newtonian fluid used in the tests has a density $\rho^f = 1.035 \times 10^{-6}$ kg/mm^3 and a kinematic viscosity $\nu^f = 3.38$ mm^2/s which is similar to the properties of blood. If we prescribe the inflow speed $\bar{U} = -50$ mm/s, this results in a Reynolds number $Re \approx 120$ based on the prescribed peak systole inflow velocity and the width of the veins which is 2 mm such that the resulting flow is within the laminar region. Parameter values for the elastic vein in the described model are as follows: The density of the upper elastic wall is $\rho^s = 1.12 \times 10^{-6}$ kg/mm^3, solid shear modulus is $\mu^s = 42.85$ kg/mms^2, Poisson ratio is $\nu^p = 0.4$, Young modulus is $E = 120$ kN/mm^2. As described before, the constitutive relations used for the materials are the incompressible Newtonian model (2) for the fluid and a hyperelastic Neo–Hooke material for the solid. This choice includes most of the typical difficulties the numerical method has to deal with, namely the incompressibility and significant deformations. From a medical point of view, the use of stents provides an efficient treatment for managing the difficult entity of intracranial aneurysms. Here, the thickness of the aneurysm wall is attenuated and the aneurysm hemodynamics changes significantly. Since the purpose of this device is to control the flux within the aneurysm in order to occlude it by a clot or rupture, the resulting flow behavior into and within the aneurysm is the main objective, particularly in view of the different stent geometries. Therefore, we decided for the 2D studies to locate the stents only in direct connection to the aneurysm. Comparing our studies with the CFD literature (see Refs. 1, 8, 22, 23 and 29), several research groups focus on CFD simulations with realistic 3D geometries, but typically assuming rigid walls. In contrast, we concentrate on the complex interaction between elastic deformations and flow perturbations induced by the stents. At the moment, we are only able to perform these simulations in 2D, however, with these studies we should be able to analyze qualitatively the influence of geometrical details onto the elastic material behavior, particularly in view of more complex blood models and constitutive equations for the structure. Therefore, the aims of our current studies can be described as follows:

(1) What is the influence of the elasticity of the walls onto the flow behavior inside of the aneurysm, particularly with regard to the resulting shape of the aneurysm?
(2) What is the influence of the geometrical details of the (2D) stents, that means shape, size, position, onto the flow behavior into and inside of the aneurysm?
(3) Do both aspects, small-scale geometrical details as well as elastic fluid-structure interaction, have to be considered simultaneously or is one of them negligible in first-order approximation?
(4) Are modern numerical methods and corresponding CFD simulations tools able to simulate qualitatively the multiphysics behavior of such biomedical configurations?

In the following, we show some corresponding results for the described prototypical aneurysm geometry, first for the steady state inflow profile, followed by nonsteady tests for the pulsatile inflow, both with rigid and elastic walls, respectively.

5.4. *Steady configurations*

Due to the given inflow profile, which is not time-dependent, and due to the low Re numbers, the flow behavior leads to a steady state which only depends on the elasticity and the shape of the stents. Moreover, for the following simulations, we only treat the aneurysm wall as elastic structure. Then, the aneurysm undergoes some slight deformations which can hardly be seen in Figs. 8 and 9. However they result in a different volume of the flow domain (see Fig. 10) and lead to a significantly different local flow behavior since the spacing between stents and elastic walls may change (see the subsequent pictures).

In the following pictures, we visualize the different flow behavior due to the velocity magnitude and by showing corresponding vector

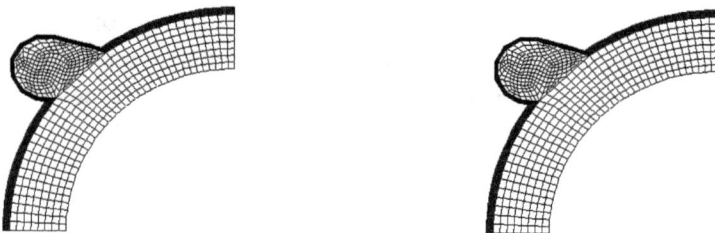

Fig. 8. Deformed mesh for steady configuration without stents, with elastic wall (left). Mesh for rigid wall (right).

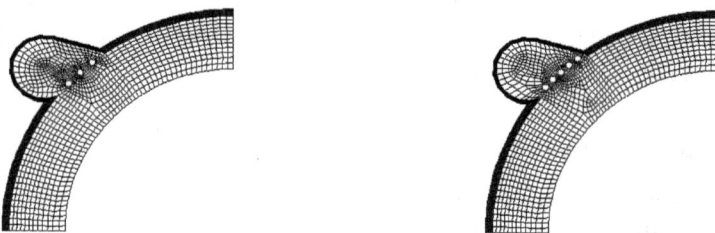

Fig. 9. Deformed mesh for steady configuration with stents: 3 stents (left) and 5 stents (right).

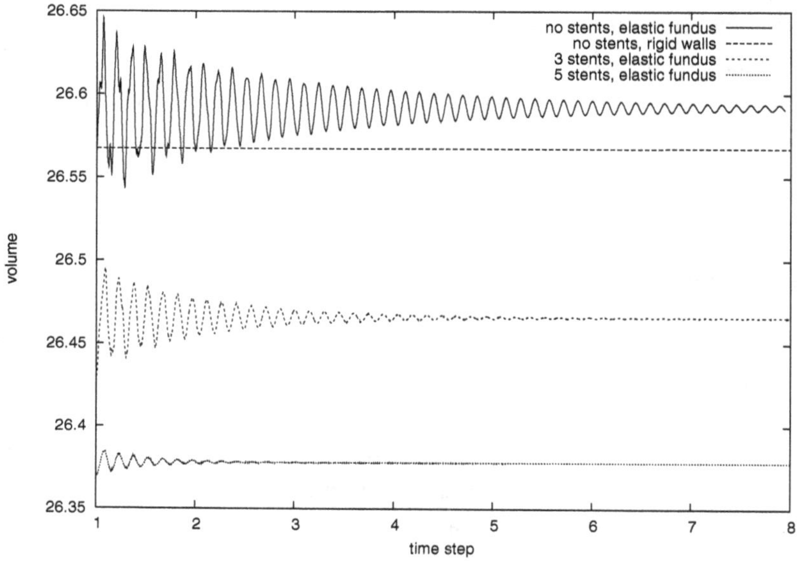

Fig. 10. Resulting volume of the fluid domain for different configurations.

plots inside of the aneurysm. Particularly the influence of the number of stents onto the complete fluid flow through the channel including the aneurysm can be clearly seen. Summarizing these results for steady inflow, the simulations show that the stent implantation across the neck of the aneurysm prevents blood penetration into the aneurysm fundus. Moreover, the elastic geometrical deformation of the wall is slightly reduced by implanting the stents while the local flow behavior inside of the aneurysm is more significantly influenced by the elastic properties of the outer wall, particularly due to different width between stents and walls of the aneurysm. In the next section, we will consider the more realistic behavior of flow configurations with time-dependent pulsatile inflow which will be analyzed for the case of elastic behavior of the aneurysm walls.

5.5. *Pulsatile configurations*

For the following pulsatile test case, we have taken again the aneurysm part as elastic while the other parts of the walls belonging to the channel are rigid. First of all, we show again (see Fig. 11) the resulting volume of the flow domain for 5, 3 and no stents. In all cases, the oscillating behavior due to the pulsative inflow is visible which also leads to different volume sizes. Looking carefully at the resulting flow behavior, we see global differences with regard to the channel flow near the aneurysm, namely due to the different flow rate into the aneurysm, and significant local differences inside of the aneurysm (see Fig. 12).

6. Summary and Future Developments

We presented a monolithic ALE formulation of fluid-structure interaction problems suitable for applications with finite deformations of the structure and laminar viscous flows, particularly arising in biomechanics. The corresponding discrete nonlinear systems result from the finite element discretization by using the high order Q_2P_1 FEM pair which are solved monolithically via discrete Newton iteration and special Krylov-multigrid approaches. While we are restricted in the presented studies to the simplified case of Newtonian fluids and small deformations, the used numerical components allow the system to be coupled with additional models of chemical and electric activation of the active response of the biological

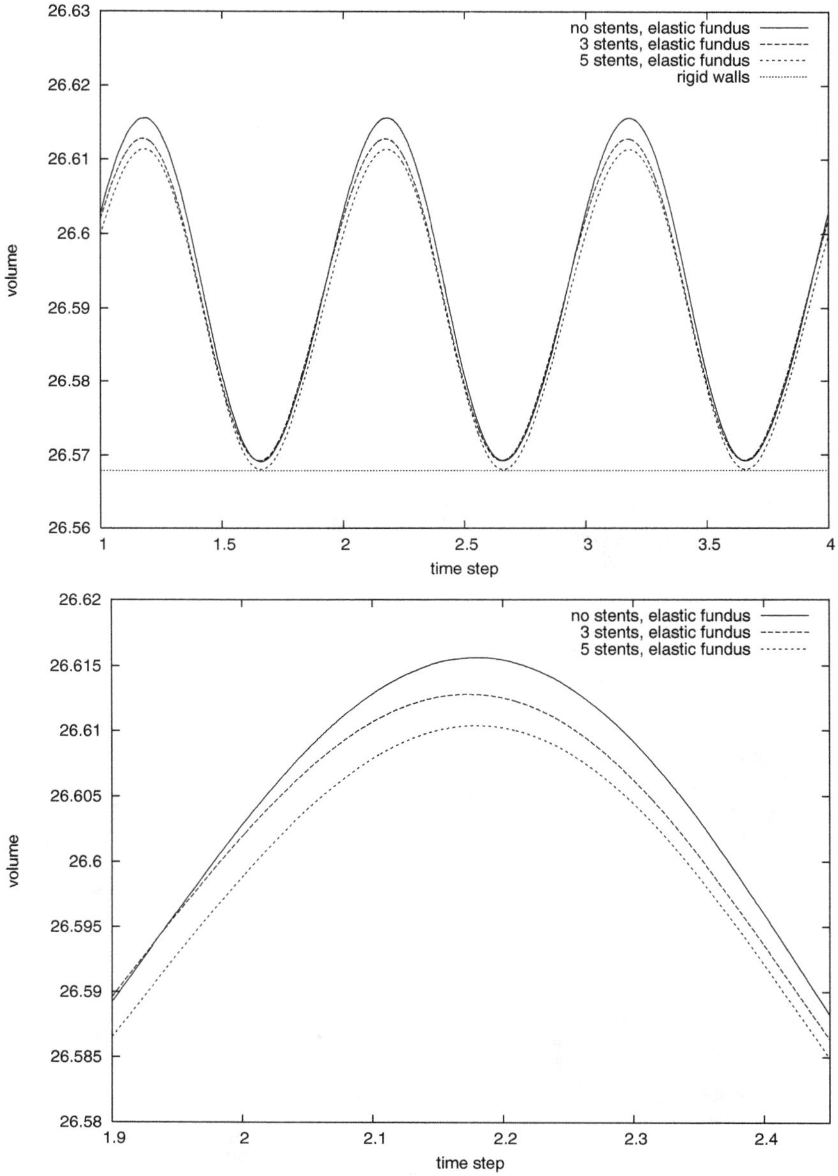

Fig. 11. Domain volume with rigid and elastic behavior of the aneurysm wall.

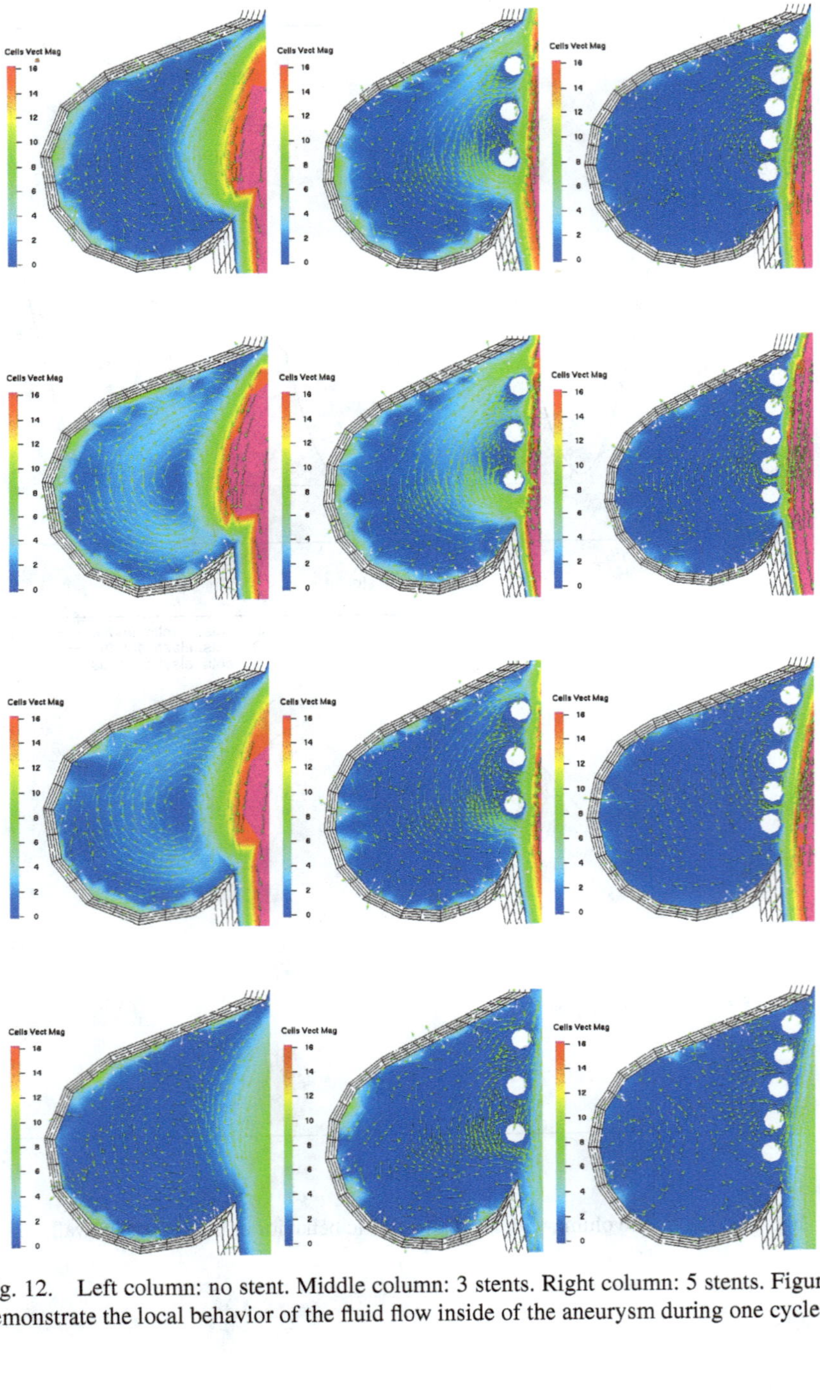

Fig. 12. Left column: no stent. Middle column: 3 stents. Right column: 5 stents. Figures demonstrate the local behavior of the fluid flow inside of the aneurysm during one cycle.

material as well as power law models used to describe the shear thinning property of blood. Further extension to viscoelastic models and coupling with mixture based models for soft tissues together with chemical and electric processes would allow to perform more realistic simulations for real applications.

In this paper, we applied the presented numerical techniques to FSI benchmarking settings ("channel flow around cylinder with attached elastic beam", see Ref. 25) which allow the validation and also evaluation of different numerical solution approaches for fluid-structure interaction problems. Moreover, we examined prototypically the influence of endovascular stent implantation onto aneurysm hemodynamics. The aim was, first of all, to study the influence of the elasticity of the walls onto the flow behavior inside of the aneurysm. Moreover, different geometrical configurations of implanted stent structures have been analyzed in 2D. These 2D results are far from providing quantitative results for such a complex multiphysics configuration, but they allow a qualitative analysis with regard to both considered components, namely the elastic behavior of the structural parts and the multiscale flow behavior due to the geometrical details of the stents. We believe that such basic studies may help towards the development of future "Virtual Flow Laboratories" which individually assist to develop personal medical tools in an individual style.

Acknowledgment

The authors want to express their gratitude to the German Research Association (DFG), funding the project as part of FOR493 and TRR30, the Jindrich Necas Center for Mathematical Modeling, project LC06052 financed by MSMT, and the Higher Education Commission (HEC) of Pakistan for their financial support of the study. The present material is also based upon work kindly supported by the Homburger Forschungsförderungsprogramm (HOMFOR) 2008.

References

1. S. Appanaboyina, F. Mut, R. Löhner, E. Scrivano, C. Miranda, P. Lylyk, C. Putman and J. Cebral, Computational modelling of blood flow in side arterial branches after stenting of cerebral aneurysm, *Int. J. Comput. Fluid Dynam.* **22** (2008):669–676.
2. D. N. Arnold, D. Boffi and R. S. Falk, Approximation by quadrilateral finite element, *Math. Comput.* **71** (2002):909–922.

3. R. Barrett, M. Berry, T. F. Chan, J. Demmel, J. Donato, J. Dongarra, V. Eijkhout, R. Pozo, C. Romine and H. Van der Vorst, *Templates for the Solution of Linear Systems: Building Blocks for Iterative Methods* (SIAM, Philadelphia, PA, 1994).
4. D. Boffi and L. Gastaldi, On the quadrilateral Q_2P_1 element for the stokes problem, *Int. J. Numer. Meth. Fluids* **39** (2002):1001–1011.
5. R. Bramley and X. Wang, *SPLIB*: A library of iterative methods for sparse linear systems, Department of Computer Science, Indiana University, Bloomington, IN (1997), http://www.cs.indiana.edu/ftp/bramley/splib.tar.gz.
6. H. Damanik, J. Hron, A. Ouazzi and S. Turek, A monolithic FEM approach for non-isothermal incompressible viscous flows, *J. Comput. Phys.* **228** (2009):3869–3881.
7. T. A. Davis and I. S. Duff, A combined unifrontal/multifrontal method for unsymmetric sparse matrices, *SACM Trans. Math. Software* **25** (1999):1–19.
8. M. A. Fernandez, J.-F. Gerbeau and V. Martin, Numerical simulation of blood flows through a porous interface, *ESAIM Math. Model. Numer. Anal.* **42** (2008):961–990.
9. Y. C. Fung, *Biomechanics: Mechanical Properties of Living Tissues* (Springer-Verlag, 1993).
10. P. M. Gresho, On the theory of semi-implicit projection methods for viscous incompressible flow and its implementation via a finite element method that also introduces a nearly consistent mass matrix, part 1: Theory, *Int. J. Numer. Meth. Fluids* **11** (1990):587–620.
11. G. A. Holzapfel, *A Continuum Approach for Engineering* (John Wiley and Sons, Chichester, UK, 2000).
12. G. A. Holzapfel, Determination of material models for arterial walls from uni-axial extension tests and histological structure, *Int. J. Numer. Meth. Fluid.* **238** (2006): 290–302.
13. J. Hron, A. Ouazzi and S. Turek, A computational comparison of two FEM solvers for nonlinear incompressible flow, in *Challenges in Scientific Computing*, ed. E. Bänsch, LNCSE:53 (Springer, 2002), pp. 87–109.
14. J. Hron and S. Turek, A monolithic FEM/multigrid solver for ALE formulation of fluid structure interaction with application in biomechanics, in *Fluid-Structure Interaction: Modelling, Simulation, Optimisation*, eds. H.-J. Bungartz and M. Schäfer, LNCSE:53 (Springer, 2006).
15. R. Löhner, J. Cebral and S. Appanaboyina, Parabolic recovery of boundary gradients, *Comm. Numer. Meth. Eng.* **24** (2008):1611–1615.
16. D. A. McDonald, *Blood Flow in Arteries*, 2nd ed (Edward Arnold, 1974).
17. R. Rannacher and S. Turek, A simple nonconforming quadrilateral stokes element, *Numer. Meth. Part. Differ. Equat.* **8** (1992):97–111.
18. M. Razzaq, Numerical techniques for solving fluid-structure interaction problems with applications to bio-engineering, Ph.D. Thesis, TU Dortmund (2009).
19. M. Razzaq, J. Hron and S. Turek, Numerical simulation of laminar incompressible fluid-structure interaction for elastic material with point constraints, in *Advances in Mathematical Fluid Mechancis — Dedicated to Giovanni paolo Galdi on the Occasion of his 60th Birthday* (Springer, 2009).
20. T. E. Tezduyar, S. Sathe, T. Cragin, B. Nanna, B. S. Conklin, J. Pausewang and M. Schwaab, Modeling of fluid structure interactions with the space time finite elements: Arterial fluid mechanics, *Int. J. Numer. Meth. Fluids* **54** (2007):901–922.

21. T. E. Tezduyar, S. Sathe, M. Schwaab and B. S. Conklin, Arterial fluid mechanics modeling with the stabilized space time fluid structure interaction technique, *Int. J. Numer. Meth. Fluids* **57** (2008):601–629.
22. R. Torri, M. Oshima, T. Kobayashi, K. Takagi and T. E. Tezduyar, Influence of wall elasticity in patient-specific hemodynamic simulations, *Comput. Fluids* **36** (2007): 160–168.
23. R. Torri, M. Oshima, T. Kobayashi, K. Takagi and T. E. Tezduyar, Numerical investigation of the effect of hypertensive blood pressure on cerebral aneurysm dependence of the effect on the aneurysm shape, *Int. J. Numer. Meth. Fluids* **54** (2007):995–1009.
24. S. Turek, *Efficient Solvers for Incompressible Flow Problems: An Algorithmic and Computational Approach* (Springer-Verlag, 1999).
25. S. Turek and J. Hron, Proposal for numerical benchmarking of fluid-structure interaction between an elastic object and laminar incompressible flow, in *Fluid-Structure Interaction: Modelling, Simulation, Optimisation*, eds. H.-J. Bungartz and M. Schäfer, LNCSE:53 (Springer, 2006).
26. S. Turek, L. Rivkind, J. Hron and R. Glowinski, Numerical study of a modified time-steeping theta-scheme for incompressible flow simulations, *J. Sci. Comput.* **28** (2006):533–547.
27. S. Turek and R. Schmachtel, Fully coupled and operator-splitting approaches for natural convection flows in enclosures, *Int. J. Numer. Meth. Fluids* **40** (2002):1109–1119.
28. S. Turek and M. Schäfer, Benchmark computations of laminar flow around cylinder, in *Flow Simulation with High-Performance Computers II*, ed. E. H. Hirschel, vol. 52 of *Notes on Numerical Fluid Mechanics* (Vieweg, 1996), pp. 547–566.
29. A. Valencia, D. Ladermann, R. Rivera, E. Bravo and M. Galvez, Blood flow dynamics and fluid–structure interaction in patient-specific bifurcating cerebral aneurysms, *Int. J. Numer. Meth. Fluids* **58** (2008):1081–1100.
30. S. P. Vanka, Implicit multigrid solutions of Navier–Stokes equations in primitive variables, *J. Comput. Phys.* **65** (1985):138–158.

CHAPTER 4

MATHEMATICAL ANALYSIS OF PARTICULATE FLOWS

Jorge San Martín

Departamento de Ingenieria Mathemática
Facultad de Ciencias Físicas y Matemáticas, Universidad de Chile
and
Centro de Modelamiento Matemático, UMR 2071 CNRS-Uchile
Casilla 170/3-Correo 3, Santiago, Chile
jorge@dim.uchile.cl

Marius Tucsnak

Nancy Université, Institut Élie Cartan, Département de Mathématiques (IECN)
B.P. 239, 54506 Vandœuvre-lès-Nancy Cedex, France
tucsnak@iecn.u-nancy.fr

This paper represents an introduction to the mathematical analysis of particulate flows. We introduce the models, which are based on the coupling of Newton's law for a collection of rigid bodies (the particles) and the equations of fluid dynamics for a liquid surrounding them. In the Introduction we briefly describe the case of an inviscid fluid. In Sec. 2 we study the existence of strong solutions in the case of a viscous incompressible fluid. We give a detailed proof of a local existence of these solutions. Moreover, we discuss some extensions and related questions, such as the global existence for small initial data. In Sec. 3 we define a concept of a weak solution and we prove their global existence up to possible contacts. A brief discussion of the existence of these contacts is also included, with updated references

1. Mathematical Models for Particulate Flows

1.1. *Introduction*

In many practical problems a fluid interacts with a solid structure, exerting stresses that may cause deformation in the structure and, thus, alter the flow of the fluid itself. These phenomena are usually called fluid-structure interactions and they occur, for instance, in aerodynamics (flow around an aircraft), medicine (blood flow in vessels), zoology (swimming of aquatic animals). The mathematical study of these problems raises several challenges, the main one being due to the fact that the domain filled by the fluid is one of the unknowns of the problem. Another difficulty which has to be tackled is that the dynamics of the system couples the ordinary differential equations modeling the solid with the partial differential equations modeling the fluid.

Within this work we focus on *particulate flows*, which design the coupled motion of a collection of rigid bodies and of a fluid surrounding them. According to the fact that the fluid ideal, viscous (compressible or incompressible) or non-Newtonian, the corresponding mathematical models are given by systems of partial and ordinary differential equations of increasing complexity. The common features of these models are the facts that the equations for the fluid and for the solids are coupled via the boundary conditions and that the equations for the fluid hold in a spatial domain which is variable with respect to time. This domain is, in most of the cases, an unknown of the problem, so that we have to tackle a free boundary value problem. However, the mathematical analysis of these problems is expected to be simpler than in the general free boundary case since for particulate flows the free boundary has a finite number of degrees of freedom, namely $6m$, where m is the number of rigid bodies (in the case of a three dimensional flow). We give below a description of the governing equations for the ideal or viscous incompressible and homogeneous fluids.

This text is meant to be an introduction into the analysis of particulate flows. We assume that the reader is familiar with the basic mathematical and numerical analysis of the Navier–Stokes or equations and we concentrate on the questions specific to the presence of rigid bodies in the fluid flow. We concentrate on the situation when the fluid-solid system feels a bounded domain. In this case the free boundary character of the system plays an important role. More precisely, if we try to write the equation in a frame connected to the solid by means of a rigid change of coordinates, the exterior boundary will be displaced, so that we still have a free boundary value problem. The situation is quite different in the case of a single body and of a fluid filling the remaining part of the space. We refer to the last subsection of Sec. 2 for some comments and references concerning the latter case.

We first give some basic equations, which are independent of the properties of the fluid. Let $\Omega \subset \mathbb{R}^3$ be an open set. We assume that, at each time $t \geq 0$, $\Omega = F(t) \cup S(t)$, where $F(t)$ is the domain filled by the fluid, whereas $S(t)$ is the domain occupied by m rigid bodies $S_1(t), \ldots, S_m(t)$. Regardless of the considered type of fluid, we know that the *Cauchy equations* hold in $F(t)$. More precisely,

$$\rho_F \left(\frac{\partial u}{\partial t} + (u \cdot \nabla) u \right) - \operatorname{div} \mathbb{T} = \rho_F b \quad (t \geq 0,\ x \in F(t)). \tag{1.1.1}$$

where the positive constant ρ_F stands for the density of the fluid, u is the Eulerian velocity field of the fluid, $\mathbb{T}$ is the Cauchy stress field in the fluid and b is the density of exterior mass forces (supposed to be known). Within this work we assume that the fluid is incompressible, which implies that

$$\operatorname{div} u = 0, \quad (t \geq 0, \quad x \in F(t)). \tag{1.1.2}$$

The equations of motion of the solids follow from Newton's laws and they can be written as

$$\begin{aligned} M_j \frac{\mathrm{d}^2 h_j}{\mathrm{d}t^2} &= -\int_{\partial S_j(t)} \mathbb{T} n \,\mathrm{d}\Gamma \\ &\quad + \int_{S_j(t)} \rho_S b \,\mathrm{d}x, \quad t \in [0, T],\ j = 1, \ldots, m, \end{aligned} \tag{1.1.3}$$

$$\begin{aligned} \frac{\mathrm{d}}{\mathrm{d}t}(J_j \omega_j) &= -\int_{\partial S_j(t)} (x - h_j) \times \mathbb{T} n \,\mathrm{d}\Gamma \\ &\quad + \int_{S_j(t)} (x - h_j) \times \rho_S b \,\mathrm{d}x, \quad t \in [0, T],\ j = 1, \ldots, m \end{aligned} \tag{1.1.4}$$

$$\frac{\mathrm{d}R_j}{\mathrm{d}t}(t) = A(\omega_j(t)) R_j(t) \quad t \in [0, T],\ j = 1, \ldots, m \tag{1.1.5}$$

where $h_j(t)$ (respectively, $\omega_j(t)$) is the position of the mass center (respectively, the angular velocity) of the rigid body $S_j(t)$. Moreover, R_j denotes the rotation tensor of the solid number j. The notation $\times$ stands for the usual vector product in $\mathbb{R}^3$ whereas n denotes the unitary normal vector field to $\partial S_j(t)$ oriented towards the interior of each solid. The skew symmetric matrix $A(\omega)$ is defined by

$$A(\omega) = \begin{pmatrix} 0 & -\omega_3 & \omega_2 \\ \omega_3 & 0 & -\omega_1 \\ -\omega_2 & \omega_1 & 0 \end{pmatrix} \quad \text{for all } \omega \in \mathbb{R}^3. \tag{1.1.6}$$

Moreover, for every $j \in \{1, \ldots, m\}$, M_j stands for the mass of S_j and $J_j(t)$ denotes the inertia matrix of $S_j(t)$ defined by

$$J_j(t)a \cdot b = \rho_j \int_{S_j(t)} [a \times (x - h_j(t))] \cdot [b \times (x - h_j(t))]\mathrm{d}x \qquad \text{for all} \quad a, b \in \mathbb{R}^3, \tag{1.1.7}$$

where ρ_j is the density of the solid S_j (supposed to be a known constant).

In order to close the system, Eqs. (1.1.1)–(1.1.5) have to be supplemented with a constitutive law for the fluid, with appropriate boundary conditions and with initial conditions.

In the case of an ideal fluid the constitutive law is

$$\mathbb{T} = -p\mathbb{I}_3, \tag{1.1.8}$$

where $\mathbb{I}_3$ is the 3×3 identity matrix and, for every $t \geq 0$, the scalar field $p(\cdot, t) : F(t) \to \mathbb{R}$ denotes the pressure field in the fluid. The fact that the fluid cannot penetrate the exterior boundary $\partial\Omega$ and the boundary of each solid $\partial S_j(t)$ can be written by using the boundary conditions

$$u(x, t) \cdot n(x) = 0 \quad (x \in \partial\Omega, \quad t \geq 0), \tag{1.1.9}$$

$$u(x, t) \cdot n(t, x) = (\dot{h}_j(t) + \omega_j(t) \times (x - h_j(t))) \cdot n(x, t) \qquad (x \in \partial S_j(t), \quad t \geq 0). \tag{1.1.10}$$

In order to complete the system we specify the initial conditions

$$u(x, 0) = u_0(x) \quad x \in F(0), \tag{1.1.11}$$

$$h_j(0) = h_j^0, \quad \dot{h}_j(0) = k_j^0, \quad R_j(0) = \mathbb{I}_3, \quad \omega_j(0) = \omega_j^0. \tag{1.1.12}$$

The constitutive law and the boundary conditions in the case of a viscous fluid will be described in Subsec. 2.1.

The outline of this work is as follows. In the next subsection we specify the equations in the case (probably the simplest) of an ideal incompressible fluid which undergoes a potential flow. Moreover, we briefly discuss some well posedness issues for both potential and non potential flows. In Sec. 2 we discuss the existence and uniqueness of strong solutions in the case of a viscous incompressible fluid. More precisely, we give a detailed proof of the local in time existence and uniqueness of solutions for smooth enough initial data. Moreover, we discuss additional questions like global existence for small data or the existence of contacts. In Sec. 3 we define a notion of weak solution in the case of a viscous incompressible fluid and we prove the existence of such solutions on arbitrary time intervals which do not include any instances of collisions of the bodies. We end by discussing some extensions and open problems.

1.2. *The case of an inviscid fluid*

We have seen in the previous subsection that the motion of a solid in an incompressible ideal fluid is governed by Eqs. (1.1.1)–(1.1.5) together with Eqs. (1.1.8)–(1.1.12). In this case, substituting Eq. (1.1.8) into Eq. (1.1.1) we obtain the classical *Euler equation for ideal fluids* which writes as

$$\rho_F\left(\frac{\partial u}{\partial t} + (u\cdot\nabla)u\right) + \nabla p = \rho_F b, \quad (t \geq 0, \quad x \in F(t)). \qquad (1.2.1)$$

The simplest mathematical model of particulate flows occurs in the case when the fluid is ideal and it undergoes a potential flow (also called an *inviscid fluid*). As has been already remarked by Kirchhoff and Kelvin, in this case, the coupled system can be reduced to a system of ordinary differential equations. The coefficients appearing in these equations are obtained by solving, at each moment t, an elliptic partial differential equation so that they cannot, in general, be expressed by simple formulas. The fact that, for particular geometries, these elliptic problems can be explicitly solved lead Kirchhoff, Kelvin, Lamb and others to remarkable analytical solutions or qualitative study of the motion of rigid bodies in an inviscid fluid. We refer to Lamb [27] and Milne-Thomson [32] for detailed results in this direction. Note that these very simple models are still used in practical problems like the guidance of ocean vehicles (see, for instance, Fossen [11] and references therein).

In the remaining part of this subsection we briefly describe the governing equations in the case of a single rigid body. The fact that the flow is potential means that for every $t \geq 0$ there exists a function $\varphi(t, \cdot)$ such that

$$u(x,t) = \nabla\varphi(x,t) \quad (t \geq 0, \quad x \in F(t)), \tag{1.2.2}$$

then Eq. (1.2.1), combined with Eq. (1.1.2), yields the well-known *Bernoulli formula*:

$$\rho_F\left(\frac{\partial\varphi}{\partial t}(x,t) + \frac{1}{2}|\nabla\varphi(x,t)|^2\right) + p(x,t) = C(t). \tag{1.2.3}$$

For the sake of simplicity, in the remaining part of this subsection we assume that the fluid lies in a bounded cavity and that it contains a single particle, denoted $S(t)$. In this case, from Eqs. (1.1.2), (1.1.9) and (1.1.10) it follows that

$$\Delta\varphi(x,t) = 0 \quad (t \geq 0, \quad x \in F(t)), \tag{1.2.4a}$$

$$\frac{\partial\varphi}{\partial n}(x,t) = 0 \quad (t \geq 0, \quad x \in \partial\Omega), \tag{1.2.4b}$$

$$\frac{\partial\varphi}{\partial n}(x,t) = [\dot{h}(t) + \omega(t) \times (x - h(t))] \cdot n(x,t) \quad (t \geq 0, \quad x \in \partial S(t)). \tag{1.2.4c}$$

From the above equations it follows that φ (and, consequently, the velocity field u) is completely determined, at each instant t, by the position and the velocity of the rigid body. Moreover from Bernoulli's formula (1.2.3) it follows that the same property is shared by the pressure field p. Therefore, p can be expressed, at each moment t as function of $h, \dot{h}, \ddot{h}$ and $R, \omega, \dot{\omega}$ so that Eq. (1.1.3)–(1.1.5) form indeed an ODE system which allows to predict the behavior of the fluid and of the particles.

The dependence of the solution φ of Eq. (1.2.4), with respect to the velocities $\dot{h}$ and ω, is clearly linear, whereas the dependence on the positions h and R is more complicated, being of a geometric nature. Therefore we do not have, in general, simple expression of the right-hand sides of Eqs. (1.1.3) and (1.1.4).

A different way of obtaining the governing equations is based on analytical mechanics. More precisely, the total kinetic energy of the system is the sum of the kinetic energy of the rigid body and of the fluid which can be written as

$$E(h(t), \dot{h}(t), R(t), \omega(t)) = \frac{1}{2} M |\dot{h}(t)|^2 + \frac{1}{2}(J(t)\omega(t)) \cdot \omega(t) + \frac{1}{2} \int_{F(t)} |\nabla \varphi(x, t)|^2 \mathrm{d}x,$$

where we have used again the fact that $\varphi(\cdot, t)$ is completely determined by $h(t)$, $\dot{h}(t)$, $R(t)$ and $\omega(t)$. In the absence of exterior forces, the system will be governed by Lagrange's equations

$$\frac{\mathrm{d}}{\mathrm{d}t} \frac{\partial E}{\partial \dot{q}_k} - \frac{\partial E}{\partial q_k} = 0 \quad (k = 1, 6), \tag{1.2.5}$$

where $q_k = h_k$ for $k \in \{1, 2, 3\}$ and (q_4, q_5, q_6) are parameters occurring in the parametrization of the orthogonal group $SO(3)$ (usually the Euler angles). The fact that Eq. (1.2.5) is equivalent with the previously described equations, based on Newton's laws, is far from being obvious from a mathematical viewpoint. For a detailed discussion of these issues we refer to the recent work of Houot and Munnier [24]. However, in order to give a flavor of the involved calculations, we derive below an expression of the force exerted by the fluid on the solid and we note that the obtained expression contains some of the terms in Eq. (1.2.5).

Proposition 1.2.1. *For every $j \in \{1, 2, 3\}$ the j-th component of the force with which the fluid acts on the solid is given by*

$$\int_{\partial S(t)} p n_j = -\frac{\rho_F}{2} \frac{\mathrm{d}}{\mathrm{d}t} \frac{\partial}{\partial \dot{h}_i} \int_{F(t)} |\nabla \varphi|^2 \mathrm{d}x - \frac{\rho_F}{2} \int_{\partial \Omega} |\nabla \varphi|^2 n_j \mathrm{d}\Gamma. \tag{1.2.6}$$

Proof. For the sake of simplicity, we take $j = 1$. From the Bernoulli formula (1.2.3) and the fact that

$$\int_{\partial S(t)} C(t) n_1 \mathrm{d}\Gamma = 0,$$

it follows that

$$\int_{\partial S(t)} p n_1 \mathrm{d}\Gamma = \rho_F (I_1 + I_2), \tag{1.2.7}$$

where

$$I_1 = -\int_{\partial S(t)} \frac{\partial \varphi}{\partial t} n_1 \mathrm{d}\Gamma,$$

$$I_2 = -\frac{1}{2}\int_{\partial S(t)} |\nabla\varphi|^2 n_1 \mathrm{d}\Gamma.$$

By using Gauss' formula it follows that

$$I_1 = -\int_{F(t)} \frac{\partial^2 \varphi}{\partial x_1 \partial t}\mathrm{d}x + \int_{\partial\Omega} \frac{\partial \varphi}{\partial t} n_1 \mathrm{d}\Gamma.$$

The above formula, combined to Reynolds' formula, Eq. (1.2.4c) and to the fact that $\partial\varphi/\partial n = 0$ on $\partial\Omega$ implies that

$$I_1 = -\frac{\mathrm{d}}{\mathrm{d}t}\int_{F(t)} \frac{\partial \varphi}{\partial x_1}\mathrm{d}x + \int_{\partial S(t)} \frac{\partial \varphi}{\partial x_1}(\dot{h} + \omega \times (x - h)) \cdot n \mathrm{d}\Gamma$$
$$+ \int_{\partial\Omega} \frac{\partial \varphi}{\partial t} n_1 \mathrm{d}\Gamma.$$

By applying the Gauss formula to the first term in the above relation we obtain

$$I_1 = -\frac{\mathrm{d}}{\mathrm{d}t}\int_{\partial S(t)} \varphi n_1 \mathrm{d}\Gamma + \int_{\partial S(t)} \frac{\partial \varphi}{\partial x_1}(\dot{h} + \omega \times (x - h)) \cdot n \mathrm{d}\Gamma.$$

From Eq. (1.2.4c) it follows that on $\partial S(t)$ we have

$$n_1 = \frac{\partial}{\partial \dot{h}_1}\frac{\partial \varphi}{\partial n},$$

so that we can use again the Gauss formula and the fact that φ is harmonic to get

$$\begin{aligned} I_1 &= -\frac{\mathrm{d}}{\mathrm{d}t}\frac{\partial}{\partial \dot{h}_1}\int_{F(t)} \mathrm{div}(\varphi\nabla\varphi)\mathrm{d}x + \int_{\partial S(t)} \frac{\partial \varphi}{\partial x_1}(\dot{h} + \omega \times (x - h)) \cdot n \mathrm{d}\Gamma \\ &= -\frac{1}{2}\frac{\mathrm{d}}{\mathrm{d}t}\frac{\partial}{\partial \dot{h}_1}\int_{F(t)} |\nabla\varphi|^2\mathrm{d}x + \int_{\partial S(t)} \frac{\partial \varphi}{\partial x_1}(\dot{h} + \omega \times (x - h)) \cdot n \mathrm{d}\Gamma. \end{aligned} \tag{1.2.8}$$

Concerning I_2, a simple integration by parts yields:

$$I_2 = -\frac{1}{2}\int_{F(t)} \frac{\partial|\nabla\varphi|^2}{\partial x_1}\mathrm{d}x + \frac{1}{2}\int_{\partial\Omega} |\nabla\varphi|^2 n_1 \mathrm{d}\Gamma. \tag{1.2.9}$$

Note that

$$\begin{aligned}\int_{F(t)} \frac{\partial|\nabla\varphi|^2}{\partial x_1}\mathrm{d}x &= 2\int_{F(t)} \left(\frac{\partial^2\varphi}{\partial x_1^2}\frac{\partial\varphi}{\partial x_1} + \frac{\partial^2\varphi}{\partial x_1\partial x_2}\frac{\partial\varphi}{\partial x_2} + \frac{\partial^2\varphi}{\partial x_1\partial x_3}\frac{\partial\varphi}{\partial x_3}\right)\mathrm{d}x \\ &= 2\int_{F(t)} \left[\frac{\partial^2\varphi}{\partial x_1^2}\frac{\partial\varphi}{\partial x_1}\mathrm{d}x + \frac{\partial}{\partial x_2}\left(\frac{\partial\varphi}{\partial x_1}\frac{\partial\varphi}{\partial x_2}\right)\right. \\ &\quad \left. + \frac{\partial}{\partial x_3}\left(\frac{\partial\varphi}{\partial x_1}\frac{\partial\varphi}{\partial x_3}\right) - \frac{\partial\varphi}{\partial x_1}\left(\frac{\partial^2\varphi}{\partial x_2^2} + \frac{\partial^2\varphi}{\partial x_3^2}\right)\right]\mathrm{d}x.\end{aligned}$$

Using the fact that φ is harmonic together with Eqs. (1.2.4b) and (1.2.4c) we obtain

$$\begin{aligned}\int_{F(t)} \frac{\partial|\nabla\varphi|^2}{\partial x_1}\mathrm{d}x &= 2\int_{F(t)} \frac{\partial}{\partial x_1}\left|\frac{\partial\varphi}{\partial x_1}\right|^2 \mathrm{d}x \\ &\quad +2\int_{F(t)} \left[\frac{\partial}{\partial x_2}\left(\frac{\partial\varphi}{\partial x_1}\frac{\partial\varphi}{\partial x_2}\right) + \frac{\partial}{\partial x_3}\left(\frac{\partial\varphi}{\partial x_1}\frac{\partial\varphi}{\partial x_3}\right)\right]\mathrm{d}x \\ &= 2\int_{\partial F(t)} \frac{\partial\varphi}{\partial x_1}\frac{\partial\varphi}{\partial n}\mathrm{d}\Gamma \\ &= 2\int_{\partial S(t)} \frac{\partial\varphi}{\partial x_1}\left[\dot{h} + \omega\times(x-h)\right]\cdot n\mathrm{d}\Gamma.\end{aligned}$$

This implies, by using Eq. (1.2.9), that

$$I_2 = -\int_{\partial S(t)} \frac{\partial\varphi}{\partial x_1}\left[\dot{h} + \omega\times(x-h)\right]\cdot n\mathrm{d}\Gamma + \frac{1}{2}\int_{\partial\Omega} |\nabla\varphi|^2 n_1 \mathrm{d}x.$$

The above formula, combined to Eq. (1.2.7) and Eq. (1.2.8) implies that Eq. (1.2.6) holds for $j = 1$. □

The fact that in the case of an inviscid fluid the governing equations are ODE's implies that the well posedness can be obtained by simply applying the Cauchy–Lipschitz theorem. However, in order to apply this theorem it is necessary to prove first that quantities like $\int_{F(t)} |\nabla\varphi|^2\mathrm{d}x$ depend smoothly of the position and of the orientation of the rigid bodies. We refer to [24] for a rigorous discussion of these issues. In the case of an ideal fluid which is not necessarily potential, the well posedness questions are quite delicate. We refer to the recent works of Ortega, Rosier and Takahashi [33, 34] for interesting results in this direction.

2. Strong Solutions in the Viscous Case

2.1. *Governing equations and main results*

In the case of a viscous incompressible fluid, Eqs. (1.1.1)–(1.1.7) are supplemented with the constitutive law of a Newtonian fluid which is

$$\mathbb{T}(x,t) = -p(x,t)\mathbb{I}_3 + 2\nu D(u),$$

where the positive constant ν is the viscosity of the fluid, $\mathbb{I}_3$ is the identity matrix and $D(u)$ is the tensor field defined by

$$D(u)_{k,l} = \frac{1}{2}\left(\frac{\partial u_k}{\partial x_l} + \frac{\partial u_l}{\partial x_k}\right),$$

and p is the pressure field in the fluid. By substituting the two above formulas in Eq. (1.1.1) and by assuming that $\rho_F = 1$ we obtain the Navier–Stokes equations

$$\frac{\partial u}{\partial t} - \nu\Delta u + (u\cdot\nabla)u + \nabla p = f \quad (t \geq 0, \quad x \in F(t)). \tag{2.1.1}$$

The incompressibility condition is of course unchanged, i.e., we still have

$$\operatorname{div} u = 0 \quad (t \geq 0, \quad x \in F(t)). \tag{2.1.2}$$

For the sake of simplicity we assume that the fluid contains a single solid $S(t)$. This means that the remaining part of the governing equations can be written

$$u(x,t) = 0 \quad (t \geq 0, \quad x \in \partial S(t)), \tag{2.1.3}$$

$$u(x,t) = \dot{h}(t) + \omega(t) \times (x - h(t)) \quad (t \geq 0) x \in \partial S(t)), \tag{2.1.4}$$

$$M\ddot{h}(t) = -\int_{\partial S(t)} \mathbb{T} n \mathrm{d}\Gamma + \rho_S \int_{S(t)} f(x,t)\,\mathrm{d}x \quad (t \geq 0), \tag{2.1.5}$$

$$\frac{\mathrm{d}}{\mathrm{d}t}(J\omega(t)) = -\int_{\partial S(t)} (x - h(t)) \times \mathbb{T} n\, \mathrm{d}\,\Gamma$$

$$+ \rho_S \int_{S(t)} (x - h(t)) \times f(x,t)\,\mathrm{d}x \quad (t \geq 0), \tag{2.1.6}$$

$$u(x,0) = u_0(x) \quad (x \in F(0)), \tag{2.1.7}$$

$$h(0) = h_0, \quad \dot{h}(0) = h_1, \quad \omega(0) = \omega_0. \tag{2.1.8}$$

In the above system the unknowns are $u(x, t)$ (the Eulerian velocity field of the fluid), $p(x, t)$ (the pressure of the fluid), $h(t)$ (the position of the mass center of the rigid body) and $\omega(t)$ (the angular velocity of the rigid body).

The constants M and ρ_S are the mass, respectively, the density, of the solid, whereas $J(t)$ is its moment of inertia at the instant t. Moreover, $f : \Omega \times [0, \infty) \to \mathbb{R}$ is the field of mass forces (acting on both the fluid and the solid). We have denoted by $\dot{w}$ and $\ddot{w}$ the derivatives of a function w depending only on the time t. If $x, y \in \mathbb{R}^3$, then $x \cdot y$ stands for the inner product of x and y and $|x|$ stands for the corresponding norm. Moreover we have denoted by $\partial S(t)$ the boundary of the rigid body at instant t and by $n(x, t)$ the unit normal to $\partial S(t)$ at the point x directed to the interior of the rigid body.

The main result in this section asserts the existence and uniqueness of strong solutions of Eqs. (2.1.1)–(2.1.8). In order to give the precise statement we will introduce several function spaces. To do this, we assume for a moment that there exists $X : \mathbb{R}^n \times [0, \infty[\to \mathbb{R}^n$ such that for each $t \geq 0$, $X|_\Omega(\cdot, t)$ is a C^∞-diffeomorphism from $F(0)$ onto $F(t)$ and from Ω onto Ω. Moreover, suppose that the mappings

$$(y, t) \mapsto D_t D_y^\alpha X(y, t), \quad \alpha \in \mathbb{N}^n,$$

exist, are continuous in $F(0)$. The detailed construction such a map X is postponed to Subsec. 2.2.

Let $u(\cdot, t)$, $t \geq 0$ be a family of functions with $u(\cdot, t) : F(t) \to \mathbb{R}^n$. Denote $v(y, t) = u(\psi(y, t), t)$, for all $t \geq 0$ and for all $y \in \Omega$. Then the functions spaces introduced above are defined, for every $T > 0$ by

$$\begin{aligned}
L^2(0, T; H^2(F(t))) &= \{u \mid v \in L^2(0, T; H^2(F(0)))\},\\
H^1(0, T; L^2(F(t))) &= \{u \mid v \in H^1(0, T; L^2(F(0)))\},\\
C([0, T], H^1(F(t))) &= \{u \mid v \in C([0, T], H^1(F(0)))\},\\
L^2(0, T; H^1(F(t))) &= \{u \mid v \in L^2(0, T; H^1(F(0)))\}.
\end{aligned}$$

We will show that there exist (strong) solutions with the velocity field in the space $\mathscr{U}(0, T; F(t))$ defined by

$$\begin{aligned}
\mathscr{U}(0, T; F(t)) = L^2(0, T; H^2(F(t))) \cap C([0, T],\\
H^1(F(t))) \cap H^1(0, T; L^2(F(t))).
\end{aligned} \tag{2.1.9}$$

Roughly speaking the functions in the above spaces are time dependent vector fields defined, at each instant t, on the variable domain $F(t)$ and which lie in classical Sobolev spaces (with respect to the space variable). Finally, for each $t \geq 0$ we introduce the homogeneous Sobolev space

$$\widehat{H}^1(F(t)) = \left\{ q \in L^2(F(t)) \mid \nabla q \in [L^2(\Omega)]^3, \quad \int_{F(t)} q \mathrm{d}x = 0 \right\}.$$

The main result in this section is:

Theorem 2.1.1. *Suppose that $\partial S(0)$ and $\partial\Omega$ are $C^{2+\mu}$-boundaries, with $\mu \in (0, 1)$. Let $f \in L^2_{\mathrm{loc}}(0, \infty; [W^{1,\infty}(\mathbb{R}^3)]^3)$ and $v_0 \in [H^1(F(0))]^3$ be such that*

$$\begin{cases} \operatorname{div} v_0 = 0, & \text{in } F(0), \\ v_0(x) = 0, & x \in \partial\Omega, \\ v_0(x) = h_1 + \omega_0 \times x, & x \in \partial F(0). \end{cases}$$

Then, there exists a time T_0 depending only on $\|v_0\|_{H^1(F(0))}$ such that the Eqs. (2.1.1)–(2.1.8) *admit a unique strong solution*

$$(v, q) \in \mathscr{U}(0, T; F(t)) \times L^2(0, T; \widehat{H}^1(F(t))),$$

for any $T \in (0, T_0)$.

Moreover, we can choose T_0 such that one the following assertions holds true:

(i) $T_0 = \infty$.
(ii) *The function $t \mapsto \|v(t)\|_{H^1(F(t))}$ is not bounded in* $[0, T_0)$.
(iii) *The distance from $S(t)$ to $\partial\Omega$ tends to zero when $t \to T_0$.*

The proof of the above theorem is quite long and it will be described in the remaining part of this section. However, for the reader's convenience, we end this subsection by an outline of the proof of Theorem 2.1.1. For the sake of simplicity we consider only the case $f \equiv 0$ and we assume that the rigid body is a ball.

The first step in the proof of Theorem 2.1.1 is to reduce Eqs. (2.1.1)–(2.1.8) to a system in which Eq. (2.1.1) is replaced by a partial differential equation holding in the cylindrical domain $F(0) \times (0, T)$. To this end, we use a change of variables, which coincides with a translation in a neighborhood of the rotating body, but it equals to the identity far from the rigid body (in particular, close to the exterior boundary). We then obtain a system equivalent to Eqs. (2.1.1)–(2.1.8), which has the form

$$
\begin{aligned}
\frac{\partial v}{\partial t} - \nu L v + M v + N v + G q &= 0, && \text{in } F(0) \times (0, T),\\
\operatorname{div} v &= 0, && \text{in } F(0) \times (0, T),\\
v(y, t) &= 0, && \text{on } \partial\Omega \times (0, T),\\
v(y, t) &= \dot{h}(t) + \omega(t) \times y, && \text{on } \partial S(0) \times (0, T),\\
M\ddot{h}(t) &= -\int_{\partial S(0)} \sigma(v, q) n \, d\Gamma, && t \in (0, T),\\
J(0)\dot{\omega}(t) = (J(0)\omega(t)) \times \omega(t) &- \int_{\partial S(0)} y \times \sigma(v, q) n d\Gamma, && t \in (0, T),\\
v(y, 0) &= u_0(y), && y \in S(0),\\
h(0) = 0, \quad \dot{h}(0) = h_1, \quad \omega(0) &= \omega_0,
\end{aligned}
$$

where

$$
\sigma(v, q) = -q\mathbb{I}_3 + 2\nu D(v).
$$

The unknowns of this system are $v(y, t)$, $q(y, t)$, $h(t)$ and $\omega(t)$. Gq is the transformed of ∇p, Lv is the transformed of Δu, Mv is a linear term in v and in ∇v, whereas Nv is a nonlinear term corresponding to $(u \cdot \nabla)u$. All the coefficients of these operators depend only on h (in a very smooth way). Moreover, Lv is close to Δv and Gq is close to ∇q for small t. Based on this fact, we prove the existence and uniqueness theorem by a fixed point argument applied to the map

$$
\mathscr{L} : \begin{pmatrix} f \\ g \end{pmatrix} \mapsto \begin{pmatrix} nu[(L - \Delta)v] - MV + (\nabla - G)q + Nv \\ (J(0)\omega) \times \omega \end{pmatrix},
$$

where (v, q, h, ω) satisfies

$$
\begin{aligned}
\frac{\partial v}{\partial t} - \nu \Delta v + \nabla q &= f, && \text{in } F(0) \times (0, T),\\
\operatorname{div} v &= 0, && \text{in } F(0) \times (0, T),\\
v(y, t) &= 0, && \text{on } \partial\Omega \times (0, T),\\
v(y, t) &= \dot{h}(t) + \omega(t) \times y, && \text{on } \partial S(0) \times (0, T),
\end{aligned}
$$

$$M\ddot{h}(t) = -\int_{\partial S(0)} \sigma n \,\mathrm{d}\Gamma, \qquad t \in (0, T),$$

$$J(0)\dot{\omega}(t) = g(t) - \int_{\partial S(0)} y \times \sigma n \mathrm{d}\Gamma, \qquad t \in (0, T),$$

$$v(y, 0) = v_0(y), \qquad y \in S(0),$$

$$h(0) = 0, \quad \dot{h}(0) = h_1, \quad \omega(0) = \omega_0.$$

For T small enough, we show that there exists a closed ball $B(0, R)$ in an appropriate Banach space such that $\mathscr{X}$ maps $B(0, R)$ into $B(0, R)$ and such that the restriction of $\mathscr{X}$ to this ball is a contraction. This clearly implies the local in time existence and uniqueness of the strong solution.

In the remaining part of this section we detail the argument briefly described above.

2.2. *A change of variables*

In this subsection we construct the change of variables which, when applied to the system (2.1.1)–(2.1.8), transforms Eq. (2.1.1) in a PDE valid, for every $t \geq 0$, in the fixed domain $F(0)$. In order to achieve this goal, this change of variables has to map $S(0)$ onto $S(t)$ and to invariate $\partial\Omega$. More precisely, our aim is to construct a map $X : \mathbb{R}^3 \times [0, \infty) \to \mathbb{R}^3$ satisfying

$$X(y, t) = y \quad (t \geq 0, y \in \partial\Omega), \tag{2.2.1}$$

$$X(y, t) = y + h(t) \quad (t \geq 0, y \in S(0)). \tag{2.2.2}$$

Moreover, in order to preserve the condition of free divergence, it will be essential to have a map X which preserves the volume element, i.e., with

$$\det J_X(y, t) = 1 \quad (y \in \mathbb{R}^3, t \geq 0), \tag{2.2.3}$$

where J_X is the Jacobian matrix of X. The remaining part of this subsection is devoted to the construction of the map X and to obtain the transformed form of Eqs. (2.1.1)–(2.1.8).

For $\mu > 0$ we denote

$$\Omega_\mu = \{x \in \Omega \mid d(x, \partial\Omega) > \mu\}.$$

We assume that h is a given smooth function defined on $[0, T]$ and with values in Ω_1 (which is supposed to be nonempty). We define the function $w : \mathbb{R}^3 \times [0, T] \to \mathbb{R}^3$ by

$$w(x, t) = \frac{1}{2}\dot{h}(t) \times x.$$

It is easily seen that

$$\operatorname{curl} w(x, t) = \dot{h}(t) \quad (t \geq 0, x \in \mathbb{R}^3). \tag{2.2.4}$$

Let $\varepsilon > 0$ and and let $\xi \in C^\infty(\mathbb{R}^3)$ be a function with compact support contained in $\Omega_{\frac{\varepsilon}{2}}$ and with $\xi \equiv 1$ on $\overline{\Omega}_\varepsilon$. We define the vector field $\Lambda : \mathbb{R}^3 \times [0, T] \to \mathbb{R}^3$ by $\Lambda = \operatorname{curl}(\xi w)$. More explicitly, for every $t \in [0, T]$ and $x \in \mathbb{R}^3$ we have

$$\Lambda(x, t) = \begin{pmatrix} \xi(x)\dot{h}_1(t) + \dfrac{\partial \xi}{\partial x_2}(x) w_3(x, t) - \dfrac{\partial \xi}{\partial x_2}(x) w_3(x, t) \\ \\ \xi(x)\dot{h}_2(t) + \dfrac{\partial \xi}{\partial x_3}(x) w_1(x, t) - \dfrac{\partial \xi}{\partial x_1}(x) w_3(x, t) \\ \\ \xi(x)\dot{h}_3(t) + \dfrac{\partial \xi}{\partial x_1}(x) w_2(x, t) - \dfrac{\partial \xi}{\partial x_1}(x) w_2(x, t) \end{pmatrix}. \tag{2.2.5}$$

It is not difficult to check that

$$\Lambda(x, t) = \begin{cases} \dot{h}(t) & \text{if} \quad x \in \overline{\Omega}_\varepsilon \supset S(t) \\ 0 & \text{if} \qquad\qquad x \notin \Omega_{\frac{\varepsilon}{2}} \end{cases}. \tag{2.2.6}$$

Next, consider the time dependent vector field $X(\cdot, t)$ satisfying

$$\begin{cases} \dfrac{\partial X}{\partial t}(y, t) = \Lambda(X(y, t), t), \quad t > 0, \\ \\ X(y, 0) = y \in \mathbb{R}^3, \end{cases} \tag{2.2.7}$$

with Λ given by Eq. (2.2.5).

Lemma 2.2.1. *For all $y \in \Omega$, the initial-value problem* (2.2.7) *admits a unique solution $X(y, \cdot) : [0, \infty) \to \Omega$. Moreover, for all $t \geq 0$, we have that the mapping $y \mapsto X(y, t)$ is a C^∞-diffeomorphism of Ω and from $F(0)$ onto $F(t)$. Moreover X satisfies* (2.2.3).

Proof. Since Λ is a C^∞ function, from the classical Cauchy–Lipschitz Theorem, it follows that there exists $\tau > 0$ such that Eq. (2.2.7) admits a unique maximal solution $X(y, \cdot)$, defined on $[0, \tau)$, which is a C^∞ function in $[0, \tau)$.

Moreover, since $\Lambda = 0$ outside Ω it clearly follows that the solution X of Eq. (2.2.7) does not blow up in finite time, so that $\tau = \infty$, and that for every $t \geq 0$ the map $y \mapsto X(y, t)$ is a C^∞ diffeomorphism of Ω (see, for instance, Hartman [18, Corollary 4.1].

On the other hand, we can check that for all $y \in S(0)$, the function

$$\widetilde{X}(y, t) = y + h(t),$$

satisfies Eq. (2.2.7). Indeed, since $\widetilde{X}(y, t) \in S(t)$ for every $y \in S(0)$ so that, by using Eq. (2.2.6), we have that $\Lambda(\widetilde{X}(y, t), t) = \dot{h}(t) = \partial\widetilde{X}/\partial t(y, t)$ for every $t \geq 0$ and $y \in S(0)$. Since the solution of Eq. (2.2.7) is unique we get that

$$X(y, t) = y + h(t) \quad (t \geq 0,\ y \in S(0)),$$

which implies that $X(\cdot, t) : S(0) \to S(t)$ is a C^∞-diffeomorphism. We have seen above that $X(\cdot, t)$ is a C^∞-diffeomorphism of Ω so we conclude that $X(\cdot, t)$ is also a C^∞-diffeomorphism from $F(0)$ onto $F(t)$.

Finally, since $\operatorname{div} \Lambda = 0$, we can apply Liouville's theorem (see, for instance, Arnold [2, p. 249]) to obtain that X satisfies Eq. (2.2.3). □

Remark 2.2.2. Note that, for every $t \geq 0$, the inverse map $Y(\cdot, t)$ of $X(\cdot, t)$ satisfies

$$\begin{aligned} &\frac{\partial Y}{\partial t}(x, t) = -\Lambda(Y(x, t)), \quad t > 0, \\ &Y(x, 0) = x \in \Omega. \end{aligned} \tag{2.2.8}$$

We are now in a position to transform Eqs. (2.1.1)–(2.1.8) into a system written in a cylindrical domain. We first define, following Inoue and Wakimoto [25], the vector field $v : F(0) \times [0, T] \to \mathbb{R}^3$ by

$$v(y, t) = J_Y(X(y, t), t)u(X(y, t), t) \quad (y \in F(0),\ t \geq 0), \tag{2.2.9}$$

where J_Y is the Jacobian matrix of the map Y from Remark 2.2.2. More explicitly, for every $k \in \{1, 2, 3\}$ we have

$$v_k(y, t) = \sum_{j=1}^{3} \frac{\partial Y_k}{\partial x_j}(X(y, t), t) v_j(X(y, t), t).$$

We also define the scalar field $q : F(0) \times [0, T] \to \mathbb{R}$ by

$$q(y, t) = p(X(y, t), t) \quad (y \in F(0),\ t \geq 0). \tag{2.2.10}$$

By using the fact that

$$J_X(y, t) J_Y(X(y, t), t) = \text{ Id} \quad (y \in F(0),\ \ t \in [0, \infty)),$$

it can be shown that:

Lemma 2.2.3. *Suppose that X and u are defined as before. Then,*

$$\operatorname{div} v(y, t) = \operatorname{div} u(X(y, t), t) \quad (y \in F(0),\ t \in [0, \infty)).$$

For a proof see, for instance, [25, Proposition 2.4]. In order to write the equations satisfied by $v(y, t)$ and $q(y, t)$ we define for each $i \in \{1, 2, 3\}$

$$\begin{aligned}(Lv)_i = &\sum_{j,k=1}^{3} \frac{\partial}{\partial y_j}\left(g^{jk}\frac{\partial v_i}{\partial y_k}\right) + 2 \sum_{j,k,l=1}^{3} g^{kl}\Gamma^i_{jk}\frac{\partial v_j}{\partial y_l} \\ &+ \sum_{j,k,l=1}^{3}\left\{\frac{\partial}{\partial y_k}(g^{kl}\Gamma^i_{jl}) + \sum_{m=1}^{n} g^{kl}\Gamma^m_{jl}\Gamma^i_{km}\right\} v_j,\end{aligned} \tag{2.2.11}$$

$$(Nv)_i = \sum_{j=1}^{3} v_j \frac{\partial v_i}{\partial y_j} + \sum_{j,k=1}^{3} \Gamma^i_{jk} v_j v_k, \tag{2.2.12}$$

$$(Mv)_i = \sum_{j=1}^{3} \frac{\partial Y_j}{\partial t}\frac{\partial v_i}{\partial y_j} + \sum_{j,k=1}^{3}\left\{\Gamma^i_{jk}\frac{\partial Y_k}{\partial t} + \frac{\partial Y_i}{\partial x_k}\frac{\partial^2 X_k}{\partial t \partial y_j}\right\} v_j, \tag{2.2.13}$$

$$(Gq)_i = \sum_{j=1}^{n} g^{ij}\frac{\partial q}{\partial y_j}, \tag{2.2.14}$$

where, for each $i,\ j,\ k \in \{1, 2, 3\}$, we have denoted (see, for instance, [9])

$$g^{ij} = \sum_{k=1}^{n} \frac{\partial Y_i}{\partial x_k}(X(y,t),t)\frac{\partial Y_j}{\partial x_k}(X(y,t),t)$$

$$\text{(metric contravariant tensor)}, \tag{2.2.15}$$

$$g_{ij} = \sum_{k=1}^{n} \frac{\partial X_i}{\partial y_k}(y,t)\frac{\partial X_j}{\partial y_k}(y,t) \quad \text{(metric covariant tensor)}, \tag{2.2.16}$$

and

$$\Gamma_{ij}^{k} = \frac{1}{2}\sum_{l=1}^{n} g^{kl}\left\{\frac{\partial g_{il}}{\partial y_j} + \frac{\partial g_{jl}}{\partial y_i} - \frac{\partial g_{ij}}{\partial y_l}\right\} \quad \text{(Christoffel's symbol)}. \tag{2.2.17}$$

With this notation, we have

Proposition 2.2.4. *The pair* (u, p) *satisfies*

$$(u,p) \in \mathscr{U}(0,T;F(t)) \times L^2(0,T;\widehat{H}^1(F(t)))$$

together with Eqs. (2.1.1)–(2.1.8) *if and only if the pair* (v, q) *defined by Eqs.* (2.2.9)–(2.2.10) *satisfies the condition*

$$(v,q) \in \mathscr{U}(0,T;F(0)) \times L^2(0,T;\widehat{H}^1(F(0))),$$

together with

$$\frac{\partial v}{\partial t} - \nu L v + M v + N v + G q = 0 \quad \text{in } F(0) \times (0,T), \tag{2.2.18}$$

$$\operatorname{div} v = 0 \quad \text{in } F(0) \times (0,T), \tag{2.2.19}$$

$$v(y,t) = \dot{h}(t) + \omega(t) \times y \quad \text{on } \partial S(0) \times (0,T), \tag{2.2.20}$$

$$M\ddot{h}(t) = -\int_{\partial S(0)} \sigma(v,q) n \, d\Gamma \quad (t \in (0,T)), \tag{2.2.21}$$

$$J(0)\dot{\omega}(t) = (J(0)\omega(t)) \times \omega(t) - \int_{\partial S(0)} y \times \sigma(v,q) n d\Gamma \quad (t \in (0,T)), \tag{2.2.22}$$

$$v(y,0) = u_0(y) \quad (y \in F(0)), \tag{2.2.23}$$

$$h(0) = h_0, \quad \dot{h}(0) = h_1, \quad \omega(0) = \omega_0. \tag{2.2.24}$$

where

$$\sigma(v, q) = -q\mathbb{I}_3 + 2\nu D(v).$$

Proof. The equivalence between Eqs. (2.1.1) and (2.2.18) has been established in Theorem 2.5 from [25]. The equivalence between Eqs. (2.1.2) and (2.2.19) follows from Lemma 2.2.3. The facts that Eq. (2.1.7) is equivalent to Eq. (2.2.23) follows directly from the change of variables.

Take next $y \in \partial S(0)$. We have already seen that corresponding solution of Eq. (2.2.7) is $X(y, t) = y + h(t)$ so that $J_X(y, t) = \mathbb{I}_3$ for every $t \geq 0$ and $y \in \partial S(0)$. This implies that

$$u(y, t) = v(X(y, t), t) \quad (t \in [0, T], \ y \in \partial S(0)),$$

so that Eq. (2.1.4) is equivalent to Eq. (2.2.20).

This concludes the proof of the proposition. □

2.3. *Estimates on the coefficients*

In this subsection T and ε are supposed to be positive and we assume that $h : [0, T] \to \Omega_{1+\varepsilon}$ is a smooth function. It is clear that the operators L, M, N and G from Eqs. (2.2.11)–(2.2.14) are completely determined by the function h. The aim of this section is to make this dependence more precise, providing estimates which play an essential role in the fixed point procedure from Subsec. 2.5.

We first notice that, from the definition (2.2.5) of Λ, it clearly follows that there exists a constant $K = K(\varepsilon, \Omega) > 0$ such that

$$\|D_x^\alpha \Lambda(\cdot, t)\|_{[L^\infty(\Omega)]^3} \leq K|\dot{h}(t)|, \quad (t \in [0, T], \ \alpha \in \mathbb{N}^3, \ |\alpha| \leq 3). \quad (2.3.1)$$

The result below yields estimates of the change of variables mappings X and Y.

Lemma 2.3.1. *There exists a positive constants $K = K(\varepsilon, \Omega)$ such that the function X defined by Eq. (2.2.7) satisfies:*

$$\|X_i\|_{L^\infty(\Omega \times (0,T))} \leq K\left(1 + \|\dot{h}\|_{L^1((0,T;\mathbb{R}^3))}\right), \quad (2.3.2)$$

$$\left\| \frac{\partial X_i}{\partial y_j} \right\|_{L^\infty(\Omega \times (0,T))} \leq \exp\left(K\|\dot{h}\|_{L^1((0,T;\mathbb{R}^3))}\right), \quad (2.3.3)$$

$$\left\| \frac{\partial^2 X_i}{\partial y_j \partial y_k} \right\|_{L^\infty(\Omega\times(0,T))} \leq KT\|\dot{h}\|_{L^1(0,T)} \exp((2+T)K\|\dot{h}\|_{L^1(0,T)}), \quad (2.3.4)$$

$$\left\| \frac{\partial^2 X_i}{\partial y_j \partial y_k \partial y_l} \right\|_{L^\infty(\Omega\times(0,T))} \leq KT(1+KT+2K\|\dot{h}\|^2_{L^1(0,T)}) \exp((4+T)K\|\dot{h}\|_{L^1(0,T)}), \quad (2.3.5)$$

for all $i,\ j,\ k,\ l \in \{1, 2, 3\}$. *Moreover, the above estimates are still valid if we replace* $\mathrm{D}^\alpha_y X$ *by* $\mathrm{D}^\alpha_x Y$, *with* $|\alpha| \leq 3$.

Proof. From Eq. (2.2.7) it follows that

$$X(y,t) = y + \int_0^t \Lambda(X(y,s),s)\mathrm{d}s \quad (t \geq 0,\ y \in \Omega).$$

By using the fact that $X(y,s) \in \Omega$ combined with Eq. (2.3.1) it follows that

$$|X(y,t)| \leq |y| + K\int_0^T |\dot{h}(s)|\mathrm{d}s \quad (t \geq 0,\ y \in \Omega),$$

which clearly implies Eq. (2.3.2).

To show that Eq. (2.3.3) we take $j \in \{1, 2, 3\}$ and we define $z_j(y,t) = \partial X/\partial y_j(y,t)$. By differentiating both sides of the Eq. (2.2.7) with respect to y_j, we have that z_j satisfies

$$\begin{aligned} \frac{\partial z_j}{\partial t} &= J(y,t)z_j(y,t), \quad t > 0, \\ z_j(y,0) &= e_j, \end{aligned} \quad (2.3.6)$$

where the matrix $J(y,t)$ is given by

$$J_{pq}(y,t) = \frac{\partial \Lambda_p}{\partial x_q}(X(y,t)) \quad (p,\ q \in \{1,2,3\}), \quad (2.3.7)$$

and $\{e_1, e_2, e_3\}$ is the canonical basis of $\mathbb{R}^3$. Consequently

$$z_j(y,t) = \exp\left(\int_0^t J(y,s)\mathrm{d}s\right) e_j,$$

which, combined with Eq. (2.3.1)), yields Eq. (2.3.3).

We next set

$$w_{jk} = \frac{\partial^2 X}{\partial y_j \partial y_k} = \frac{\partial z_j}{\partial y_k}.$$

Derivating Eq. (2.3.6) with respect to y_k it follows that

$$\begin{aligned} \frac{\partial w_{jk}}{\partial t}(y,t) &= \frac{\partial J}{\partial y_k}(y,t) z_j(y,t) + J(y,t) w_{jk}(y,t), \quad t > 0, \\ w_{jk}(y,0) &= 0, \end{aligned} \tag{2.3.8}$$

which implies that

$$w_{jk}(y,t) = \int_0^t \frac{\partial J}{\partial y_k}(y,s) z_j(y,s) \mathrm{d}s + \int_0^t J(y,s) w_{jk}(y,s) \mathrm{d}s. \tag{2.3.9}$$

On the other hand from Eq. (2.3.7) it follows that

$$\frac{\partial J_{pq}}{\partial x_k}(y,t) = \sum_{r=1}^{3} \frac{\partial^2 \Lambda_p}{\partial x_r \partial x_q}(X(y,t)) \frac{\partial X}{\partial y_r}(y,t). \tag{2.3.10}$$

The above formula, combined with Eqs. (2.3.1) and (2.3.3) yields that

$$\left| \frac{\partial J}{\partial y_k}(y,t) z_j(y,t) \right| \leq K \|\dot{h}\|_{L^1(0,T)} \exp(2K \|\dot{h}\|_{L^1(0,T)}) \qquad (t \in (0,T),\ y \in \Omega). \tag{2.3.11}$$

Note also that Eqs. (2.3.3) and (2.3.7) imply that

$$|J(y,t) w_{jk}(y,t)| \leq K \|\dot{h}\|_{L^1(0,T)} |w_{jk}(y,t)| \quad (t \in (0,T),\ y \in \Omega).$$

The above inequality, combined with Eqs. (2.3.9) and (2.3.11), implies that

$$\begin{aligned} |w_{jk}(y,t)| \leq{}& KT \|\dot{h}\|_{L^1(0,T)} \exp(2K \|\dot{h}\|_{L^1(0,T)}) \\ &+ K \|\dot{h}\|_{L^1(0,T)} \int_0^t |w_{jk}(y,s)| \mathrm{d}s \quad (t \in (0,T),\ y \in \Omega). \end{aligned}$$

By applying Gronwall's inequality it follows that

$$|w_{jk}(y,t)| \leq KT \|\dot{h}\|_{L^1(0,T)} \exp((2+T)K \|\dot{h}\|_{L^1(0,T)}) \quad (t \in (0,T),\ y \in \Omega),$$

which clearly implies Eq. (2.3.4).

We next set

$$\eta_{jkl} = \frac{\partial^3 X}{\partial y_j \partial y_k \partial y_l} = \frac{\partial w_{jk}}{\partial y_l}.$$

Derivating Eq. (2.3.8) with respect to y_l we obtain

$$\frac{\partial \eta_{jkl}}{\partial t}(y,t) = \frac{\partial^2 J}{\partial y_k \partial y_l}(y,t) z_j(y,t) + \frac{\partial J}{\partial y_k}(y,t) w_{jl}(y,t) + \frac{\partial J}{\partial y_l}(y,t) w_{jk}(y,t) \\ + J(y,t)\eta_{jkl}(y,t) \quad (t \in (0,T), \quad y \in \Omega), \tag{2.3.12}$$

$$\eta_{jkl}(y,0) = 0. \tag{2.3.13}$$

On the other hand, derivating Eq. (2.3.10) with respect to y_l we have

$$\frac{\partial^2 J_{pq}}{\partial x_k \partial y_l}(y,t) = \sum_{r,s=1}^{3} \frac{\partial^2 \Lambda_p}{\partial x_s \partial x_r \partial x_q}(X(y,t)) \frac{\partial X}{\partial y_s}(y,t) \frac{\partial X}{\partial y_r}(y,t) \\ + \sum_{r=1}^{3} \frac{\partial^2 \Lambda_p}{\partial x_r \partial x_q}(X(y,t)) \frac{\partial^2 X}{\partial y_l \partial y_r}(y,t) \quad (t \in [0,T], \ y \in \Omega).$$

The above relation, combined with Eqs. (2.3.1), (2.3.3) and (2.3.4), implies that

$$\left\| \frac{\partial^2 J}{\partial y_k \partial y_l}(y,t) z_j \right\|_{L^\infty(\Omega \times (0,T))} \leq K(1+KT) \exp\left((3+T)K\|\dot{h}\|_{L^1(0,T)}\right). \tag{2.3.14}$$

Integrating Eq. (2.3.12) with respect to time and using Eqs. (2.3.1), (2.3.4), (2.3.7), (2.3.10) and (2.3.14) it follows that

$$|\eta_{jkl}(y,t)| \leq KT(1+KT) \exp\left((3+T)K\|\dot{h}\|_{L^1(0,T)}\right) \\ + 2TK^2 \|\dot{h}\|^2_{L^1(0,T)} \exp\left((3+T)K\|\dot{h}\|_{L^1(0,T)}\right) \\ + K\|\dot{h}\|_{L^1(0,T)} |\eta_{jkl}(y,t)| \quad (t \in (0,T), \ y \in \Omega).$$

By applying Gronwall's inequality we obtain Eq. (2.3.5).

The similar estimates for $D_x^\alpha Y$ can be checked in the same way, by using the fact that Y satisfies Eq. (2.2.8). □

Corollary 2.3.2. *There exists a constant $K = K(\varepsilon, \Omega) > 0$ such that for all $m, l \in \{1, 2, 3\}$ we have*

$$\left\| \frac{\partial X_m}{\partial y_l} - \delta_{ml} \right\|_{L^\infty(\Omega\times(0,T))} \leq KT\|\dot{h}\|_{L^1(0,T)} \exp(K\|\dot{h}\|_{L^1(0,T)}), \quad (2.3.15)$$

$$\|g_{ml} - \delta_{ml}\|_{L^\infty(\Omega\times(0,T))} \leq KT\|\dot{h}\|_{L^1(0,T)} \exp(K\|\dot{h}\|_{L^1(0,T)}), \quad (2.3.16)$$

where δ_{ml} denotes the Kronecker's symbol. Moreover, the same estimates above are still valid if we replace $\partial X_m/\partial y_l$ by $\partial Y_m/\partial x_l$ and g_{ml} by g^{ml}, respectively.

Proof. From the definition of the function $X(y, t)$, we have that for each $m, l \in \{1, \ldots, n\}$

$$\frac{\partial X_m}{\partial y_l}(y, 0) = \delta_{ml}.$$

Therefore, from the mean value theorem and the fact that X is of class C^2, we get that for any $t \in (0, T)$, there exists $\xi \in (0, t)$ such that

$$\frac{\partial X_m}{\partial y_l}(y, t) - \delta_{ml} = \frac{\partial^2 X_m}{\partial t \partial y_l}(y, \xi)(t - 0) = \frac{\partial \Lambda_m}{\partial y_l}(X(y, \xi))t$$

$$= t \sum_{k=1}^{n} \frac{\partial \Lambda_m}{\partial x_k}(X(y, \xi)) \frac{\partial X_k}{\partial y_l}(y, \xi).$$

The above formula, combined to Eqs. (2.3.1) and (2.3.1) yields estimates (2.3.15).

In order to prove Eq. (2.3.15) it suffices to remark that, from the definition of g_{ml}, we have that

$$|g_{ml} - \delta_{ml}| = \left| \sum_{k=1}^{n} \left(\frac{\partial X_m}{\partial y_k} - \delta_{mk} \right) \frac{\partial X_l}{\partial y_k} + \sum_{k=1}^{n} \delta_{mk} \left(\frac{\partial X_l}{\partial y_k} - \delta_{kl} \right) \right|,$$

and to use Eqs. (2.3.3) and (2.3.15). The similar estimates for the first derivatives of Y and g^{ml} can be proved in the same manner. □

Corollary 2.3.3. *Let (g^{ij}) be the metric covariant tensor and let (Γ^k_{ij}) be the Christoffel's symbols defined in Eqs. (2.2.15) and (2.2.17), respectively.*

Then there exists a constant $K = K(\varepsilon, \Omega)$ *such that*

$$\|g^{jk}\|_{L^\infty(\Omega\times(0,T))} \leq 3\exp(2KT), \tag{2.3.17}$$

$$\|\Gamma_{ij}^k\|_{L^\infty(\Omega\times(0,T))} \leq KT\exp(KT), \tag{2.3.18}$$

for all $i,\ j,\ k \in \{1, 2, 3\}$.

Proof. Estimate (2.3.17) follows easily from Lemma 2.3.1 so we skip the proof. Moreover, from Eq. (2.2.15) it follows that

$$\frac{\partial g^{jk}}{\partial y_l} = \sum_{m=1}^{3}\sum_{p=1}^{3} \frac{\partial X_p}{\partial y_l}\left(\frac{\partial^2 Y_j}{\partial x_p \partial x_m}\frac{\partial Y_k}{\partial x_m} + \frac{\partial^2 Y_k}{\partial x_p \partial x_m}\frac{\partial Y_j}{\partial x_m}\right).$$

The above formula, combined with Eq. (2.2.17) and to Lemma 2.3.1 yields Eq. (2.3.18). □

Let $v \in \mathscr{U}(0, T; F(t))$ and $q \in L^2(0, T); \widehat{H}^1(\Omega))$ be given functions, where the space $\mathscr{U} = \mathscr{U}(0, T; F(t))$ has been defined in Eq. (2.1.9) and where the variable domain $F(t)$ is associated to the given trajectory $h(t)$ of the mass center of the solid. Denote

$$R_1 = \|v\|_{\mathscr{U}} + \|\dot{h}\|_{L^2(0,T)}, \tag{2.3.19}$$

$$R_2 = \|\nabla q\|_{L^2((0,T;[L^2(F(0))]^3)} + \|\dot{h}\|_{L^2(0,T)}. \tag{2.3.20}$$

By combining Lemma 2.3.1, Corollaries 2.3.2 and 2.3.3 it is not difficult to prove the result below.

Lemma 2.3.4. *With the above notation, there exist positive constants* $K,\ K_1$, *depending only on* Ω *and* ε, *such that*

1. $\|\nu(L - \Delta)v\|_{L^2((0,T;[L^2(F(0))]^3)} \leq KTR_1 \exp(K_1T)$,
2. $\|Mv\|_{L^2(0,T;[L^2(F(0))]^3)} \leq KT^{1/2}R_1 \exp(K_1T)$,
3. $\|(\nabla - G)q\|_{L^2(0,T;[L^2(F(0))]^3)} \leq KTR_2 \exp(K_1T)$.

Roughly speaking, the above result says that, for T small enough, the operators L, G defined in Eqs. (2.2.11) and (2.2.14) are close to the operators Δ and ∇, respectively, and that the operator M defined in Eq. (2.2.13) is small in an appropriate sense. In order to estimate the nonlinear operator N defined in Eq. (2.2.12) we need the following simple result which is proved, for instance, [37].

Lemma 2.3.5. *With the notation in Lemma* 2.3.4, *for any* $v, w \in \mathscr{U}(0, T; F(t))$, *we have that* $(w \cdot \nabla)v \in L^{5/2}(0, T; L^2(F(t))$, *and for all* $i, j \in \{1, 2, 3\}$, *we have that* $w_i v_j \in L^\infty(0, T; L^2(F(t)))$. *Moreover, there exists a constant* $C > 0$, *depending only on* Ω, *such that*:

$$\|(w \cdot \nabla)v\|_{L^{5/2}(0,T;L^2)} \leq C\|w\|_{L^\infty(0,T;H^1)}\|v\|^{1/5}_{L^\infty(0,T;H^1)}\|v\|^{4/5}_{L^2(0,T;H^2)}, \tag{2.3.21}$$

$$\|w_i v_j\|_{L^\infty(0,T;L^2)} \leq C\|w\|_{L^\infty(0,T;H^1)}\|v\|_{L^\infty(0,T;H^1)}. \tag{2.3.22}$$

We are now in a position to estimate the operator N.

Corollary 2.3.6. *There exist constants* K_1, *depending only on* Ω *and* ε *and* $C = C(\Omega)$ *such that for every* $w \in \mathscr{U}(0, T; F(t))$ *we have*

$$\|Nw\|_{L^2(0,T;L^2)} \leq CT^{1/10} \exp(K_1 T)\|w\|^2_{\mathscr{U}}.$$

Proof. From Eq. (2.3.21) it follows that

$$\|(w \cdot \nabla)w\|_{L^{5/2}(0,T;L^2)} \leq C\|w\|^{6/5}_{L^\infty(0,T;H^1)}\|w\|^{4/5}_{L^2(0,T;H^2)} \leq C\|w\|^2_{\mathscr{U}}.$$

Thus, by the Hölder inequalities, we deduce that

$$\|(w \cdot \nabla)w\|_{L^2(0,T;L^2)} \leq T^{1/10}\|(w \cdot \nabla)w\|_{L^{5/2}(0,T;L^2)} \leq CT^{1/10}\|w\|^2_{\mathscr{U}}.$$

To conclude, we use the above estimate together with (2.3.18) and (2.3.22). □

2.4. *Analysis of the linearized problem*

We have seen in the previous section that the operators L, G defined in Eqs. (2.2.11) and (2.2.14) are close to the operators Δ and ∇, respectively. Therefore it seems natural to tackle the initial and boundary value problem (2.2.18)–(2.2.24) as a perturbation of the system obtained by replacing Eq. (2.2.18) by the evolution Stokes equation. This is why this section is devoted to the study of the initial and boundary value problem

$$\frac{\partial v}{\partial t}(x, t) - \nu\Delta v(x, t) + \nabla q(x, t) = f(x, t), \quad \text{in } F(0) \times [0, T], \tag{2.4.1}$$

$$\operatorname{div} u = 0, \quad \text{in } F(0) \times [0, T], \tag{2.4.2}$$

$$v(y,t) = \dot{h}(t) + \omega(t) \times y, \qquad y \in \partial S(0),\ t \in [0,T], \tag{2.4.3}$$

$$M\ddot{h}(t) = -\int_{\partial S(0)} \sigma(v,q) n \mathrm{d}\Gamma, \qquad t \in [0,T], \tag{2.4.4}$$

$$J(0)\dot{\omega}(t) = -\int_{\partial S(0)} y \times \sigma(v,q) \mathrm{d}y + g(t), \qquad t \in [0,T], \tag{2.4.5}$$

$$v(x,0) = u_0(x), \qquad x \in S(0), \tag{2.4.6}$$

$$h(0) = h_0,\ \dot{h}(0) = h_1,\ \omega(0) = \omega_0, \tag{2.4.7}$$

where

$$\sigma(v,q) = -q\mathbb{I}_3 + 2\nu D(v).$$

A simple and very useful idea is to extend v to a function defined on the whole of Ω by putting

$$v(y,t) = \dot{h}(t) + \omega(t) \times y \quad (y \in S(0)).$$

This naturally leads to the introduction of the function spaces :

$$\mathscr{H} = \{\phi \in [L^2(\Omega)]^3 \mid \operatorname{div}\phi = 0 \text{ in } \Omega,\ D(\phi) = 0 \text{ in } S(0)\}, \tag{2.4.8}$$

$$\mathscr{V} = \{\phi \in [H_0^1(\Omega)]^3 \mid \operatorname{div}\phi = 0 \text{ in } \Omega,\ D(\phi) = 0 \text{ in } S(0)\}. \tag{2.4.9}$$

According to a classical result (see, for instance, [39, Lemma 1.1]), for every $\phi \in \mathscr{H}$, there exists $v_\phi, \omega_\phi \in \mathbb{R}^3$ such that

$$\phi(y) = v_\phi + \omega_\phi \times y, \quad (y \in S(0)).$$

Define an inner product on $[L^2(\Omega)]^3$ by

$$\langle \psi, \phi \rangle = \int_{F(0)} \psi \cdot \phi \mathrm{d}y + \rho_S \int_{S(0)} \psi \cdot \phi \mathrm{d}y, \tag{2.4.10}$$

where $\rho_S > 0$ is the density of the rigid body. The corresponding norm is clearly equivalent to the usual norm in $[L^2(\Omega)]^3$. For $\psi,\ \phi \in \mathscr{H}$ we can use the definition (1.1.7) of the inertia tensor to obtain that

$$\langle \psi, \phi \rangle = \int_{F(0)} \psi \cdot \phi \mathrm{d}y + M v_\phi \cdot v_\psi + J(0)\omega_\phi \cdot \omega_\psi. \tag{2.4.11}$$

In order to solve Eqs. (2.4.1)–(2.4.7) we use a semi-group approach. We first define the operator $\mathscr{A} : \mathscr{D}(\mathscr{A}) \to [L^2(\Omega)]^3$ by

$$\mathscr{D}(\mathscr{A}) = \{\phi \in \mathscr{V} \| \ \phi_{|F(0)} \in [H^2(F(0))]^3\},$$

$$\mathscr{A}\phi = \begin{cases} -\nu\Delta\phi & \text{in } \Omega, \\ \frac{2\nu}{M}\int_{\partial B} D(\phi)n d\Gamma + \left[2\nu[J(0)]^{-1}\int_{\partial S(0)} y \times D(\phi)n d\Gamma\right] \times y & \text{in } S(0), \end{cases} \tag{2.4.12}$$

for every $\phi \in \mathscr{D}(\mathscr{A})$. We show below that the operator $A : \mathscr{D}(A) \to \mathscr{H}$ determining the dynamics of Eqs. (2.4.1)–(2.4.7) is defined by

$$\mathscr{D}(A) = \mathscr{D}(\mathscr{A}), \tag{2.4.13}$$

$$A\phi = \mathbb{P}\mathscr{A}\phi \quad (\phi \in \mathscr{D}(A)), \tag{2.4.14}$$

where $\mathbb{P}$ is the orthogonal projector from $[L^2(\Omega)]^3$ onto $\mathscr{H}$ ($\mathscr{H}$ is clearly a closed subspace of $[L^2(\Omega)]^3$) and where, in the expression of $\mathscr{A}\phi$, $D(\phi)$ on $\partial S(0)$ is the trace of the restriction of $D(\phi)$ to $F(0)$. The result below, which can be seen as a very simple version of Korn's inequality, will be used several times in the sequel.

Lemma 2.4.1. *For every* $\phi \in [H_0^1(\Omega)]^3$ *with* $\operatorname{div}\phi = 0$ *we have*

$$\|\nabla\phi\|^2_{[L^2(\Omega)]^9} = 2\|D(\phi)\|^2_{[L^2(\Omega)]^9}.$$

Proof. Let $\phi \in [\mathscr{D}(\Omega)]^3$ with $\operatorname{div}\phi = 0$. After some simple calculations, we get

$$\nabla\phi : \nabla\phi - 2D(\phi) : D(\phi) = \operatorname{div}((\nabla\phi)^*\phi - (\operatorname{div}\phi)\phi) + (\operatorname{div}\phi)^2.$$

Integration the above formula on Ω and using the fact that ϕ vanishes on $\partial\Omega$ we obtain

$$\|\nabla\phi\|^2_{[L^2(\Omega)]^9} = 2\|D(\phi)\|^2_{[L^2(\Omega)]^9} + \|\operatorname{div}\phi\|^2_{L^2(\Omega)} = 2\|D(\phi)\|^2_{[L^2(\Omega)]^9}.$$

By density, the above formula holds for every $\phi \in [H_0^1(\Omega)]^3$ with $\operatorname{div}\phi = 0$. □

Proposition 2.4.2. *The operator A defined by Eqs.* (2.4.13) *and* (2.4.14) *is self-adjoint and positive. Moreover, we have that*

$$\mathscr{D}(A^{\frac{1}{2}}) = \mathscr{V}, \tag{2.4.15}$$

$$\|\phi\|_{[H^2(F(0))]^3} \leq C\|A\phi\|_{[L^2(\Omega)]^3} \quad (\phi \in \mathscr{D}(A)). \tag{2.4.16}$$

Proof. Let $\phi, \psi \in \mathscr{D}(A)$. According to Eqs. (2.4.11) and (2.4.12) and to the definition (2.4.10) of the inner product in $[L^2(\Omega)]^3$, we have

$$\begin{aligned}\langle A\phi, \psi\rangle &= \langle \mathscr{A}\phi, \psi\rangle \\ &= -\nu \int_{F(0)} \Delta\phi \cdot \psi \mathrm{d}y + \left[\frac{2\nu}{M}\int_{\partial S(0)} D(\phi)n\, \mathrm{d}\Gamma\right] \cdot \left(\rho_s \int_{S(0)} \psi \mathrm{d}y\right) \\ &\quad + 2\nu\rho_S \int_{S(0)} \left\{[J(0)]^{-1}\left[\int_{\partial S(0)} y \times D(\phi)n\, \mathrm{d}\Gamma\right] \times y\right\} \cdot \psi \mathrm{d}y.\end{aligned} \tag{2.4.17}$$

Since $\psi \in \mathscr{D}(A)$, there exist $v_\psi, \omega_\psi \in \mathbb{R}^3$ such that

$$\psi(y) = v_\psi + \omega_\psi \times y \quad (y \in S(0)), \tag{2.4.18}$$

so that

$$\rho_s \int_{S(0)} \psi \mathrm{d}y = M v_\psi, \tag{2.4.19}$$

$$\begin{aligned}&2\nu\rho_S \int_{S(0)} \left\{[J(0)]^{-1}\left[\int_{\partial S(0)} y \times D(\phi)n\, \mathrm{d}\Gamma\right] \times y\right\} \cdot \psi \mathrm{d}y \\ &\quad = 2\nu\rho_S \int_{S(0)} \left\{[J(0)]^{-1}\left[\int_{\partial S(0)} y \times D(\phi)n\, \mathrm{d}\Gamma\right] \times y\right\} \cdot (\omega \times y)\mathrm{d}y.\end{aligned}$$

The above formula and Eq. (1.1.7) yield

$$\begin{aligned}&2\nu\rho_S \int_{S(0)} \left\{[J(0)]^{-1}\left[\int_{\partial S(0)} y \times D(\phi)n\, \mathrm{d}\Gamma\right] \times y\right\} \cdot \psi \mathrm{d}y \\ &\quad = 2\nu \left[\int_{\partial S(0)} y \times D(\phi)n\, \mathrm{d}\Gamma y\right] \cdot \omega.\end{aligned} \tag{2.4.20}$$

By combining Eqs. (2.4.17), (2.4.19) and (2.4.20) it follows that

$$\langle A\phi, \psi\rangle = -\nu \int_{F(0)} \Delta\phi \cdot \psi \mathrm{d}y + 2\nu v_\psi \cdot \int_{\partial S(0)} D(\phi) n \,\mathrm{d}\Gamma + 2\nu\omega_\psi \cdot \left[\int_{\partial S(0)} y \times D(\phi) n \,\mathrm{d}\Gamma\right]. \tag{2.4.21}$$

On the other hand,

$$\Delta\phi \cdot \psi = 2\mathrm{div}\,(D(\phi)) \cdot \psi = 2\mathrm{div}\,(D(\phi)\psi) - 2D(\phi) : D(\psi)$$

so that Eq. (2.4.21) yields that for every $\phi, \psi \in \mathscr{D}(A)$, we have

$$\langle A\phi, \psi\rangle = 2\nu \int_\Omega D(\phi) : D(\psi)\mathrm{d}y - 2\nu \int_{\partial S(0)} D(\phi)\psi \cdot n\mathrm{d}\Gamma + 2\nu v_\psi \cdot \int_{\partial S(0)} D(\psi) n \,\mathrm{d}\Gamma + 2\nu\omega_\psi \cdot \left[\int_{\partial S(0)} y \times D(\phi) n \,\mathrm{d}\Gamma\right].$$

The above formula and Eq. (2.4.18) imply that

$$\langle A\phi, \psi\rangle = 2\nu \int_\Omega D(\phi) : D(\psi)\mathrm{d}y = \langle \phi, A\psi\rangle \quad (\phi,\ \psi \in \mathscr{D}(A)), \tag{2.4.22}$$

so that A is symmetric. Moreover, by taking $\phi = \psi$ in Eq. (2.4.22), it follows that we have

$$\langle A\phi, \phi\rangle = 2\nu \int_\Omega |D(\phi)|^2 \mathrm{d}y = \nu \int_\Omega |\nabla\phi|^2 \mathrm{d}y \quad (\phi \in \mathscr{D}(A)), \tag{2.4.23}$$

so that A is strictly positive. In order to prove A is self-adjoint, it suffices to check that $A : \mathscr{D}(A) \to \mathscr{H}$ is onto. We have this to show that for every $f \in \mathscr{H}$ there exists $\phi \in \mathscr{D}(A)$ such that

$$A\phi = f, \tag{2.4.24}$$

which is equivalent to

$$\langle A\phi, \eta\rangle = \langle f, \eta\rangle \quad (\eta \in \mathscr{V}).$$

By using Eq. (2.4.22) we see the above equality is equivalent to

$$2\nu \int_{\Omega} D(\phi) : D(\eta) \mathrm{d}y = \langle f, \eta \rangle \quad (\eta \in \mathscr{V}). \tag{2.4.25}$$

Consider the bilinear form $a : \mathscr{V} \times \mathscr{V} \to \mathbb{R}$ defined by

$$a(\phi, \eta) = 2\nu \int_{\Omega} D(\phi) : D(\eta) \mathrm{d}y \quad (\phi,\ \eta \in \mathscr{V}).$$

We clearly have that a is bounded. Moreover, from Eq. (2.4.22) and Poincaré's inequality it follows that a is elliptic in the sense that there exists $C > 0$ such that

$$a(\phi, \phi) \geq C \|\phi\|^2_{[H^1(\Omega)]^3}, \quad (\phi \in \mathscr{V}).$$

On the other hand, the mapping

$$\eta \mapsto \langle f, \eta \rangle$$

is a linear and continuous form on $\mathscr{V}$, so that, by the Lax–Milgram theorem, there exists a unique $\phi \in \mathscr{V}$ satisfying Eq. (2.4.25). In particular, Eq. (2.4.25) holds for $\eta \in \mathscr{D}(F(0))$ with $\operatorname{div} \eta = 0$, so that, according to Propositions 1.1 and 1.2 from Temam [40], there exists $p \in \mathscr{D}'(F(0))$ such that

$$-\nu \Delta \phi + \nabla p = f, \quad \text{in } \mathscr{D}'(F(0)).$$

Since $\phi \in \mathscr{V}$, we also have that

$$\begin{aligned} \operatorname{div} \phi &= 0 && \text{in } F(0), \\ \phi(y) &= v_\phi + \omega_\phi \times y && \text{on } \partial S(0), \\ \phi(y) &= 0 && \text{on } \partial\Omega, \end{aligned}$$

with $v_\phi,\ \omega_\phi \in \mathbb{R}^3$. A standard regularity result for the Stokes system (see Cattabriga [3]) implies that $\phi \in [H^2(F(0))]^3$. We already knew that $\phi \in \mathscr{V}$ so that we have $\phi \in \mathscr{D}(A)$ and $A\phi = f$. Consequently A is onto so that it is self adjoint. The fact that it is strictly positive and it satisfies Eqs. (2.4.15) follows from Eq. (2.4.23). Finally, Eq. (2.4.16) follows easily from the already mentioned regularity result for the Stokes system. □

A direct consequence of Proposition 2.4.2 is:

Corollary 2.4.3. *Let* $f \in L^2(0, T; [L^2(F(0))]^3)$, $g \in [L^2(0, T)]^3$ *and* $u_0 \in [H^1(\Omega)]^2$ *such that*

$$\begin{aligned} \operatorname{div} u_0 &= 0, && \text{in } F(0), \\ u_0(y) &= h_1 + \omega_0 \times y, && y \in \partial S(0), \\ u_0(y) &= 0, && y \in \partial\Omega \end{aligned}$$

Then the system (2.4.1)–(2.4.7) *admits a unique solution* (v, q, h, ω) *with*

$$\begin{aligned} &v \in L^2(0, T; [H^2(F(0))]^3) \cap C([0, T]; \\ &[H^1(F(0))]^3) \cap H^1(0, T; [L^2(F(0))]^3), \end{aligned}$$

$$q \in L^2(0, T; \widehat{H}^1(F(0))), \quad h \in H^2(0, T; \mathbb{R}^3), \ \omega \in H^1(0, T; \mathbb{R}^3).$$

Moreover, there exists a positive constant K such that

$$\begin{aligned} &\|v\|_{L^2(0,T;[H^2(F(0))]^3)} + \|v\|_{L^\infty(0,T;[H^1(F(0))]^3)} + \|v\|_{H^1(0,T;[L^2(F(0))]^3)} \\ &\quad + \|\nabla q\|_{[L^2(0,T;L^2(F(0)))]^3} + \|\dot{h}\|_{H^1(0,T;\mathbb{R}^3)} + \|\omega\|_{H^1(0,T;\mathbb{R}^3)} \\ &\quad \le K\big(\|u_0\|_{[H^1(F(0))]^3} + |h_1| + |\omega_0| + \|f\|_{L^2(0,T;[L^2(F(0))]^3)} \\ &\quad + \|g\|_{[L^2(0,T)]^3}\big). \end{aligned} \tag{2.4.26}$$

The constant K depends only on Ω *and on T and it is non-decreasing with respect to T.*

2.5. *Proof of the main result*

Denote

$$E = L^2(0, T; [L^2(F(0))]^3) \times L^2(0, T; \mathbb{R}^3),$$

and consider the mapping

$$\mathscr{L} : E \to E, \tag{2.5.1}$$

$$(f, g) \mapsto (\nu(L - \Delta)v - Mv + (\nabla - G)q + Nv, \ J(0)\omega \times \omega), \tag{2.5.2}$$

where (v, q, h, ω) is the solution of Eqs. (2.4.1)–(2.4.7). It is easily seen that a quadruple (v, q, h, ω), with

$$h \in [H^2(0,T)]^3, \quad \omega \in [H^1(0,T)]^3,$$
$$v \in \mathscr{U}(0;T;F(0))$$

is a solution of Eqs. (2.1.1)–(2.1.8) if and only if (v, q, h, ω) is a fixed point of $\mathscr{L}$. This equivalence and Proposition 2.2.4 show that in order to prove Theorem 2.1.1 it suffices to show the existence and uniqueness of a fixed point of the mapping $\mathscr{L}$ defined by Eqs. (2.5.1) and (2.5.2). We first show that $\mathscr{L}$ invariates an appropriate ball of the Hilbert space E.

Proposition 2.5.1. *For $R > 0$ we denote by $B(0, R)$ the closed ball centered at the origin and of radius R of the Hilbert space E. Then, for every $R > 0$ there exists $T > 0$ such that $\mathscr{L}(B(0,R)) \subset B(0,R)$.*

Proof. According to Corollary 2.4.3 we have

$$\|v\|_{\mathscr{U}} + \|\nabla q\|_{[L^2(0,T;L^2(F(0)))]^3} + \|\dot{h}\|_{H^1(0,T;\mathbb{R}^3)} + \|\omega\|_{H^1(0,T;\mathbb{R}^3)} + \|\omega\|_{L^\infty(0,T;\mathbb{R}^3)} \leq C\left(\|u_0\|_{[H^1(F(0))]^3} + |h_1| + |\omega_0| + R\right), \quad (2.5.3)$$

where C depends only on Ω and on T and it is non-decreasing with respect to T. By applying Lemma 2.3.4 it follows that

$$\|\nu(L-\Delta)v - Mv + (\nabla - G)q\|_{L^2(0,T;[L^2(\Omega)]^3)} \leq KC(T+\sqrt{T})\exp(K_1 T)(\|u_0\|_{[H^1(F(0))]^3} + |h_1| + |\omega_0| + R), \quad (2.5.4)$$

with K and K_1 depending only on Ω and on ε. Moreover, by combining Eq. (2.5.3) and Corollary 2.3.6, there exist constants K_1, depending only on Ω and ε and $C = C(\Omega)$, such that

$$\|Nv\|_{L^2(0,T;L^2)} \leq CT^{1/10}\exp(K_1 T)\big(\|u_0\|_{[H^1(F(0))]^3} + |h_1| + |\omega_0| + R\big). \quad (2.5.5)$$

On the other hand, estimate (2.5.3) easily implies that

$$\|J(0)\omega \times \omega\|_{L^2(0,T;\mathbb{R}^3)} \leq CT^{\frac{1}{2}}\left(\|u_0\|_{[H^1(F(0))]^3} + |h_1| + |\omega_0| + R\right).$$

The above estimate, (2.5.3) and (2.5.4) imply that for every $T \in (0, 1)$ we have

$$\|\mathscr{L}(f, g)\|_E \leq KT^{1/10} \exp(K_1) \left(\|u_0\|_{[H^1(F(0))]^3} + |h_1| + |\omega_0| + R\right),$$

which clearly implies the conclusion. □

We are now in a position to prove the main result of this section.

Proof of Theorem 2.1.1. The first step consists in proving that, for every $R > 0$, we have that $\mathscr{L}|_{B(0,R)}$ is a contraction for T small enough. This can be done by calculations very similar with those in the proof of Proposition 2.5.1 so that we omit them. In this way we obtain the existence and uniqueness of a local in time solution defined on $[0, T_0)$.

The second step consists in remarking that the conditions on T_0 obtained at the previous step depend only on $\|u_0\|_{H^1}$ and on ε (recall that ε is, roughly speaking, the distance from $\partial S(0)$ to $\partial\Omega$). Therefore if

$$\sup_{t\in[0,T_0)} \|v(t)\|_{H^1} < \infty \quad \text{and} \quad \inf_{t\in[0,T_0)} d(h(t), \partial\Omega) > 1,$$

then the local in time solution can be extended to a larger interval. This fact implies that one of the assertions *(i)*, *(ii)* or *(iii)* holds true. □

2.6. *Remarks and bibliographical notes on Sec. 2*

The material in this section is essentially based on Takahashi [37]. Since we assumed that the rigid bodies are balls, the fixed point procedure is slightly simplified, following Takahashi and Tucsnak [38]. Note that [37] also provides global existence for small initial data. As far as we know, the first existence and uniqueness result of strong solutions for the problem considered in this section has been given in Grandmont and Maday [16]. This result, local in time, assumes that the inertia of the solid is large with respect to the inertia of the fluid and it uses the passage to Lagrangian coordinates.

A problem related to the questions studied in this section is the problem of the motion of a single solid in a fluid which fills the remaining part of $\mathbb{R}^3$. In this case there is a natural change of variables such that the Navier–Stokes equation in the unknown time-dependent domain is replaced by an equation which holds in the fixed and known domain $F(0)$. This change of variable is very simple, just a rigid displacement. However, this apparently very simple change of variables leads to a new difficulty. More precisely, the transformed equation contains the term

$$[(\omega \times y) \cdot \nabla]v \quad ((y \in \mathbb{R}^3 \setminus F(0)),$$

which does not define a bounded operator on standard spaces (due to its behavior when $|y| \to \infty$). Tackling this difficulty requires refined techniques and we refer to Galdi [12, 13], Galdi and Silvestre [14] for results in this direction. The same difficulty has to be solved in the study of the Navier–Stokes flow around an obstacle rotating with a *given* angular velocity. In this case we do not have a free boundary problem but we still have to tackle the linear unbounded term $[(\omega \times y) \cdot \nabla]v$. The study of this question leads to interesting results, often connected to semigroup theory (see, for instance, Hishida [21], Galdi and Silvestre [15] or Cumsille and Tucsnak [5]).

Another related problem concerns the two dimensional case, i.e., in which the solids are supposed to be cylinders of infinite lengths. In this case it has been shown in [37] and [38] that the strong solutions are global in time, up to possible contacts.

Finally, let us mention the interesting works of Hillairet [20] and Hesla [19]. In these works it is shown that, in two space dimensions and for particular forms of the solids, the contact between two solids never occurs. In other words, in the above context any strong solution is global in time.

3. Weak Solutions in the Viscous Case

3.1. *Notation and preliminaries*

In this section we consider the initial and boundary value problem modeling the motion of rigid bodies in a viscous incompressible fluid from a different perspective. More precisely, we are interested in defining a concept of weak solution and in proving the existence of such solution. This existence result will be global in time, provided that there is no contact between the solids or between a solid and the exterior boundary. We first recall the governing equation and then we introduce some notation.

As in the previous sections, we limit to the case of one rigid body and we denote by $\Omega \subset \mathbb{R}^3$ the open bounded set representing the domain occupied by the fluid and the solid. Let S_0 be a given open subset of Ω representing the initial position of the rigid body (the particle). We denote by $h(t)$ the position vector of the mass center of the rigid body at the moment t, by $R(t)$ the orthogonal matrix giving its angular orientation and by $\omega(t)$ its angular velocity. With this notation, the domain $S(t)$ occupied by the rigid body at instant t is given by

$$S(t) = \{R(t)(y - h(0)) + h(t), \quad y \in S_0\}. \tag{3.1.1}$$

We denote by $F(t) = \Omega \setminus \overline{S(t)}$ the domain occupied by the fluid at instant t. Recall that the full system of equations modeling the motion of the fluid and of the rigid body can be written as

$$\rho_F\left(\frac{\partial u}{\partial t} + (u \cdot \nabla)u\right) - \nu\Delta u + \nabla p = \rho_F g, \qquad x \in F(t), t \in [0, T], \tag{3.1.2}$$

$$\operatorname{div} u = 0, \quad x \in F(t), t \in [0, T], \tag{3.1.3}$$

$$u = 0, \quad x \in \partial\Omega, \ t \in [0, T], \tag{3.1.4}$$

$$u = \frac{\mathrm{d}h}{\mathrm{d}t} + \omega \times (x - h), \quad x \in \partial S(t), \ t \in [0, T], \tag{3.1.5}$$

$$M\frac{\mathrm{d}^2 h}{\mathrm{d}t^2} = -\int_{\partial S(t)} \mathbb{T}\vec{n}d\Gamma + \int_{S(t)} \rho_S g dx, \tag{3.1.6}$$

$$\frac{\mathrm{d}}{\mathrm{d}t}(J\omega) = -\int_{\partial S(t)} (x - h) \times \mathbb{T}\vec{n}d\Gamma + \int_{S(t)} (x - h) \times \rho_S g dx, \quad t \in [0, T], \tag{3.1.7}$$

with the initial conditions

$$u(x, 0) = u_0(x), \quad x \in F(0), \tag{3.1.8}$$

$$h(0) = h_0 \in \mathbb{R}^3, \ \frac{\mathrm{d}h}{\mathrm{d}t}(0) = h_1 \in \mathbb{R}^3, \ \omega(0) = \omega_0 \in \mathbb{R}^3. \tag{3.1.9}$$

In the above system the unknowns are $u(x, t)$ (the Eulerian velocity field of the fluid), $h(t)$ (the position of the mass center of the rigid body) and $\omega(t)$ (the angular velocity of the rigid bodies). The fluid is supposed to be homogeneous of density ρ_F whereas the density of the solid is supposed to be constant and it is denoted by ρ_S.

Moreover we have denoted by $\partial\Omega$ the boundary of Ω, by $\partial S(t)$ the boundary of the rigid body at instant t and by $n(x,t)$ the outwards unit vector field normal to $\partial F(t)$ and by $g(x,t)$ the applied body forces (per unit mass). The constant $\nu > 0$ stands for the viscosity of the fluid. Further, we have denoted by M (respectively, by J) the mass (respectively, the inertia moment related to the mass center) of the rigid body and by $\mathbb{T}$ the Cauchy stress tensor field in the fluid. The components $(T_{kl})_{k,l\in\{1,\dots,3\}}$ of $\mathbb{T}$ are related to the velocity field u by

$$T_{kl}(x,t) = -p(x,t)\delta_{kl} + \nu\left(\frac{\partial u_k}{\partial x_l} + \frac{\partial u_l}{\partial x_k}\right), \quad k,l = 1,\dots,3. \quad (3.1.10)$$

We next introduce some notation and we recall some simple results which will be useful in the remaining part of this section.

We suppose that the boundary of Ω is of class $\mathscr{C}^2$. For $\sigma > 0$ and $G \subset \mathbb{R}^3$ an open set we denote by G_σ the σ-neighborhood of G, i.e.,

$$G_\sigma = \{x \in \mathbb{R}^3 : d(x,G) < \sigma\}, \quad (3.1.11)$$

and we define the function space

$$K_\sigma(\chi) = \{u \in [H_0^1(\Omega)]^3 \mid \operatorname{div} u = 0 \quad \text{and } \mathrm{D}(u) = 0 \text{ in } S_\sigma(\chi)\}. \quad (3.1.12)$$

Moreover, we denote by $K_0(\chi)$ the closure of $\cup_{\sigma>0} K_\sigma(\chi)$ in $H^1(\Omega)$.

Let us now go back to the notations in problem (3.1.2)–(3.1.11). We suppose that the set $S_0(t)$, representing the region occupied by the solid body at instant t, is open and that the boundary of $S(t)$, is of class $\mathscr{C}^2$. Due to the regularity assumptions above, it can be easily checked that the following result holds.

Proposition 3.1.1. *There exists $\delta > 0$ such that for all $x \in S_0(0)$ (respectively for all $x \in \mathbb{R}^3 \setminus \overline{\Omega}$) there exists a open disk B of radius δ included in $S_0(0)$ (respectively in $\mathbb{R}^3 \setminus \overline{\Omega}$) and containing x.*

Throughout this section we fix $\delta > 0$ satisfying the conditions in the above proposition. In the particular case when we choose, in Eq. (3.1.12),

$\sigma = \delta$ we denote by G_{ext} the set G_δ. More precisely we put

$$G_{\text{ext}} = \{x \in \mathbb{R}^3 | d(x, G) < \delta\},$$

and we denote by G_{int} the "δ-kernel" of G defined by

$$G_{\text{int}} = \{x \in \mathbb{R}^3 | B(x, \delta) \subset G\}.$$

We remark that, due to Proposition 3.1.1, the δ-neighbourhood of the G_{int} and the "δ-kernel" of G_{ext} are equal to G.

For $f \in L^1_{loc}(\mathbb{R}^3, \mathbb{R}^3)$ we denote by $\overline{f}$ the convolution of f by a radially symmetric regularizing kernel supported in $B(0, \delta)$. More precisely, we put

$$\overline{f} = w_\delta * f = \int_{\mathbb{R}^3} w_\delta(x - y) f(y) dy, \tag{3.1.13}$$

with $w_\delta \in \mathcal{D}(\mathbb{R}^3)$, $w_\delta(x) = \tilde{w}_\delta(|x|)$, $\int_{\mathbb{R}^3} w_\delta(x) dx = 1$ and $\operatorname{supp} w_\delta \subset B(0, \delta)$.

The remark below, which can be checked by a simple calculation, plays an important role in the remaining part of this work.

Remark 3.1.2. If u is a rigid velocity field in the set G (i.e., $\mathrm{D}(u) = 0$ in G) then $\overline{u}(x) = u(x)$ for all $x \in G_{\text{int}}$.

We denote by $\varphi(\cdot, t)$ the characteristic function of $S_0(t)$. We denote by $\psi(\cdot, t)$ the characteristic function of the "δ-kernel" of $S_0(t)$, i.e.,

$$S(\psi) = (S_0(t))_{\text{int}}.$$

The outline of the remaining part of this section is as follows: In Subsec. 3.2 we introduce some function spaces, we give the weak form of the governing equations and we state the main result. In Subsec. 3.3 we introduce a penalized problem and describe the main steps of the proof of the existence result. In Subsec. 3.4 we apply classical results of DiPerna and Lions in order to pass to the limit in the transport equation of the density. In Subsec. 3.5 we derive several technical results which are then used, in Subsec. 3.6, to prove the compactness of the sequence of approximated velocity fields. The main result is proved in Subsec. 3.7.

3.2. *Weak formulation of the governing equations*

In this section we give a weak formulations of the governing Eqs. (3.1.2)–(3.1.11) and we state the main existence result.

Let $G \subset \mathbb{R}^3$ be an open bounded set with a $\mathscr{C}^2$ boundary. We first recall some spaces which are classical in the theory of the Navier–Stokes equations and we refer, for instance, to Temam [40] for their detailed study. More precisely, we set

$$\mathscr{V}(G) = \{v \in \mathscr{C}_0^\infty(G; \mathbb{R}^3) | \mathrm{div}\, v = 0\},$$

and denote by $V(G)$, respectively by $H(G)$, its closure in $[H^1(G)]^3$, respectively, in $[L^2(G)]^3$. It is known that

$$V(G) = \{v \in [H_0^1(G)]^3 | \mathrm{div}\, v = 0\},$$
$$H(G) = \{v \in [L^2(G)]^3 | \mathrm{div}\, v = 0,\ v \cdot n = 0 \text{ in } \mathrm{H}^{-1/2}(\partial G)\}.$$

For $v \in L^2(G, \mathbb{R}^3)$ we denote by $\mathrm{D}(v)$ the tensor field defined by

$$\mathrm{D}_{ij}(v) = \frac{1}{2}\left(\frac{\partial v_i}{\partial x_j} + \frac{\partial v_j}{\partial x_i}\right), \quad i, j = 1, \ldots, 3,$$

where the derivatives are calculated in the distributions sense, i.e., in $\mathscr{D}'(G)$. We say that $v \in L^2(G, \mathbb{R}^3)$ is a rigid velocity field if $\mathrm{D}_{ij}(v) = 0$, in $\mathscr{D}'(G)$, for $i, j \in \{1, \ldots, 3\}$. We next introduce some function spaces specific to our problem. Let $T > 0$ and let Q be the cylinder $Q = \Omega \times [0, T]$. We denote

$$\mathrm{Char}(Q) = \{g : Q \to \{0, 1\}\}, \quad \mathrm{Char}(\Omega) = \{g : \Omega \to \{0, 1\}\},$$

i.e., $\psi \in \mathrm{Char}(Q)$ if and only if ψ is the characteristic function of some subset of Q. For $\chi \in \mathrm{Char}(\Omega)$ we denote

$$K(\chi) = \{v \in V(\Omega) : \chi \mathrm{D}(v) = 0 \text{ in } L^2(\Omega)\}, \tag{3.2.1}$$
$$S(\chi) = \{x \in \Omega : \chi(x) = 1\}.$$

The space $K(\chi)$ is clearly a closed subspace of $V(\Omega)$. According to Lemma 1.1 in Temam [39], if $S(\chi)$ is an open connected subset of Ω then, for every $v \in K(\chi)$ there exist a vector k_v and a constant ℓ_v such that

$$v(x) = k_v + \ell_v \times x, \ \forall x \in S(\chi). \tag{3.2.2}$$

We give below, for later use, some properties of the space $K(\chi)$, in relation to the spaces $K_\sigma(\chi)$ which have been defined in Eq. (3.1.13). For the proof of the first one we refer to San Martin, Starovoitov and Tucsnak [35].

Proposition 3.2.1. *For any $\xi \in K(\chi)$ there exists a sequence of functions $\{\xi_\sigma\}_{\sigma>0} \subset K(\chi)$ satisfying the conditions: $\xi_\sigma \in K_\sigma(\chi), \ \forall \sigma > 0$ and $\xi_\sigma \to \xi$ in $H^1(\Omega)$ as $\sigma \to 0$.*

As a consequence of the result above we obtain

Corollary 3.2.1. *The spaces $K(\chi)$ and $K_0(\chi)$ coincide.*

Proof. For any $\sigma > 0$ we have that $K_\sigma(\chi) \subset K(\chi)$. Hence

$$\bigcup_{\sigma>0} K_\sigma(\chi) \subset K(\chi).$$

Since $K(\chi)$ is a closed subspace of H_0^1 then

$$K_0(\chi) = \overline{\bigcup_{\sigma>0} K_\sigma(\chi)} \subset K(\chi).$$

The opposite inclusion follows directly from Proposition 3.2.1, thus the result is proved. □

For $\psi \in \mathrm{Char}(Q)$, we denote by $L^p(0, T; K(\psi))$ the space of functions $v \in L^p(0, T; V(\Omega))$ such that $v(t) \in K(\psi(\cdot, t))$ for almost all $t \in [0, T]$.

In order to define weak solutions of Eqs. (3.1.2)–(3.1.11) we follow the ideas in [6] and [22]. This weak formulation is global in the sense that the unknown functions are defined on the whole domain Ω. More precisely, instead of considering separately the velocity (respectively, the density) fields of the fluid and the rigid body, we consider only one velocity field u defined on $\Omega \times [0, T]$ such that the restriction of $u(\cdot, t)$ to $S(t)$, is a rigid velocity field and we introduce a new unknown function $\varphi(x, t)$, which is the characteristic function of $S(t)$. More precisely, let (u, p, h, ω) be a classical solution of Eqs. (3.1.2)–(3.1.11). For every $t \geq 0$ and every $x \in S(t)$ we denote

$$u(x,t) = \frac{\mathrm{d}h}{\mathrm{d}t}(t) + \omega(t) \times (x - h(t)). \tag{3.2.3}$$

It is not difficult to check that

$$\frac{\partial \varphi}{\partial t} + \operatorname{div}(\varphi u) = 0, \tag{3.2.4}$$

in $\mathscr{D}'(Q)$. This fact suggests to include the above equation in the weak form of our system.

Moreover, we consider a global density field, determined by the position of the solid at each instant t defined by

$$\rho(x,t) = \rho_F(1 - \varphi(x,t)) + \rho_S \varphi(x,t) = \begin{cases} \rho_F & \text{if } x \in F(t) \\ \rho_S & \text{if } x \in S(t). \end{cases}$$

Weak solutions of our problem can be defined as follows:

Definition 3.2.2. Let $u_0 \in H(\Omega)$ and let φ_0 be the characteristic function of S_0. A set of functions $\{u, \varphi\}$ such that

$$u \in L^\infty(0,T; H(\Omega)) \cap L^2(0,T; K(\varphi)), \tag{3.2.5}$$

$$\varphi \in \mathrm{Char}(Q) \cap \mathscr{C}^{0,1/p}(0,T; L^p(\Omega)),\ 1 \leq p < \infty, \tag{3.2.6}$$

is said to be a weak solution of Eqs. (3.1.2)–(3.1.11) if it satisfies the condition

1. The equalities

$$\int_Q (\rho u(\xi_t + (u \cdot \nabla)\xi) - \nu \mathrm{D}(u) : \mathrm{D}(\xi))\mathrm{d}x\mathrm{d}t$$
$$= -\int_\Omega \rho(x,0) u_0 \cdot \xi(x,0)\mathrm{d}x - \int_Q \rho g \cdot \xi \mathrm{d}x\mathrm{d}t, \tag{3.2.7}$$

$$\int_Q \varphi(\eta_t + (u \cdot \nabla)\eta)\mathrm{d}x\mathrm{d}t = -\int_\Omega \varphi_0 \cdot \eta(x,0)\mathrm{d}x, \tag{3.2.8}$$

hold for any functions $\xi \in H^1(Q) \cap L^2(0,T; K(\varphi))$, $\xi(T) = 0$, $\eta \in \mathscr{C}^1(Q)$, $\eta(T) = 0$.

2. There exists a family of isometries $\{\mathscr{A}_{s,t}\}_{s,t\in[0,T]}$ of $\mathbb{R}^3$ such that

$$S(\varphi(t)) = \mathscr{A}_{s,t}(S(\varphi(s))), \quad \forall s, t \in [0, T], \tag{3.2.9}$$

and $\mathscr{A}_{s,t}$ are Lipschitz-continuous with respect to s and t.

Remark 3.2.2. Equation (3.2.7) is quite similar to the classical weak formulation of the Navier–Stokes system. Indeed, the only difference is the fact that the solution u and the test function φ are in $K(\varphi)$ instead of $V(\Omega)$. This approach has the advantage that u is by definition a rigid velocity field in $S(t)$ and the disadvantage that the space of test functions depends on the solution by the intermediate of φ.

Equation (3.2.8) is nothing else than the weak form of transport Eq. (3.2.4).

Finally, condition (3.2.9) says that the obtained solution is regular enough to retrieve the rigid displacement from this weak formulation.

Remark 3.2.3. It is easy to check that if (u, p, h, ω) is a strong solution of Eqs. (3.1.2)–(3.1.11) then (u, φ), where φ is the characteristic function of $S(t)$ defined in Eq. (3.1.1) and u is extended on all of Q by using Eq. (3.2.3), is a weak solution of Eqs. (3.1.2)–(3.1.11), in the sense of Definition 3.2.2.

Conversely if (u, φ) is smooth enough weak solution of Eqs. (3.1.2)–(3.1.11), in the sense of Definition 3.2.2,

The main result of this section is

Theorem 3.2.3. *Assume that* $u^0 \in H(\Omega)$, $g \in L^2(Q)$, $\rho^0 \in L^\infty(\Omega)$, $\rho^0 \geq m_0 > 0$ *for some constant* m_0 *and that the boundaries* $\partial\Omega$, $\partial S_0(0)$, *are of class* $\mathscr{C}^2$. *Then there exists* $T > 0$ *such that Eqs.* (3.1.2)–(3.1.11) *admit at least one weak solution on* $[0, T)$. *Moreover, this solution satisfies the energy estimate*

$$\int_\Omega \rho|u|^2 dx + \int_Q \nu\,|D(u)|^2\, dxdt \leq C\left\{\int_\Omega \rho^0|u^0|^2 dx + \|g\|^2_{L^2(Q)}\right\}, \tag{3.2.10}$$

for some constant $C > 0$. *We have either* $T = +\infty$ *(i.e., the solution is global) or there is a contact in finite time.*

Finally, there exists a family of isometries $\{\mathscr{A}_{s,t}\}_{s,t\in[0,T]}$ *of* $\mathbb{R}^3$ *such that*

$$S(\varphi(t)) = \mathscr{A}_{s,t}(S(\varphi(s))), \quad \forall s,t \in [0,T] \tag{3.2.11}$$

and $\mathscr{A}_{s,t}$ *is Lipschitz-continuous with respect to* s *and* t.

3.3. *Main steps of the proof Theorem 3.2.3*

The first step in the proof of Theorem 3.2.3 is to approximate the rigid body by a very viscous fluid. In this way we introduce a penalized problem. More precisely, for given $n \in \mathbb{N}$, $u^0 \in H(\Omega)$, $\rho^0 \in L^\infty(\Omega)$, and $\psi^0 \in L^\infty(\Omega) \cap \mathrm{Char}(\Omega)$, we consider the following penalized problem.

Find a set of functions $\{u_n,\ \rho_n,\ \varphi_n,\ \varphi_n,\ \psi_n\}$ such that

$$u_n \in L^\infty(0,T;H(\Omega)) \cap L^2(0,T;V(\Omega)), \tag{3.3.1}$$

$$\psi_n,\ \varphi_n \in \mathrm{Char}(Q) \cap \mathscr{C}^{0,1/p}(0,T;L^p(\Omega)), \tag{3.3.2}$$

$$\rho_n \in L^\infty(Q), \tag{3.3.3}$$

$$S(\varphi_n) = (S(\psi_n))_{\mathrm{ext}} \tag{3.3.4}$$

and that relations

$$\int_Q (\rho_n u_n(\xi_t + (u_n\cdot\nabla)\xi) - (\nu + n\varphi_n)\mathrm{D}(u_n) : \mathrm{D}(\xi))dxdt = \tag{3.3.5}$$

$$= -\textstyle\int_\Omega \rho^0 u^0 \cdot \xi(\cdot,0)dx - \int_Q \rho_n g\cdot \xi dxdt,$$

$$\int_Q \rho_n(\eta_t + (u_n\cdot\nabla)\eta)dxdt = -\int_\Omega \rho^0\cdot\eta(\cdot,0)dx, \tag{3.3.6}$$

$$\int_0^T\int_{\Omega_{\mathrm{ext}}} \psi_n(\gamma_t + (\overline{u_n}\cdot\nabla)\gamma)dxdt = -\int_{\Omega_{\mathrm{ext}}} \psi^0\cdot\gamma(\cdot,0)dx, \tag{3.3.7}$$

hold for any functions $\xi \in H^1(Q) \cap L^2(0,T;V(\Omega))$, $\xi(\cdot,T) = 0$, $\eta \in \mathscr{C}^1(Q)$, $\eta(\cdot,T) = 0$, $\gamma \in \mathscr{C}^1((0,T)\times\Omega_{\mathrm{ext}})$, $\gamma(\cdot,T) = 0$.

The function $\overline{u_n}(\cdot,t)$ in Eq. (3.3.7) is defined as in Eq. (3.1.14) after extending u by zero outside Ω. The replacement of u_n by $\overline{u_n}$ in Eq. (3.3.7), which is much smoother, allows the application of some standard results on ordinary differential equations and on characteristics of transport equations. Moreover, due to Remark 3.1.2 we will obtain a rigid motion when $n\to\infty$, without passing to the limit with respect to δ.

The result below asserts the existence of weak solutions for Eqs. (3.3.1)–(3.3.7). This result can be proved following step by step the classical methods of investigation of the Navier–Stokes equations for non-homogeneous fluids (see [1] or [30]). This is why we omit the proof.

Theorem 3.3.1. *For any* $n \in \mathbb{N}$, $u^0 \in H(\Omega)$, $\rho^0 \in L^\infty(\Omega)$, $\psi^0 \in L^\infty(\Omega) \cap$ $Char(\Omega)$ *there exists at least a solution of the penalized problem* (3.3.1)–(3.3.7). *This solution has the following properties*:

$$\int_\Omega \rho_n |u_n|^2 dx + \int_Q (\nu + n\varphi_n)|D(u_n)|^2 dxdt$$

$$\leq C\left\{\int_\Omega \rho^0|u^0|^2 dx + \|g\|^2_{L^2(Q)}\right\}, \tag{3.3.8}$$

for some constant $C > 0$,

$$\|\rho_n(t)\|_{L^p(\Omega)} = \left\|\rho^0\right\|_{L^p(\Omega)}, \quad 1 \leq p \leq \infty, \tag{3.3.9}$$

$$\|\psi_n(t)\|_{L^p(\Omega_{ext})} = \left\|\psi^0\right\|_{L^p(\Omega_{ext})}, \quad 1 \leq p \leq \infty. \tag{3.3.10}$$

Moreover, for all $t \in [0, T]$ *the function* $\psi_n(\cdot, t)$ *take, a.e. in* Ω, *only two values*: 0 *and* 1.

According to Theorem 3.3.1 the sequences $\{u_n\}$, $\{\rho_n\}$, $\{\psi_n\}$ have subsequences (which we also denote by $\{u_n\}$, $\{\rho_n\}$, $\{\psi_n\}$) such that

$$u_n \to u \text{ in } L^2(0, T; V(\Omega)) \text{ weakly and in } L^\infty(0, T; H(\Omega)) \; * -\text{weakly}, \tag{3.3.11}$$

$$\rho_n \to \rho \text{ in } L^\infty(Q) \; * -\text{weakly}, \tag{3.3.12}$$

$$\psi_n \to \psi \text{ in } L^\infty(0, T, L^\infty(\Omega_{ext})) \; * -\text{weakly}. \tag{3.3.13}$$

Moreover, denote by φ the characteristic functions of $(S(\psi))_{ext}$.

The second step of the proof consists in showing that the weak limits defined above satisfy the transport equations. More precisely, we will show that the following result holds true.

Proposition 3.3.2. *The functions* u, ρ *and* φ *defined above satisfy relations* (3.2.5)–(3.2.6) *and* (3.2.8).

The proof of this result, which is based on the results of R. DiPerna and P.-L. Lions (see [8] and [30]), is given in Subsec. 3.4.

The third and the most technical step of the proof consists in proving the following result, which is proved in Subsec. 3.6.

Theorem 3.3.3. *The sequence* $\{u_n\}$ *in Eq.* (3.3.11) *converges strongly to* u *in* $L^2(Q)$.

The last step consists in combining Proposition 3.3.2 and Theorem 3.3.3 in order to prove our main existence theorem.

3.4. *Compactness of the density field*

3.4.1. *Some background on the transport equation*

In this subsection we gather, for easy reference, some basic facts about transport equations and in particular concerning compactness of weak solutions. We do not give proofs, we only refer to the relevant literature.

Let us consider the problem of finding $\psi \in L^\infty(Q)$ such that

$$\frac{\partial \psi}{\partial t} + \text{div}\,(\psi v) = 0, \qquad \text{in } \mathscr{D}'(Q), \tag{3.4.1}$$

$$\psi(x, 0) = \psi_0(x), \qquad \text{in } L^\infty(\Omega), \tag{3.4.2}$$

where v is a given vector field $v \in L^2(0, T; V(\Omega))$ and $\psi_0 \in L^\infty(\Omega)$. We recall the following result of DiPerna and Lions (see [8]).

Proposition 3.4.1. *The problem* (3.4.1) *and* (3.4.2) *has a unique weak solution* $\psi \in L^\infty(Q) \cap \mathscr{C}\left([0, T]; L^1(\Omega)\right)$, *in the sense that there exists a unique* $\psi \in L^\infty(Q) \cap \mathscr{C}([0, T]; L^1(\Omega))$ *such that*

$$\int_Q \psi(\eta_t + (v\cdot\nabla)\eta)dxdt = -\int_\Omega \psi_0\eta(\cdot, 0)dx, \quad \forall \eta \in \mathscr{C}^1(Q),\ \eta(\cdot, T) = 0.$$

Furthermore, if the data satisfies $\psi_0(x) \in \{0, 1\}$ *a.e. in* Ω *then* $\psi(x, t) \in \{0, 1\}$ *a.e. in* Q.

For a proof of Proposition 3.4.1, we refer to [8]. Let us only point out that the previous problem was not complemented by boundary conditions because the velocity field v vanishes on $\partial\Omega$.

We will essentially use the following compactness result, also due to DiPerna and Lions (see for instance, [30]).

Theorem 3.4.2. *Let* $\{\psi_n\}_{n>0}$ *and* $\{v_n\}_{n>0}$ *be two sequences such that*

$$\{\psi_n\} \subset \mathscr{C}([0,T]; L^1(B_R)) \quad \textit{for all } R > 0,$$
$$\{v_n\} \subset L^2(0,T; V(\Omega)).$$

If the sequence $\{\psi_n\}$ *is bounded in* $L^\infty(Q)$, *the sequence* $\{v_n\}$ *is bounded in* $L^2(0,T;V(\Omega))$ *and*

$$\frac{\partial \psi_n}{\partial t} + div\,(\psi_n v_n) = 0 \quad \textit{in } \mathscr{D}'(Q),$$
$$\psi_n(0) \to \psi_0 \quad \textit{in } L^1(\Omega),$$
$$v_n \rightharpoonup v \quad \textit{weakly in } L^2(0,T;V(\Omega)),$$

for some $\psi_0 \in L^\infty(\Omega)$, $\psi_0 \geq 0$ *a.e. then* $\{\psi_n\}$ *converges strongly in* $\mathscr{C}([0,T];L^p(\Omega))$ *for all* $1 \leq p < \infty$ *to the unique solution* $\psi \in L^\infty(Q) \cap \mathscr{C}([0,T];L^1(\Omega))$ *of the problem*

$$\frac{\partial \psi}{\partial t} + div\,(\psi v) = 0 \quad \textit{in } \mathscr{D}'(Q)$$
$$\psi(x,0) = \psi_0(x) \quad \textit{a.e. in } \Omega$$

3.4.2. *Passage to the limit in the transport equations*

In this subsection we apply the results in the previous subsection to the sequences of solutions of the penalized problem (3.3.1)–(3.3.7).

In order to prove Proposition 3.3.2 we first notice that, by Theorem 3.4.2, we have the following result.

Lemma 3.4.3. *The sequences* $\{\rho_n\}$, $\{\psi_n\}$ *contain subsequences (which we also denote by* $\{\rho_n\}$, $\{\psi_n\}$*) such that*

$\rho_n \to \rho$ *strongly in* $\mathscr{C}([0,T];L^p(\Omega))$, $(1 \leq p < \infty)$,
$\psi_n \to \psi$ *strongly in* $L^p(\Omega_{ext} \times]0,T[)$, $(1 \leq p < \infty)$.

Corollary 3.4.4. *The corresponding subsequence of* $\{\varphi_n\}$ *(which we also denote by* $\{\varphi_n\}$*) converge to* φ *strongly in* $L^p(\Omega_{ext} \times]0,T[)$, $(1 \leq p < \infty)$.

We can obtain more information about the convergence of ψ_n, φ_n by using the regularity of the vector field $\overline{u_n}$. In order to obtain this information we recall some classical notions on ordinary differential equations and characteristics of transport equations.

Let us consider the following Cauchy problem:

$$\begin{cases} \dfrac{dX(t)}{dt} = \overline{u_n}(X, t), \\ X(s) = y, \end{cases} \tag{3.4.3}$$

where $y \in \Omega_{\text{ext}}$ and $s \in [0, T]$ are given. Since for almost all $t \in [0, T]$, $\overline{u_n}(\cdot, t) \in \mathscr{D}(\mathbb{R}^3)$ and $\overline{u_n}(x, \cdot) \in L^\infty(0, T; \mathbb{R}^3)$ for all $x \in \Omega_{\text{ext}}$, it follows from classical results (see for instance [31, Sec. 68] that Eq. (3.4.3) admits a unique solution defined in $[0, T]$. Moreover, since $\overline{u_n}|_{\partial\Omega_{\text{ext}}} = 0$, it follows that $X(t) \in \Omega_{\text{ext}}$ for all $t \in [0, T]$. Let us denote by $\mathscr{M}^n_{s,t}(y)$ this unique solution.

The properties of the family of mappings $\mathscr{M}^n_{s,t}(y)$ can be summarized by the following result.

Lemma 3.4.5.

(a) *The set of functions*

$$y \to \mathscr{M}^n_{s,t}(y),$$

is bounded in $\mathscr{C}^2(\Omega_{ext}; \mathbb{R}^3)$, *uniformly with respect to* $s, t \in [0, T]$ *and* $n > 0$.

(b) *The set of functions*

$$s \to \mathscr{M}^n_{s,t}(y),$$

is bounded in $W^{1,\infty}(0, T; \mathbb{R}^3)$, *uniformly with respect to* $t \in [0, T]$, $y \in \Omega_{ext}$ *and* $n > 0$.

Moreover, the set of functions

$$t \to \mathscr{M}^n_{s,t}(y),$$

is bounded in $W^{1,\infty}(0, T; \mathbb{R}^3)$, *uniformly with respect to* $t \in [0, T]$, $y \in \Omega_{ext}$ *and* $n > 0$.

(c) $\det\left(\frac{\partial \mathscr{M}^n_{s,t}(y)}{\partial y}\right) = 1$ *for any* $y \in \Omega_{ext}$, $s, t \in [0, T]$, $n > 0$.

Proof. The boundedness of $\mathcal{M}_{s,t}(\cdot)$ in $\mathcal{C}(\Omega_{\text{ext}})$ is a direct consequence of the fact that $\mathcal{M}_{s,t}(y) \in \Omega_{\text{ext}}$ for all $y \in \Omega_{\text{ext}}$. Moreover, according to Theorem 1A in [28, Sec. 57] (see also [31, Sec. 69]) for each fixed $(s, t) \in [0, T] \times [0, T]$ the function $\mathcal{M}_{s,t}(\cdot)$ is $\mathcal{C}^1(\Omega_{\text{ext}})$ and the functions $t \to \partial\mathcal{M}_{s,t}(y)/\partial y_i$, $i = 1, \ldots, 3$, are absolutely continuous in t and they satisfy the linear initial value problem

$$\frac{d}{dt}\left(\frac{\partial\mathcal{M}_{s,t}(y)}{\partial y_i}\right) = \nabla_x\overline{u_n}(\mathcal{M}_{s,t}(y), t)\frac{\partial\mathcal{M}_{s,t}(y)}{\partial y_i} \text{a.e. in } [0, T], \tag{3.4.4}$$

$$\frac{\partial\mathcal{M}_{s,s}(y)}{\partial y_1} = \begin{pmatrix}1\\0\\0\end{pmatrix}, \quad \frac{\partial\mathcal{M}_{s,s}(y)}{\partial y_2} = \begin{pmatrix}0\\1\\0\end{pmatrix}, \quad \frac{\partial\mathcal{M}_{s,s}(y)}{\partial y_3} = \begin{pmatrix}0\\0\\1\end{pmatrix}. \tag{3.4.5}$$

Since $\overline{u_n}$ is bounded in $L^\infty(0, T; \mathcal{C}^2(\Omega_{\text{ext}}))$ relations (3.4.4) and (3.4.5) above imply that assertion (a) of the lemma is true.

In order to prove assertion (b) we first notice that the boundedness of the functions $t \to \mathcal{M}_{s,t}(y)$ and $s \to \mathcal{M}_{s,t}(y)$ in $\mathcal{C}(\Omega_{\text{ext}})$ is a direct consequence of the fact that $\mathcal{M}_{s,t}(y) \in \Omega_{\text{ext}}$ for all $y \in \Omega_{\text{ext}}$ and $t, s \in [0, T]$. Moreover, according to the Eq. (3.4.3), it is clear that $t \to \partial\mathcal{M}_{s,t}(y)/\partial t$ is bounded in $L^\infty(0, T; \mathbb{R}^3)$. For the function $s \to \partial\mathcal{M}_{s,t}(y)/\partial s$ we notice that it is absolutely continuous in t and satisfies the linear initial value problem

$$\frac{d}{dt}\left(\frac{\partial\mathcal{M}_{s,t}(y)}{\partial s}\right) = \nabla_x\overline{u_n}(\mathcal{M}_{s,t}(y), t)\frac{\partial\mathcal{M}_{s,t}(y)}{\partial s} \quad \text{a.e. in } [0, T], \tag{3.4.6}$$

$$\left.\frac{\partial\mathcal{M}_{s,t}(y)}{\partial s}\right|_{t=s} = -\overline{u_n}(y, s). \tag{3.4.7}$$

Since $\overline{u_n}$ is bounded in $L^\infty(0, T; \mathcal{C}^1(\Omega_{\text{ext}}))$, relations (3.4.6) and (3.4.7) above imply that the function $s \to \mathcal{M}_{s,t}(y)$ is bounded in $W^{1,\infty}(0, T; \mathbb{R}^3)$. This ends the proof of assertion (b).

In order to prove assertion (c) it suffices to notice that relations (3.4.4) and (3.4.5) above and the classical Liouville theorem imply that

$$\text{Det}\left(\frac{\partial\mathcal{M}_{s,t}(y)}{\partial y}\right) = \exp\int_s^t \text{div}\left(\overline{u_n}(\mathcal{M}_{s,\eta}(y), \eta)\right) d\eta \tag{3.4.8}$$

and to use the fact that $\text{div}\overline{u_n} = 0$. □

From the lemma above we can conclude the following corollaries.

Corollary 3.4.6. *The sequence* $\{\mathscr{M}^n\}$ *converges to* $\mathscr{M}$ *in* $\mathscr{C}^{0,\alpha}([0,T]\times[0,T];\mathscr{C}^1(\Omega_{ext}))$, $\alpha<1$, *as* $n\to\infty$, *where* $\mathscr{M}_{s,t}(y)$ *is the unique solution of the Cauchy problem*

$$\frac{dX(t)}{dt}=\overline{u}(X,t),$$
$$X(s)=y\in\Omega_{ext}.$$

Simple calculations show that the solution of the transport Eq. (3.3.7) is

$$\psi_n^i(x,t)=\psi^{i,0}(\mathscr{M}_{t,0}^n(x)). \tag{3.4.9}$$

The relation above, Corollary 3.4.6 and the dominated convergence theorem imply the following result

Corollary 3.4.7. *The function* ψ *in Lemma* 3.4.3 *satisfies the condition*

$$\psi(x,t)=\psi^0(\mathscr{M}_{t,0}(x)),\quad \forall x\in\Omega_{ext},\ \forall t\in[0,T]. \tag{3.4.10}$$

Proof of Proposition 3.3.2. From Eqs. (3.3.8) and (3.4.12) we conclude that $\varphi \mathrm{D}(u)=0$. This fact implies that there exist a rigid function v such that $v(x,t)=u(x,t)$ for $x\in S(\varphi(t))$. It follows from Remark 3.1.2 that

$$\overline{u}(x,t)=v(x,t),\ \forall x\in S(\psi(t)). \tag{3.4.11}$$

Let us define $\mathscr{A}_{s,t}(y)$ as the unique solution of the problem:

$$\begin{cases}\dfrac{dX(t)}{dt}=v(X,t),\\[2mm] X(s)=y\in\mathbb{R}^3.\end{cases} \tag{3.4.12}$$

If $y\in S(\psi(s))$ then, by Eq. (3.4.11), we have that $\mathscr{A}_{s,t}(y)=\mathscr{M}_{s,t}(y)$. So, relation (3.4.10) can be rewritten as

$$\psi(x,t)=\psi^0\left(\mathscr{A}_{t,0}(x)\right),\quad \forall x\in\Omega_{\text{ext}},\ \forall t\in[0,T]. \tag{3.4.13}$$

Since $\mathscr{A}_{t,0}(x)$ is a rigid displacement, relation (3.4.13) implies that

$$\varphi(x,t)=\varphi^0(\mathscr{A}_{t,0}(x)),\quad \forall x\in\Omega_{\text{ext}},\ \forall t\in[0,T]. \tag{3.4.14}$$

The relation above implies that

$$\frac{\partial\varphi(x,t)}{\partial t} + \operatorname{div}\,(\varphi(x,t)v(x,t)) = 0 \text{ in } \mathscr{D}'(\Omega_{\text{ext}} \times [0,T)).$$

In other terms we showed that

$$\int_0^T \int_{\Omega_{\text{ext}}} \varphi(\eta_t + (v\cdot\nabla)\eta)dxdt = -\int_{\Omega_{\text{ext}}} \varphi^0\eta_0 dx$$

for all $\eta \in \mathscr{C}^1(Q)$, $\eta(T) = 0$.

Moreover since $\varphi v = \varphi u$ and $u(x,t) = 0$ for $x \in \Omega_{\text{ext}} \setminus \Omega$

$$\int_0^T \int_{\Omega} \varphi(\eta_t + (u\cdot\nabla)\eta)dxdt = -\int_{\Omega} \varphi^0\eta_0 dx \qquad (3.4.15)$$

for all $\eta \in \mathscr{C}^1(Q)$, $\eta(T) = 0$.

We have thus proved that φ and u satisfy Eq. (3.2.8). Moreover, by Proposition 3.4.1 we have that $\varphi \in \text{Char}(Q)$. □

3.5. *Some technical results*

In this section we give several technical results which are an essential ingredient of the proof of the compactness of the velocity field.

If $\sigma > 0$ and $G \subset \mathbb{R}^3$ we denote by G_σ the σ-neighbourhood of G, i.e.,

$$G_\sigma = \{x \in \mathbb{R}^3 : d(x,G) < \sigma\}.$$

Let us notice that if we take $\sigma = \delta$, where $\delta > 0$ is the number fixed in Subsec. 3.1, then $G_\delta = G_{\text{ext}}$.

Let $\chi\colon \Omega \to \{0,1\}$ be the characteristic function of a subset $S(\chi)$ of Ω. We suppose that the boundary of $S(\chi)$ is of class $\mathscr{C}^2$. Let us introduce the functions spaces

$V^s(\Omega)$ the closure of $\mathscr{V}(\Omega)$ in $H^s(\Omega)$, $0 < s \le 1$,
$K^s(\chi)$ the closure of $K(\chi)$ in $H^s(\Omega)$, $0 \le s \le 1$,

where the spaces $\mathscr{V}(\Omega)$ and $K(\chi)$ were introduced in Subsec. 3.2. We note that $V^1(\Omega) = V(\Omega)$ and $K^1(\chi) = K(\chi)$ where the space $V(\Omega)$ was also introduced in Subsec. 3.2.

Moreover we define several projection operators.

First we denote by $P^s(\chi)$, the orthogonal projector of $H^s(\Omega)$ onto $K^s(\chi), 0 \le s \le 1$.

If $\sigma > 0$ we denote by $P^s_\sigma(\chi)$ the orthogonal projector of $H^s(\Omega)$ onto the space of functions which are rigid velocity fields in a σ-neighbourhood of $S(\chi)$. More precisely for $0 \le s < 1$ we set $P^s_\sigma(\chi) = P^s(\mathbb{I}_{S_\sigma(\chi)})$ where $\mathbb{I}_{S_\sigma(\chi)}$ is the characteristic function of $S_\sigma(\chi)$.

Lemma 3.5.1. *For any $\sigma > 0$ there exists $n_0 > 0$ (depending only on σ) such that*

$$S(\varphi_n(t)) \subset S_\sigma(\varphi(t)) \quad \textit{and} \quad S(\varphi(t)) \subset S_\sigma(\varphi_n(t))$$

for all $n > n_0$, for all $t \in [0, T]$ and for all $i = 1, \dots, N$.

Proof. According to Corollary 3.4.6, we have that $\mathscr{M}^n \to \mathscr{M}$ in $\mathscr{C}([0, T] \times [0, T] \times \Omega)$. This fact, combined to Eqs. (3.4.9) and (3.4.10) implies that, for any $\sigma > 0$, there exists $n_0 > 0$ such that

$$S(\psi_n(t)) \subset S_\sigma(\psi(t)) \quad \text{and} \quad S(\psi(t)) \subset S_\sigma(\psi_n(t)) \tag{3.5.1}$$

for all $n > n_0$, $t \in [0, T]$, $i = 1, \dots, N$. By considering the δ-neighborhood of the sets above and by using Eq. (3.3.4) we obtain that for all $\sigma > 0$ we have the relations

$$\begin{aligned} S_\delta(\psi_n(t)) &= S(\varphi_n), \quad S_{\sigma+\delta}(\psi) = S_\sigma(\varphi), \\ S_\delta(\psi(t)) &= S(\varphi) \quad \text{and} \quad S_{\sigma+\delta}(\psi_n) = S_\sigma(\varphi_n). \end{aligned} \tag{3.5.2}$$

Relations (3.5.1) and (3.5.2) implies the conclusion of the lemma. □

Proposition 3.5.2. *Let u and φ be the functions considered in Proposition 3.3.2 then*

$$\lim_{\sigma\to 0} \int_0^T \left\| P^s_\sigma(\varphi(\cdot, t))u(\cdot, t) - u(\cdot, t) \right\|^2_{L^2(\Omega)} dt = 0. \tag{3.5.3}$$

Proof. For almost every $t \in [0, T]$ we have that $u(t) \in K(\varphi(\cdot, t))$. Then, by Proposition 3.2.1 there exists a sequence $\{u_\sigma\}_{\sigma>0}$ that converges to $u(t)$ in $K(\varphi(\cdot, t))$ and such that $u_\sigma \in K_\sigma(\varphi(\cdot, t))$, for all $\sigma > 0$.

Then we have

$$\left\| P_\sigma^s(\varphi(\cdot,t))u(\cdot,t) - u(\cdot,t) \right\|_{L^2(\Omega)} \leq \left\| P_\sigma^s(\varphi(\cdot,t))u(\cdot,t) - u(\cdot,t) \right\|_{V^s(\Omega)} \leq \|u_\sigma - u(\cdot,t)\|_{H_0^1(\Omega)}.$$

We conclude that the sequence of functions $f_\sigma(t) = \| P_\sigma^s(\varphi(\cdot,t))u(\cdot,t) - u(\cdot,t)\|^2_{L^2(\Omega)}$ converges to zero for a.e. $t \in [0,T]$. Since $\{f_\sigma\}$ is bounded from above by the function $g \in L^1(0,T)$ defined by $g(t) = \|u(\cdot,t)\|^2_{H_0^1(\Omega)}$, by using the Lebesgue dominated convergence theorem we conclude that assertion (3.5.3) holds true. □

The main result of this subsection is

Proposition 3.5.3. *For all $s \in [0,1)$ we have that*

$$\lim_{\sigma\to 0}\lim_{n\to\infty} \left\| P_\sigma^s(\varphi(\cdot,t))u_n - u_n \right\|_{L^2(0,T];V^s(\Omega))} = 0. \tag{3.5.4}$$

Proof. Let us first suppose that, for an arbitrary $\sigma_0 > 0$, there exists a family of functions $(u_n^\sigma(\cdot,t))_{n,\sigma}$ such that $u_n^\sigma(\cdot,t) \in K_\sigma^s(\varphi(\cdot,t))$ and

$$\lim_{\sigma\to 0}\lim_{n\to\infty} \left\| u_n^\sigma(\cdot,t) - u_n(\cdot,t) \right\|_{V^s(\Omega)} = 0. \tag{3.5.5}$$

for almost all $t \in [0,T]$. Since

$$\left\| P_\sigma^s(\varphi(\cdot,t))u_n(\cdot,t) - u_n(\cdot,t) \right\|_{V^s(\Omega)} \leq \left\| u_n^\sigma(\cdot,t) - u_n(\cdot,t) \right\|_{V^s(\Omega)},$$

relation (3.5.5) still holds if we replace $u_n^\sigma(\cdot,t)$ by $P_\sigma^s(\varphi(\cdot,t))u_n(\cdot,t)$. Moreover, the function $t \to \left\| w_n^\sigma(\cdot,t) \right\|_{V^s(\Omega)}$, where

$$w_n^\sigma(\cdot,t) = P_\sigma^s(\varphi(\cdot,t))u_n(\cdot,t) - u_n(\cdot,t),$$

is measurable, and

$$\int_0^T \left\| w_n^\sigma(\cdot,t) \right\|^{2/s}_{V^s(\Omega)} dt \leq \int_0^T \|u_n(\cdot,t)\|^{2/s}_{V^s(\Omega)} dt \tag{3.5.6}$$

$$\leq C\int_0^T \|u_n(\cdot,t)\|^{2(1-s)/s}_{L^2(\Omega)} \|u_n(\cdot,t)\|^2_{V^1(\Omega)} dt \leq C\int_0^T \|u_n(\cdot,t)\|^2_{V^1(\Omega)} dt \leq C$$

due to the energy estimate (3.3.8).

Relations (3.5.5) and (3.5.6) yield Eq. (3.5.4).

To conclude the proof it suffices to consider, following [35], a family of functions $(u_n^\sigma(\cdot, t))_{n,\sigma}$ such that $u_n^\sigma(\cdot, t) \in K_\sigma^s(\varphi(\cdot, t))$, for all $n \geq 1$ and which satisfies Eq. (3.5.5). □

3.6. *Proof of Theorem 3.3.3*

In order to prove Theorem 3.3.3 we need two results. The first one is

Proposition 3.6.1. *For any $s \in (0, 1)$ there exists a $\sigma_0 > 0$ such that, for any $\sigma \in (0, \sigma_0)$ we have*

$$\lim_{n\to\infty} \int_Q \rho_n u_n P_\sigma^s(\varphi(\cdot, t))(u_n)dxdt = \int_Q \rho u P_\sigma^s(\varphi(\cdot, t))(u)dxdt. \quad (3.6.1)$$

Proof. Consider an arbitrary $\sigma > 0$. Due to Lemma 3.5.1, there exists $n_0 > 0$ such that

$$S(\varphi_n(t)) \subset S_{\sigma/2}(\varphi(t)) \quad \forall t \in [0, T]$$

for all $n > n_0$.

Moreover, if we divide the interval $[0, T]$ in N_T subintervals $I_1 = [0, \tau]$, $I_2 = [\tau, 2\tau], \ldots, I_{N_T} = [(N_T - 1)\tau, N_T\tau]$, where $\tau = T/N_T$, then the regularity of the function $t \to \mathscr{A}_{0,t}(y)$ implies that there exists $\tau > 0$ (depending on σ) such that

$$S_{\sigma/2}(\varphi(t)) \subset S_\sigma(\varphi(k\tau)), \quad (3.6.2)$$

$$S_{\sigma/2}(\varphi(k\tau)) \subset S_\sigma(\varphi(t)) \quad (3.6.3)$$

for all $t \in I_k$, and for all $k = 1, \ldots, N_T$.

More precisely, if L is the Lipschitz constant of the function $t \to \mathscr{A}_{0,t}(y)$, then there exist

$$\tau \in \left[\frac{\sigma}{\sigma_0/T + 2(L+1)}, \frac{\sigma}{2(L+1)}\right],$$

satisfying Eqs. (3.6.2) and (3.6.3). In particular, it follows that there exists a constant $C > 0$ such that for all $\sigma \in (0, \sigma_0)$ there exist $\tau \geq C\sigma$ satisfying Eqs. (3.6.2) and (3.6.3).

Let us take one of the intervals I_k, $k = 1, \ldots, N_T$. In Eq. (3.3.5) we consider a test function ξ, which is equal to zero if $t \notin I_k$ and such that $\xi(\cdot, t) \in K_{\sigma/2}(\varphi(\cdot, k\tau))$ for all $t \in I_k$. In this case relation (3.3.5) implies, by using classical estimates on the Navier–Stokes equations (see, for instance [29, pp. 70–71]), that there exists a constant $C > 0$ such that

$$\left| \int_{I_k} \int_{\Omega} \rho_n u_n \xi_t dx dt \right| \leq C \|\xi\|_{L^2(I_k; V(\Omega))}, \quad \forall n > n_0.$$

The relation above implies that the sequence $\{d/dt(P^0_{\sigma/2}(\varphi(\cdot, k\tau))(\rho_n u_n))\}$ is bounded in $L^2(I_k; [K_{\sigma/2}(\varphi(\cdot, k\tau))]^*)$, where $[K_{\sigma/2}(\varphi(\cdot, k\tau))]^*$ is the dual space of $K_{\sigma/2}(\varphi(\cdot, k\tau))$ with respect to the pivot space $K^0_{\sigma/2}(\varphi(\cdot, k\tau))$. Moreover, from Eqs. (3.3.8) and (3.3.9) it follows that the sequence $\{\rho_n u_n\}$ is bounded in $L^2(I_k \times \Omega)$, so the sequence $\{P^0_{\sigma/2}(\varphi(\cdot, k\tau))(\rho_n u_n)\}$ is also bounded in $L^2(I_k; K^0_{\sigma/2}(\varphi(\cdot, k\tau)))$.

Since, for all $s > 0$, the inclusion $K^0_{\sigma/2}(\varphi(\cdot, k\tau)) \subset [K^s_{\sigma/2}(\varphi(\cdot, k\tau))]^*$ is compact, it follows from Aubin's theorem that the sequence $\{P^0_{\sigma/2}(\varphi(\cdot, k\tau))(\rho_n u_n)\}$ is relatively compact in $L^2(I_k; [K^s_{\sigma/2}(\varphi(\cdot, k\tau))]^*)$. Moreover, by Lemma 3.4.3 and relation (3.3.11) we have that $\rho_n u_n \rightharpoonup \rho u$ weakly in $L^2(I_k; L^2(\Omega))$. It follows that

$$\begin{aligned} P^0_{\sigma/2}(\varphi(\cdot, k\tau))(\rho_n u_n) &\to P^0_{\sigma/2}(\varphi(\cdot, k\tau))(\rho u) \\ &\quad \text{in } L^2(I_k; [K^s_{\sigma/2}(\varphi(\cdot, k\tau))]^*) \text{ strongly}, \end{aligned} \tag{3.6.4}$$

for all $s > 0$.

On the other hand, by relation (3.6.3) we have also that

$$P^0_{\sigma/2}(\varphi(\cdot, k\tau)) P^s_{\sigma}(\varphi(\cdot, t)) = P^s_{\sigma}(\varphi(\cdot, t)), \quad \forall t \in I_k \text{ and } \forall s \geq 0.$$

Using the relation above and the fact that $P^0_{\sigma/2}(\varphi(\cdot, k\tau))$ is self adjoint in $L^2(\Omega)$ we obtain that

$$\begin{aligned} &\int_{I_k} \langle \rho_n u_n, P^s_{\sigma}(\varphi(\cdot, t))(u_n) \rangle_{L^2(\Omega)} dt \\ &\quad = \int_{I_k} \langle P^0_{\sigma/2}(\varphi(\cdot, k\tau))(\rho_n u_n), P^s_{\sigma}(\varphi(\cdot, t))(u_n) \rangle_{L^2(\Omega)} dt \\ &\quad = \int_{I_k} \langle P^0_{\sigma/2}(\varphi(\cdot, k\tau))(\rho_n u_n), P^s_{\sigma}(\varphi(\cdot, t))(u_n) \rangle_{[K^s_{\sigma/2}]^*, K^s_{\sigma/2}} dt. \end{aligned}$$

By using Eq. (3.6.4) it follows that

$$\lim_{n\to\infty}\int_{I_k}\langle \rho_n u_n,\, P^s_\sigma(\varphi(\cdot,t))(u_n)\rangle_{L^2(\Omega)}dt$$
$$=\int_{I_k}\langle P^0_{\sigma/2}(\varphi(\cdot,k\tau))(\rho u),\, P^s_\sigma(\varphi(\cdot,t))(u)\rangle_{L^2(\Omega)}dt$$
$$=\int_{I_k}\langle \rho u,\, P^s_\sigma(\varphi(\cdot,t))(u)\rangle_{L^2(\Omega)}dt,\quad \forall k=1,\ldots,N_T.$$

By summing up the relations above, from $k=1$ to $k=N_T$, we obtain the assertion of Proposition 3.6.1. □

We also need the following result.

Proposition 3.6.2. *The sequences* $\{\rho_n\}$, $\{u_n\}$ *defined above satisfy the relation*

$$\lim_{n\to\infty}\int_Q \rho_n u_n^2 = \int_Q \rho u^2.$$

Proof. We clearly have that

$$\begin{aligned}\int_Q \rho_n u_n^2 - \int_Q \rho u^2 = &\int_0^T\int_\Omega \big(\rho_n u_n\cdot P^s_\sigma(\varphi(\cdot,t))[u_n]\\ &- \rho u\cdot P^s_\sigma(\varphi(\cdot,t))[u]\big)\,dxdt\, +\\ &+\int_0^T\int_\Omega \rho_n u_n\cdot\big(u_n - P^s_\sigma(\varphi(\cdot,t))[u_n]\big)\,dxdt\\ &+\int_0^T\int_\Omega \rho u\cdot\big(P^s_\sigma(\varphi(\cdot,t))[u]-u\big)\,dxdt.\end{aligned}\tag{3.6.5}$$

In order to estimate the last integral in the right hand side of Eq. (3.6.5) we notice that

$$\left|\int_Q \rho u\cdot\big(P^s_\sigma(\varphi(\cdot,t))[u]-u\big)\right| \le C\int_0^T \big\|P^s_\sigma(\varphi(\cdot,t))[u]-u\big\|_{L^2(\Omega)}dt,\tag{3.6.6}$$

where $C=\|\rho u\|_{L^\infty(0,T;L^2(\Omega))}$.

On the other hand, by Propositions 3.5.2 and 3.5.3, for every $\gamma > 0$ there exists $\sigma_0 > 0$ such that for every $\sigma \in (0, \sigma_0)$ we have:

$$\lim_{n\to\infty} \left\| P^s_\sigma(\varphi(\cdot, t))u_n - u_n \right\|_{L^2([0,T];L^2(\Omega))} \le \gamma, \tag{3.6.7}$$

$$\left\| P^s_\sigma(\varphi(\cdot, t))u - u \right\|_{L^2([0,T];L^2(\Omega))} \le \gamma. \tag{3.6.8}$$

For σ satisfying the conditions above, relations (3.6.6) and (3.6.8) imply that

$$\left| \int_Q \rho u \cdot \left(P^s_\sigma(\varphi)[u] - u\right) \right| \le$$

$$C \int_{[0,T]} \left\| P^s_\sigma(\varphi)[u] - u \right\|_{L^2(\Omega)} dt \le C\gamma.$$

The second integral in the right hand side of Eq. (3.6.5) can be estimated in a completely similar manner, by using Eq. (3.6.7) instead of Eq. (3.6.8). Moreover, by Proposition 3.6.1 the first integral in the right hand side of Eq. (3.6.5) tends to zero when $n \to \infty$. Since $\gamma > 0$ is arbitrary we obtain the conclusion of the proposition. □

Proof of Theorem 3.3.3 We first notice that

$$\left| \int_Q \rho \left(u_n^2 - u^2\right) \right| \le \left| \int_Q \left(\rho_n u_n^2 - \rho u^2\right) \right| + \left| \int_Q \left(\rho_n - \rho\right) u_n^2 \right|. \tag{3.6.9}$$

Since u_n is bounded in $L^\infty(0, T; L^2(\Omega))$ and in $L^2(0, T; H^1(\Omega))$ we can easily deduce that u_n is bounded in $L^4(Q)$. Moreover, by Lemma 3.4.3, we have that $\rho_n \to \rho$ strongly in $L^2(Q)$, so the second term in the right hand side of Eq. (3.6.9) tends to zero when $n \to \infty$. Since, by Proposition 3.6.2, the first term in the right hand side of Eq. (3.6.9) also tends to zero we conclude that

$$\lim_{n\to\infty} \int_Q \rho(u_n^2 - u^2) = 0. \tag{3.6.10}$$

Moreover,

$$\int_Q |u_n - u|^2 \le \frac{1}{m_0}\left(\int_Q \rho(u_n^2 - u^2) + \int_Q 2\rho u \cdot (u - u_n)\right), \tag{3.6.11}$$

where m_0 is defined in Theorem 3.2.3. The right hand side of Eq. (3.6.11) tends to zero by Eq. (3.6.10) and from the fact that $u_n \rightarrow u$ in $L^2(Q)$ weakly. We have thus proved the strong convergence of u_n to u in $L^2(Q)$.

□

3.7. *Proof of the main results*

3.7.1. *Proof of Theorem 3.2.3.*

By Proposition 3.3.2 (proved in the Subsec. 3.4), the functions u, ρ, φ satisfy relations (3.2.5), (3.2.6) and (3.2.8). So, in order to prove the existence of at least one weak solution of Eq. (3.1.2)–(3.1.11) we have only to prove that relation (3.2.7) is also satisfied. Due to Theorem 3.3.3 and to Lemma 3.4.3 we have that, up to the extraction of a subsequence,

$$u_n \overset{n\to\infty}{\longrightarrow} u, \quad \text{in } L^2(Q) \text{ strongly}, \tag{3.7.1}$$

$$\rho_n \overset{n\to\infty}{\longrightarrow} \rho, \quad \text{in } \mathscr{C}([0,T]; L^p(\Omega)), \quad (1 \le p < \infty) \text{ strongly}. \tag{3.7.2}$$

Let σ be an arbitrary positive number. We choose the test function ξ in Eq. (3.3.5) such that $\xi \in H^1(Q) \cap L^2(0,T; K_\sigma(\varphi))$. Then, due to Lemma 3.5.1, there exists $n_0 > 0$ (depending only on σ) such that

$$\varphi_n \mathrm{D}(\xi) = 0, \quad \text{in } L^2(Q), \quad \forall n > n_0$$

and, consequently,

$$\int_Q \varphi_n \mathrm{D}(u_n) : \mathrm{D}(\xi) dxdt = 0, \quad \forall n > n_0.$$

By using Eqs. (3.7.1) and (3.7.2) we can pass to the limit in the relation (3.3.5) to get that relation (3.2.7) holds for any $\xi \in H^1(Q) \cap L^2(0,T; K_\sigma(\varphi))$. By using Proposition 3.2.1 and Corollary 3.2.1, it follows that Eq. (3.2.7) holds for any $\xi \in H^1(Q) \cap L^2(0,T; K(\varphi))$.

We still have to show the regularity property (3.2.6). Let us consider the function $\varphi \in C(0,T; L^p(\Omega)) \cap Char(Q)$, $1 \le p \le \infty$. Since $S(\varphi(t)) = \mathscr{A}_{s,t}(S(\varphi(s)))$ and $\mathscr{A}_{s,t}$ is Lipschitz-continuous with respect to s and t, there exists a constant C such that for all $s, t \in [0,T]$ we have

$$S(\varphi(t)) \subset S_\gamma(\varphi(s)), \quad S(\varphi(s)) \subset S_\gamma(\varphi(t)),$$

where $\gamma = C|t-s|$.

Therefore,

$$\begin{aligned}
&\max\left\{\mu(S(\varphi(t))\setminus S(\varphi(s))),\quad \mu(S(\varphi(s))\setminus S(\varphi(t)))\right\}\\
&\quad\le \max\left\{\mu(S_\gamma(\varphi(s))\setminus S(\varphi(s))),\quad \mu(S_\gamma(\varphi(t))\setminus S(\varphi(t)))\right\}\\
&\quad\le C\,|\partial S(\varphi(t))|\,|t-s|,
\end{aligned}$$

where $|\partial S(\varphi(t))| = |\partial S(\varphi(0))|$ is the length of the boundary of the rigid body. This quantity is bounded since $\partial S(\varphi(0))$ is of the class C^2. Hence we have

$$\begin{aligned}
\|\varphi(t)-\varphi(s)\|_{L_p(\Omega)} &= \mu\big(S_\gamma(\varphi(t))\triangle S(\varphi(s))\big)^{1/p}\\
&\le C\,|t-s|^{1/p},\quad 1\le p<\infty,
\end{aligned}$$

where $A\triangle B = (A\cup B)\setminus(A\cap B)$.

This ends the proof of the existence of at least one weak solution of Eqs. (3.1.2)–(3.1.11). The energy estimate (3.2.10) follows directly from Eq. (3.3.8).

Finally, representation (3.2.11) is already obtained in the proof of Proposition 3.3.2 (see relation (3.4.14)).

The theorem is entirely proved.

3.8. *Remarks and bibliographical notes on Sec. 3*

The presentation in this section is essentially based on [35], which also contains a study of possible collisions in the two-dimensional case. For the sake of completeness we mention that, as far as we know, the first results on the existence of weak solutions of Eqs. (3.1.2)–(3.1.11) have been proved in Judakov [26] and Serre [36]. In the above mentioned references the authors assume that the fluid-solid system fills the whole space. This allows a simple change of variables reducing the problem to a fixed domain (see also the corresponding comments on strong solutions at the end of Sec. 2). The problem in a bounded domain with several rigid bodies has been studied much later in Hoffmann and Starovoitov [22], [23], Desjardins and Esteban [6], [7], Conca, San Martin and Tucsnak [4] and in Gunzburger, Lee and Seregin.[17] In [6] and [7], the authors prove global existence up to collisions in the 2D and in the 3D cases, for both incompressible and compressible fluid. In [22] and [23], the authors show global existence for one rigid body in the presence of eventual collisions. The same type of results is obtained in [4] and in [17] by different methods. The methods in the papers quoted above do not seem to be applicable in the case of several rigid body with eventual collisions. Another paper tackling specifically the three-dimensional case and proposing a modeling of collisions in this case is Feireisl [10].

References

1. S. N. Antontsev, A. V. Kazhikhov and V. N. Monakhov, *Boundary Value Problems in Mechanics of Nonhomogeneous Fluids* (North-Holland Publishing Co., Amsterdam, 1990). Translated from Russian.
2. V. Arnold, *Ordinary Differential Equations* (Springer-Verlag, Berlin, 1992). Translated from Russian.
3. L. Cattabriga, Su un problema al contorno relativo al sistema di equazioni di Stokes, *Math. J. Univ. Padova* **31** (1961):308–340.
4. C. Conca, J. San Martín and M. Tucsnak, Existence of solutions for the equations modelling the motion of a rigid body in a viscous fluid, *Comm. Part. Differ. Equat.* **25** (2000):1019–1042.
5. P. Cumsille and M. Tucsnak, Wellposedness for the Navier–Stokes flow in the exterior of a rotating obstacle, *Math. Methods Appl. Sci.* **29** (2006):595–623.
6. B. Desjardins and M. Esteban, Existence of weak solutions for the motion of rigid bodies in a viscous fluid, *Arch. Ration. Mech. Anal.* **146** (1999):59–71.
7. B. Desjardins and M. Esteban, On weak solutions for fluid-rigid structure interaction: compressible and incompressible models, *Comm. Part. Differ. Equat.* **25** (2000):1399–1413.
8. R. J. DiPerna and P.-L. Lions, Ordinary differential equations, transport theory and Sobolev spaces, *Invent. Math.* **98** (1989):511–547.
9. L. Eisenhart, *Riemannian Geometry* (Princeton University Press, Princeton, N. J., 1949).
10. E. Feireisl, On the motion of rigid bodies in a viscous incompressible fluid, *J. Evol. Equat.* **3** (2003):419–441.
11. T. I. Fossen, *Guidance and Control of Ocean Vehicles* (John Wiley, Chichester, 1994).
12. G. Galdi, On the steady self-propelled motion of a body in a viscous incompressible fluid, *Arch. Ration. Mech. Anal.* **148** (1999):53–88.
13. G. P. Galdi, On the motion of a rigid body in a viscous liquid: A mathematical analysis with applications, in *Handbook of Mathematical Fluid Dynamics.* Vol. 1 (North-Holland, Amsterdam, 2002), pp. 653–791.
14. G. P. Galdi and A. L. Silvestre, Strong solutions to the problem of motion of a rigid body in a Navier–Stokes liquid under the action of prescribed forces and torques, in *Nonlinear Problems in Mathematical Physics and Related Topics, I* (Kluwer/Plenum, New York, 2002), pp. 121–144.
15. G. P. Galdi and A. L. Silvestre, Strong solutions to the Navier–Stokes equations around a rotating obstacle, *Arch. Ration. Mech. Anal.* **176** (2005):331–350.

16. C. Grandmont and Y. Maday, Existence for an unsteady fluid-structure interaction problem, *M2AN Math. Model. Numer. Anal.* **34** (2000):609–636.
17. M. Gunzburger, H.-C. Lee and G. Seregin, Global existence of weak solutions for viscous incompressible flows around a moving rigid body in three dimensions, *J. Math. Fluid Mech.* **2** (2000):219–266.
18. P. Hartman, *Ordinary Differential Equations*, 2nd ed. (Birkhäuser Boston, Mass., 1982).
19. T. Hesla, *Collisions of smooth bodies in viscous fluids: A mathematical investigation*, Ph.D. dissertation, University of Minnesota, Minneapolis (2004).
20. M. Hillairet, Lack of collision between solid bodies in a 2D incompressible viscous flow, *Comm. Part. Differ. Equat.* **32** (2007):1345–1371.
21. T. Hishida, An existence theorem for the Navier–Stokes flow in the exterior of a rotating obstacle, *Arch. Ration. Mech. Anal.* **150** (1999):307–348.
22. K.-H. Hoffmann and V. Starovoitov, On a motion of a solid body in a viscous fluid. Two-dimensional case, *Adv. Math. Sci. Appl.* **9** (1999):633–648.
23. Zur Bewegung einer Kugel in einer zähen Flüssigkeit, *Doc. Math.* **5** (2000):15–21 (electronic).
24. J. Houot and A. Munnier, On the motion and collisions of rigid bodies in an ideal fluid, *Asymptot. Anal.* **56** (2008):125–158.
25. A. Inoue and M. Wakimoto, On existence of solutions of the Navier–Stokes equation in a time dependent domain, *J. Fac. Sci. Univ. Tokyo Sect. IA Math.* **24** (1977): 303–319.
26. N. Judakov, The solvability of the problem of the motion of a rigid body in a viscous incompressible fluid, *Dinamika Splosn. Sredy* (1974), pp. 249–253, 255.
27. H. Lamb, *Hydrodynamics*, 6th ed. (Cambridge Mathematical Library, Cambridge University Press, Cambridge, 1993), with a foreword by R. A. Caflisch.
28. E. Lee and L. Markus, *Foundations of Optimal Control Theory* (John Wiley & Sons, New York, 1967).
29. J.-L. Lions, *Quelques Méthodes de Résolution des Problèmes aux Limites non Linéaires* (Dunod, 1969).
30. P.-L. Lions, *Mathematical Topics in Fluid Mechanics* (Oxford Science Publications, 1996).
31. E. J. McShane, *Integration* (Princeton University press, 1944).
32. L. M. Milne-Thomson, *Theoretical Hydrodynamics*, 4th ed (The Macmillan Co., New York, 1960).
33. J. Ortega, L. Rosier and T. Takahashi, On the motion of a rigid body immersed in a bidimensional incompressible perfect fluid, *Ann. Inst. H. Poincaré Anal. Non Linéaire* **24** (2007):139–165.
34. J. H. Ortega, L. Rosier and T. Takahashi, Classical solutions for the equations modelling the motion of a ball in a bidimensional incompressible perfect fluid, *M2AN Math. Model. Numer. Anal.* **39** (2005):79–108.
35. J. A. San Martín, V. Starovoitov and M. Tucsnak, Global weak solutions for the two-dimensional motion of several rigid bodies in an incompressible viscous fluid, *Arch. Ration. Mech. Anal.* **161** (2002):113–147.
36. D. Serre, Chute libre d'un solide dans un fluide visqueux incompressible. Existence, *Japan J. Appl. Math.* **4** (1987):99–110.
37. T. Takahashi, Analysis of strong solutions for the equations modeling the motion of a rigid-fluid system in a bounded domain, *Adv. Differ. Equat.* **8** (2003):1499–1532.

38. T. Takahashi and M. Tucsnak, Global strong solutions for the two-dimensional motion of an infinite cylinder in a viscous fluid, *J. Math. Fluid Mech.* **6** (2004) 53–77.
39. R. Temam, Problèmes Mathématiques en Plasticité (Gauthier-Villars, Montrouge, 1983).
40. *Navier–Stokes Equations*, 3rd ed. (North-Holland Publishing Co., Amsterdam, 1984), with an appendix by F. Thomasset.

CHAPTER 5

FLUID-STRUCTURE INTERACTION BETWEEN BLOOD AND ARTERIAL WALLS

A. Quarteroni

CMCS-MATHICSE-SB, Ecole Polytechnique Fédérale, Lausanne
MOX, Dipartimento di Matematica "F. Brioschi", Politecnico di Milano, Milan

Mathematical models of the cardiovascular system, followed by the use of efficient and accurate numerical algorithms, have allowed applied mathematicians to make substantial progresses in the computer simulation and interpretation of the circulatory system functionality. In this work we will address some of the most basic models that are used to describe blood flow dynamics in local arterial environments and to predict the vessel wall deformation in compliant arteries. After deriving the equations that describe the blood flow in the arterial lumen and the deformation of the vessel walls, we will address their coupling and comment on the way this complex problem can be solved efficiently by iterative fluid-structure interaction algorithms.

1. Introduction

In recent years, the application of mathematical models of the cardiovascular system, seconded by the use of efficient and accurate numerical algorithms, has made impressive progress in the interpretation of the circulatory system functionality. Models have been used to describe both physiological and pathological situations, as well as in the perspective of providing patient specific design indications to surgical planning. The main impulse to develop this field of study is the increasing demand from the medical community for scientifically rigorous and quantitative investigations of cardiovascular diseases, which are responsible today for a great amount of deaths in industrialized societies.

The vascular system is highly complex and able to regulate itself: an excessive decrease in blood pressure will cause the smaller arteries, the arterioles, to contract and the heart rate to increase, whereas an excessive blood pressure is counter-reacted by a relaxation of the arterioles (which causes a reduction of the periphery resistance to the flow) and a decrease of the heart beat. Yet, it may happen that some pathological conditions develop. For example, the arterial wall may become more rigid, due to illness or unhealthy living habits, or fat may accumulate in the arterial walls causing a reduction of the vessel section (a stenosis). The way these pathologies affect the blood field, as well as the possible outcome of a surgical intervention to cure them, may be studied by numerical simulations. The latter are less invasive than *in-vivo* investigation, and far more accurate and flexible than *in-vitro* experiments. Numerical models require patient's data that can be generated by radiological acquisition through, e.g., computer tomography, magnetic resonance, doppler anemometry, etc. These data will be used to generate the geometrical shape of the computational domain, and to provide initial and boundary conditions for the system of partial differential equations that constitutes the mathematical model.

In this work we will address some of the most basic models that are used to describe blood flow dynamics in local arterial environments and to predict the vessel wall deformation in compliant arteries. After deriving the equations that describe the blood flow in the arterial lumen and the deformation of the vessel walls, we will address their coupling and comment on the way this complex problem can be solved efficiently by iterative (FSI) (fluid-structure interaction) algorithms.

Modeling FSI in the blood circulatory system is a broad and complex mathematical subject that cannot be addressed exhaustively in few pages. Here we sketch some of the main topics. This can be considered as a synthetic presentation aimed at stimulating the potentially interested reader. For a more thorough and extensive analysis we refer to the Chaps. 3, 8 and 9 of Ref. 10.

2. Kinematics and Dynamics of Continuous Media

Let $\widehat{\Omega} \subset \mathbb{R}^3$ be a bounded, open and simply connected subset of $\mathbb{R}^3$, with smooth boundary, filled by a *continuous medium*. We shall refer to $\widehat{\Omega}$ as the *reference configuration* of the medium under consideration.

A *deformation* of $\widehat{\Omega}$ is a smooth one-to-one mapping

$$\hat{\varphi} : \widehat{\Omega} \longrightarrow \Omega, \quad \widehat{\boldsymbol{x}} \longrightarrow \boldsymbol{x} = \hat{\varphi}(\widehat{\boldsymbol{x}}),$$

which carries each point $\widehat{\boldsymbol{x}} = (\widehat{x}_1, \widehat{x}_2, \widehat{x}_3)$ of $\widehat{\Omega}$ into a new position $\boldsymbol{x} = \hat{\varphi}(\widehat{\boldsymbol{x}})$ of $\Omega \subset \mathbb{R}^3$. The *displacement* of the *material point* $\widehat{\boldsymbol{x}}$ is given by the vector

$$\hat{\eta}(\widehat{\boldsymbol{x}}) = \hat{\varphi}(\widehat{\boldsymbol{x}}) - \widehat{\boldsymbol{x}}. \tag{1}$$

The local deformation is linked to the *deformation gradient*, defined as

$$\widehat{\boldsymbol{F}}(\widehat{\boldsymbol{x}}) = \nabla_{\hat{x}}\hat{\varphi}. \tag{2}$$

The symbol $\nabla_{\hat{x}}$ indicates the gradient with respect to the $\widehat{\boldsymbol{x}}$ coordinates. Sometimes we will omit the suffix when it is clear from the context which coordinate system we are adopting. The deformation gradient is a second-order tensor field, that is $\widehat{\boldsymbol{F}} : \widehat{\Omega} \to \mathbb{R}^{3\times3}$ being $\mathbb{R}^{3\times3}$ the space of three-dimensional matrices; its value is given by the 3×3 matrix of components

$$\widehat{F}_{ij} = \frac{\partial x_i}{\partial \widehat{x}_j}, \quad i, j = 1, 2, 3.$$

We also assume that its determinant

$$\widehat{J} = \det \widehat{\boldsymbol{F}}, \tag{3}$$

called the *Jacobian* of the deformation, is everywhere strictly positive, whence the mapping is *orientation preserving*.

To relate differential operators acting on the two configurations, the following result, which is obtained by applying the usual rules for the gradient of composite functions, is useful:

If $\hat{f} : \widehat{\Omega} \to \mathbb{R}$ *is a regular function and* $f : \Omega \to \mathbb{R}$ *is defined as* $f(\boldsymbol{x}) = \hat{f}(\widehat{\boldsymbol{\phi}}^{-1}(\widehat{\boldsymbol{x}}))$, *then*

$$\nabla_{\widehat{x}}\widehat{f} = \widehat{\boldsymbol{F}}\nabla f.$$

Let us assume that we have a sufficiently regular second-order tensor field $\boldsymbol{T} : \Omega \longrightarrow \mathbb{R}^{3\times3}$, defined on the deformed configuration. The *Piola transformation* of $\boldsymbol{T}$ associated to the given deformation $\hat{\varphi}$ is the second-order tensor field $\widehat{\mathbf{\Pi}} = \mathcal{P}_{\hat{\varphi}}(\boldsymbol{T}) : \widehat{\Omega} \to \mathbb{R}^{3\times 3}$ given by

$$\widehat{\mathbf{\Pi}} = \widehat{J}(\widehat{\boldsymbol{x}})\boldsymbol{T}(\hat{\boldsymbol{\varphi}}(\widehat{\boldsymbol{x}}))\widehat{\boldsymbol{F}}^{-T}(\widehat{\boldsymbol{x}}), \tag{4}$$

for all $\widehat{\boldsymbol{x}} \in \widehat{\Omega}$. Using a short-hand notation we may write $\mathcal{P}_{\hat{\varphi}}(\boldsymbol{T}) = \widehat{J}\widehat{\boldsymbol{T}}\widehat{\boldsymbol{F}}^{-T}$.

The *inverse Piola transformation* of $\widehat{\mathbf{\Pi}}$ returns the tensor $\boldsymbol{T}(\boldsymbol{x})$ according to

$$\boldsymbol{T}(\boldsymbol{x}) = \widehat{J}^{-1}(\hat{\boldsymbol{\varphi}}^{-1}(\boldsymbol{x}))\widehat{\mathbf{\Pi}}(\hat{\boldsymbol{\varphi}}^{-1}(\boldsymbol{x}))\widehat{\boldsymbol{F}}^{T}(\hat{\boldsymbol{\varphi}}^{-1}(\boldsymbol{x})), \tag{5}$$

or, more simply, $\boldsymbol{T} = J^{-1}\widehat{\mathbf{\Pi}}\boldsymbol{F}^T$.

If $\boldsymbol{T}$ is a regular tensor field in Ω and $\widehat{\mathbf{\Pi}}$ its Piola transformation, we have

$$\mathbf{div}_{\hat{x}}\widehat{\mathbf{\Pi}} = J\ \mathbf{div}\ \boldsymbol{T}, \tag{6}$$

where $\mathbf{div}_{\hat{x}}$ is the divergence with respect to the $\widehat{\boldsymbol{x}}$ coordinates and the equality has to be understood on corresponding points in $\widehat{\Omega}$ and Ω, respectively (see Ref. 7 for a proof).

As a result, by the application of the divergence theorem, we have

$$\int_{\partial\widehat{\Omega}} \widehat{\mathbf{\Pi}}\widehat{\boldsymbol{n}}\, \mathrm{d}\hat{\gamma} = \int_{\partial\Omega} \boldsymbol{T}\boldsymbol{n}\, \mathrm{d}\gamma, \tag{7}$$

whenever $\widehat{\mathbf{\Pi}}$ and $\boldsymbol{T}$ are related by Eq. (4).

In general, if $F = F(\boldsymbol{x}, t)$, t being the time variable, we will usually indicate with F_t the function of the space variable only, defined as $F_t(\boldsymbol{x}) = F(\boldsymbol{x}, t)$, at any fixed time t. Any smooth map

$$\hat{\boldsymbol{\varphi}} : \widehat{\Omega} \times \mathbb{R}^+ \longrightarrow \mathbb{R}^3, \quad (\widehat{\boldsymbol{x}}, t) \longrightarrow \boldsymbol{x} = \hat{\boldsymbol{\varphi}}(\widehat{\boldsymbol{x}}, t),$$

such that at any $t \geq 0$ $\hat{\boldsymbol{\varphi}}_t$ is a deformation, is called a *motion*. Thus, a motion is a one-parameter family of deformations, the parameter t being the time. Without loss of generality we have assumed here that the motion starts at $t = 0$ (initial time). The reference configuration $\widehat{\Omega}$ is in principle arbitrary, yet often it coincides with the initial configuration, i.e., $\widehat{\Omega} = \Omega(0)$. When not otherwise stated, we will implicitly make this assumption.

The point $\boldsymbol{x} = \hat{\boldsymbol{\varphi}}(\widehat{\boldsymbol{x}}, t)$ is the position at time t of the *material point* (also called *material particle*) identified by $\widehat{\boldsymbol{x}}$, while $\Omega(t) = \hat{\boldsymbol{\varphi}}(\widehat{\Omega}, t)$ denotes the *current configuration* at time t.

In this context, the displacement is now also function of time, $\hat{\eta}(\widehat{\boldsymbol{x}}, t) = \hat{\varphi}(\widehat{\boldsymbol{x}}, t) - \widehat{\boldsymbol{x}}$ being the displacement at time t.

All the kinematic quantities above can be extended to a motion. In particular, $\widehat{\boldsymbol{F}}$ and $\widehat{J}$ still indicate the deformation gradient and Jacobian, respectively, yet are now function also of time. For instance, $\widehat{\boldsymbol{F}}(\widehat{\boldsymbol{x}}, t) = \nabla_{\widehat{\boldsymbol{x}}}\hat{\varphi}(\widehat{\boldsymbol{x}}, t)$.

Given a subdomain $\widehat{V}$ of the domain $\widehat{\Omega}$, the set $V(t) = \{\boldsymbol{x} \in \Omega(t) : \boldsymbol{x} = \hat{\varphi}(\widehat{\boldsymbol{x}}, t), \ \widehat{\boldsymbol{x}} \in \widehat{V}\}$ is formed by the same material particles as $\widehat{V}$ and is called a *material* (*sub*)*domain*, or also *material volume*. Its size (volume) is

$$|V| = \int_V d\boldsymbol{x} = \int_{\widehat{V}} \widehat{J}(\widehat{x}, t) d\widehat{\boldsymbol{x}}, \tag{8}$$

then if $\widehat{J}$ is constant in time (i.e., $\partial\widehat{J}/\partial t = 0$) the material subdomain does not change its measure during motion.

The *velocity* is a major kinematic quantity and is the time derivative of the displacement:

$$\widehat{\boldsymbol{u}}(\widehat{\boldsymbol{x}}, t) = \frac{\partial}{\partial t}\hat{\eta}(\widehat{\boldsymbol{x}}, t) = \frac{\partial}{\partial t}\hat{\varphi}(\widehat{\boldsymbol{x}}, t), \tag{9}$$

the last equality is obtained by using the definition (1), now referred to time t.

3. Lagrangian, Eulerian and ALE Formulations

All physical quantities can be defined either on the reference or on the current configuration. For instance the field $\widehat{\rho} : \widehat{\Omega} \times \mathbb{R}^+ \to \mathbb{R}^+$ indicates the density, i.e., $\widehat{\rho}(\widehat{\boldsymbol{x}}, t)$ is the density at time t in the material point $\widehat{\boldsymbol{x}}$. Yet, the invertibility of the mapping allows us to refer the same quantity to the current configuration: for all $t > 0$

$$\rho(\boldsymbol{x}, t) = \widehat{\rho}(\hat{\varphi}^{-1}(\boldsymbol{x}, t), t), \quad \boldsymbol{x} \in \Omega(t)$$

is the density at the physical point $\boldsymbol{x} \in \Omega(t)$ at time t occupied by the material particle $\widehat{\boldsymbol{x}}$.

Using the reference or the current variables is a matter of convenience. The independent variables $(\widehat{\boldsymbol{x}}, t)$ are adopted in the *Lagrangian formulation*, whereas the $(\boldsymbol{x}, t)$ pair is employed in the *Eulerian formulation*. In the Lagrangian formulation we focus on the material particle $\widehat{\boldsymbol{x}}$ and its evolution; in the Eulerian formulation we observe what happens at a given point $\boldsymbol{x}$ in the physical space. When a field is expressed in the Eulerian coordinates it is also referred to as an *Eulerian field*, while a *Lagrangian field*, also called *material field*, is a field expressed in Lagrangian coordinates.

We will adopt the same symbol for a given physical quantity. Yet, the superscript $\widehat{\ }$ will denote a Lagrangian field. To summarize, for a quantity q we have

$$\widehat{q}(\widehat{\boldsymbol{x}}, t) = q(\boldsymbol{x}, t), \quad \text{with } \boldsymbol{x} = \hat{\boldsymbol{\varphi}}(\widehat{\boldsymbol{x}}, t),\ \widehat{\boldsymbol{x}} \in \widehat{\Omega},\ t > 0. \tag{10}$$

We will also make use of the composition operator: $\widehat{q}(\cdot, t) = q(\cdot, t) \circ \hat{\boldsymbol{\varphi}}$. Conversely,

$$q(\boldsymbol{x}, t) = \widehat{q}(\widehat{\boldsymbol{x}}, t), \quad \text{with } \widehat{\boldsymbol{x}} = \hat{\boldsymbol{\varphi}}^{-1}(\boldsymbol{x}, t),\ \boldsymbol{x} \in \Omega(t),\ t > 0, \tag{11}$$

or, more simply, $q(\cdot, t) = \widehat{q}(\cdot, t) \circ \hat{\boldsymbol{\varphi}}^{-1}$ (see Fig. 1).

To solve the differential equations governing the motion of a fluid or a solid we need to identify the *computational domain* where we want to solve the equations, on the boundary of which we need to provide suitable boundary conditions. In a solid, where the displacements are often relatively small, the computational domain is often taken to be a subset of $\widehat{\Omega}$ and the Lagrangian formulation is thus preferred.

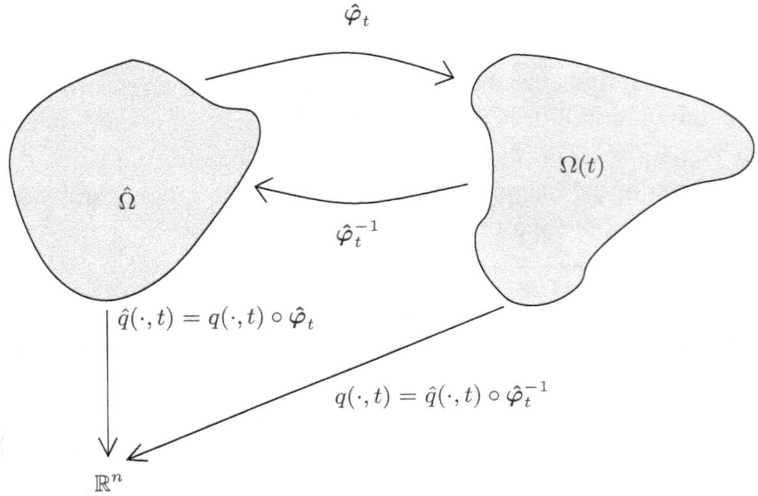

Fig. 1. Eulerian and Lagrangian description of a scalar field.

In a fluid the situation is rather different. The displacements are extremely large and, moreover, usually irrelevant, since when solving for a fluid we are normally interested in the velocity field, or other related quantities, rather than on the displacement itself. Therefore, the computational domain is normally taken as a fixed region $\Omega \subset \mathbb{R}^3$, and located where we are interested to compute the solution. No special requirements is made on Ω apart that it should be "filled by the fluid", that is $\Omega \subset \Omega(t)$ for all times t we are observing the motion. Although for the sake of simplicity we have set as the time interval for our equations the whole positive real line, yet in practical computations the time interval of interest is obviously finite. The Eulerian framework is then here preferable. However, the Lagrangian frame is useful as a tool to formally derive the equations from fundamental principles. Yet, in many situations of practical interest in haemodynamics, such as blood flowing in a compliant artery, the computational domain for the fluid cannot be fixed in time, as it has to follow the displacements of the fluid-wall interface. The Lagrangian frame is not of help here, since certainly we do not wish to follow the evolution of the blood particles as they circulate along the whole cardiovascular system! We usually wish to compute the flow field in a domain confined in the area of interest, yet following the movement of the wall interface (see, for instance, Fig. 2).

The computational domain, which we will indicate in the following with $\omega(t)$, is neither fixed nor a material subdomain, since its time evolution is not the fluid motion. To describe its evolution it turns out to be useful to introduce another, intermediate, frame of reference, called *Arbitrary Lagrangian Eulerian* (ALE).

Typically what is given is the evolution of the boundary of $\omega(t)$. We will show further on that it is possible to build from this information the auxiliary motion

$$\widetilde{\mathcal{A}} : \widetilde{\omega} \times \mathbb{R}^+ \rightarrow \mathbb{R}^3, \quad (\widetilde{\boldsymbol{x}}, t) \rightarrow \boldsymbol{x} = \widetilde{\mathcal{A}}(\widetilde{\boldsymbol{x}}, t),$$

Fig. 2. The ALE computational domain for the fluid in a compliant artery with the pressure distribution at time 6 s (left) and 12 s (right).The domain deforms to follow the arterial wall movement, yet the axial position of its proximal and distal boundary is kept fixed.

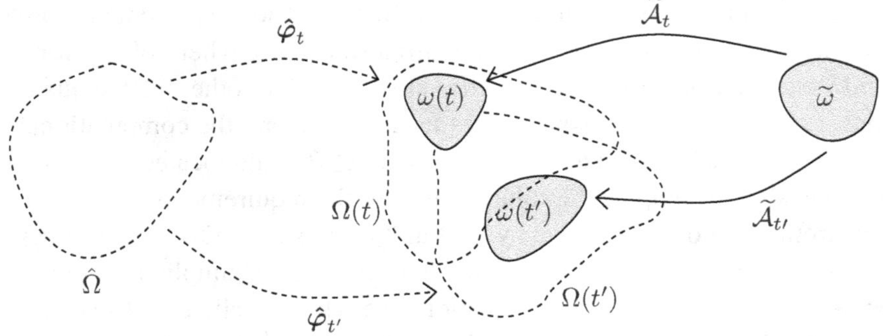

Fig. 3. The moving computational domain $\omega(t)$ and the ALE map. Here, for generality, we show an arbitrary reference computational domain $\widetilde{\omega}$. Most of the times, however, it is chosen to coincide with $\omega(0)$.

such that $\omega(t) = \widetilde{\mathcal{A}}(\widetilde{\omega}, t)$, for all $t > 0$, see Fig. 3. Here, $\widetilde{\omega} \subset \mathbb{R}^3$ is the reference configuration for the computational domain, which in general corresponds to the initial position at $t = 0$, i.e., $\omega(0)$. Figure 3 gives a sketch of the situation.

In the ALE formulation we have then the interplay of (at least) two motions: the one of the medium under consideration and that of the computational domain. The former is governed by physical laws, the latter is rather arbitrary, provided that the domain boundary movement is respected.

We can define the *computational domain velocity*, also called *ALE velocity*, as

$$\widetilde{\boldsymbol{w}}(\widetilde{\boldsymbol{x}}, t) = \frac{\partial \widetilde{\mathcal{A}}}{\partial t}(\widetilde{\boldsymbol{x}}, t), \quad \forall \widetilde{\boldsymbol{x}} \in \widetilde{\omega}, \tag{12}$$

which can be mapped to the Eulerian frame, in short hand notation $\boldsymbol{w}(\cdot, t) = \widetilde{\boldsymbol{w}}(\cdot, t) \circ \widetilde{\mathcal{A}}^{-1}$.

In general, $\boldsymbol{w}(\boldsymbol{x}, t) \neq \boldsymbol{u}(\boldsymbol{x}, t)$. However, we can note two particular cases:

(i) $\boldsymbol{w} = \mathbf{0}$: the computational domain is fixed as $\omega(t) = \omega(0)$ for all times; we recover the Eulerian formulation;
(ii) $\boldsymbol{w} = \boldsymbol{u}$: the computational domain $\omega(t)$ is now a material domain; we recover the Lagrangian formulation.

For a given scalar Eulerian field q (the discussion applies also to vector or tensor fields), we define the *Eulerian time-derivative* as simply

$$\frac{\partial q}{\partial t}(\boldsymbol{x}, t), \qquad \boldsymbol{x} \in \Omega(t). \tag{13}$$

In other words, we look at the rate of change of q at a fixed point $\boldsymbol{x}$ in the physical space, where the current configuration lives. It is nothing else than the classical partial derivative.

Let now $\widehat{q}$ be the Lagrangian description of q. We define the *material time-derivative* of q as the Eulerian description of the time derivative of the Lagrangian field $\partial\widehat{q}/\partial t$, namely

$$\frac{Dq}{Dt}(\cdot, t) = \frac{\partial \widehat{q}}{\partial t}(\cdot, t) \circ \hat{\boldsymbol{\varphi}}_t^{-1}. \tag{14}$$

Therefore, using Eq. (14),

$$\frac{D}{Dt}q(\boldsymbol{x}, t) = \frac{d}{dt}q(\hat{\boldsymbol{\varphi}}(\widehat{\boldsymbol{x}}, t), t), \quad \text{with } \boldsymbol{x} = \hat{\boldsymbol{\varphi}}(\widehat{\boldsymbol{x}}, t). \tag{15}$$

The material derivative of q at $(\boldsymbol{x}, t)$ is thus the rate of variation in time of q perceived by an observer which moves with the particle $\widehat{\boldsymbol{x}}$ located at time t in the point $\boldsymbol{x}$.

Standard application of the chain rule for the composition of functions in Eq. (15) yields the following identity for any given Eulerian field q:

$$\frac{Dq}{Dt} = \boldsymbol{u} \cdot \nabla q + \frac{\partial q}{\partial t}. \tag{16}$$

The same type of considerations may be extended to the ALE formulation. In particular the *ALE time-derivative* $\partial q/\partial t_{|\tilde{\mathcal{A}}}$ of a field q may be defined in a way analogous to the material derivative. In particular, for each $\boldsymbol{x} \in \omega(t)$ and $t > 0$ we have

$$\frac{\partial q}{\partial t}_{|\widetilde{\mathcal{A}}} = \frac{\mathrm{d}}{\mathrm{d}t}q(\widetilde{\mathcal{A}}(\widetilde{\boldsymbol{x}}, t), t), \quad \text{with } \boldsymbol{x} = \widetilde{\mathcal{A}}(\widetilde{\boldsymbol{x}}, t). \tag{17}$$

In other words, we look at the rate of change of q in a point that moves with the computational domain. This relation is of utmost utility in the context of the numerical discretisation. When computing numerically a solution in a moving domain we are usually interested in the variation of quantities collocated at the nodes of a computational mesh, the latter necessarily

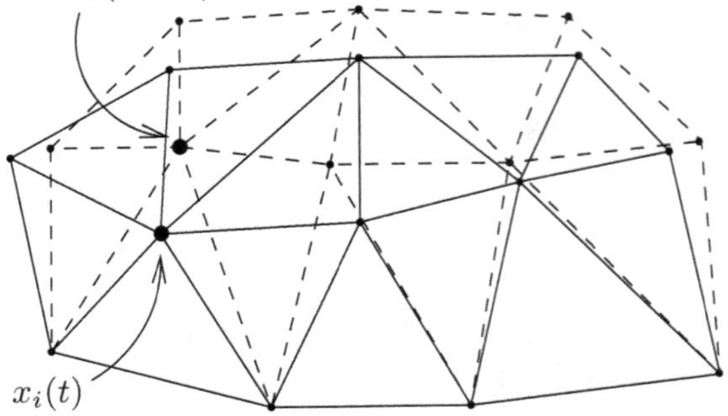

Fig. 4. Example of a moving mesh.

follows the evolution of the computational domain. In Fig. 4 we show one of such a node at two different times, namely $\boldsymbol{x}_i(t)$ and $\boldsymbol{x}_i(t+\delta t)$, where i is the node index.

The following identity holds:

$$\frac{\partial q}{\partial t}_{|\widetilde{\mathcal{A}}} = \boldsymbol{w} \cdot \nabla q + \frac{\partial q}{\partial t}. \tag{18}$$

The transport term $\boldsymbol{w} \cdot \nabla q$ accounts for the variations of q caused by the motion of the computational domain. It is clearly zero if the domain is fixed, while it coincides with the transport term in the material derivative (16) if $\boldsymbol{w} = \boldsymbol{u}$.

4. Mass and Momentum Conservation Principles

The *mass* of an arbitrary material domain $V(t)$ at time t is given by

$$\int_{V(t)} \rho \, \mathrm{d}\boldsymbol{x}, \tag{19}$$

being ρ the density (or volume mass) of the continuous medium. The units of measurement of density are $[\rho] = \mathrm{kg/m^3}$.

In classical mechanics the mass of a body does not change during the motion, a principle known as the *mass conservation*. Therefore,

$$\frac{\mathrm{d}}{\mathrm{d}t} \int_{V(t)} \rho \, \mathrm{d}\boldsymbol{x} = 0, \tag{20}$$

holds true for any $V(t)$ at any time. This is an integral statement, we want to express it "point-wise". To this aim, we use the Reynolds transport formula to obtain

$$\frac{\mathrm{d}}{\mathrm{d}t}\int_{V(t)} \rho \, \mathrm{d}\boldsymbol{x} = \int_{V(t)} \left(\frac{\partial \rho}{\partial t} + \mathrm{div}(\rho \boldsymbol{u}) \right) \mathrm{d}\boldsymbol{x},$$

by which, due to the arbitrariness of $V(t)$, we get the following

Proposition 4.1 (Continuity equation). *If ρ indicates the density of a continuous medium, mass conservation implies*

$$\frac{\partial \rho}{\partial t} + \mathrm{div}(\rho \boldsymbol{u}) = 0, \quad \text{in } \Omega(t), \tag{21}$$

for all $t > 0$, that is

$$\frac{\partial \rho}{\partial t} + \sum_{i=1}^{3} \frac{\partial}{\partial x_i}(\rho u_i) = 0.$$

If the fluid has *constant density* then Eq. (21) implies the well known *incompressibility equation*

$$\mathrm{div}\, \boldsymbol{u} = 0 \quad \text{in } \Omega(t), \quad t > 0. \tag{22}$$

The well known Newton's law yields the *conservation of linear momentum*. The rate of change of the momentum of a material domain $V(t)$, given by $\int_{V(t)} \rho \boldsymbol{u} \, \mathrm{d}\boldsymbol{x}$, is equal to the resultant of the external forces acting on it, that is

$$\frac{\mathrm{d}}{\mathrm{d}t}\int_{V(t)} \rho \boldsymbol{u} \, \mathrm{d}\boldsymbol{x} = \boldsymbol{F} = \boldsymbol{F}_v + \boldsymbol{F}_s.$$

The force $\boldsymbol{F}$ is the composition of two terms: a *volume force* $\boldsymbol{F}_v$, and a *surface force* $\boldsymbol{F}_s$. The former acts on each particle of $V(t)$ (like the force of gravity) and is expressed as the integral of the density times a *specific force* (i.e., force per unit of weight) $\boldsymbol{f}$ which has the dimension of an acceleration, $[\boldsymbol{f}] = \mathrm{m/s^2}$. The latter is instead responsible for the mutual interaction between the material contained in $V(t)$ and the exterior, through the boundary $\partial V(t)$. More precisely, $\boldsymbol{F}_s$ is equal to the surface integral of the so called *Cauchy stress* $\boldsymbol{t}$, which has the dimension of force per unit area, $[\boldsymbol{t}] = \mathrm{N/m^2}$, that is $\boldsymbol{F}_s = \int_{\partial V(t)} \boldsymbol{t} \, d\boldsymbol{\gamma}$.

According to the Cauchy postulate, $\boldsymbol{t}$ can be computed by applying to the normal $\boldsymbol{n}$ of $\partial V(t)$ a symmetric second-order tensor $\boldsymbol{\sigma} : \Omega(t) \to \mathbb{R}^{3\times 3}$, called the *Cauchy stress tensor*, i.e.,

$$\boldsymbol{t} = \boldsymbol{\sigma}\boldsymbol{n} \quad \text{on } \partial V(t), \quad \text{or, componentwise, } t_i = \sum_{j=1}^{3} \sigma_{ij} n_j. \tag{23}$$

The symmetry is in fact an implication of the conservation of angular momentum. The Cauchy postulate implies that the dependence of $\boldsymbol{t}$ on the geometry of $\partial V(t)$ is only through its normal. This holds true in most situations.

The *momentum conservation law* can then be expressed by the following equation,

$$\begin{aligned} \frac{\mathrm{d}}{\mathrm{d}t} \int_{V(t)} \rho \boldsymbol{u}\, \mathrm{d}\boldsymbol{x} &= \int_{V(t)} \rho \boldsymbol{f}\, \mathrm{d}\boldsymbol{x} + \int_{\partial V(t)} \boldsymbol{\sigma}\boldsymbol{n}\, \mathrm{d}\gamma \\ &= \int_{V(t)} \rho \boldsymbol{f}\, \mathrm{d}\boldsymbol{x} + \int_{V(t)} \mathbf{div}\, \boldsymbol{\sigma}\, d\boldsymbol{x}, \end{aligned} \tag{24}$$

valid for all material domains $V(t)$. To obtain the last equality we have used the divergence theorem. Finally, by exploiting the Reynolds transport formula, we obtain

Proposition 4.2 (Momentum conservation). *Assume that* (20) *holds. Then* (24) *is equivalent to*

$$\rho \frac{\partial \boldsymbol{u}}{\partial t} + \rho(\boldsymbol{u} \cdot \nabla)\boldsymbol{u} - \mathbf{div}\, \boldsymbol{\sigma} = \rho \boldsymbol{f}, \quad \text{in } \Omega(t), \quad t > 0. \tag{25}$$

Componentwise, we have

$$\rho \frac{\partial u_i}{\partial t} + \rho \sum_{j=1}^{3} u_j \frac{\partial u_i}{\partial x_j} - \rho \sum_{j=1}^{3} \frac{\partial \sigma_{ij}}{\partial x_j} = f_i, \quad i = 1, 2, 3.$$

The equations may be written in conservative form as

$$\frac{\partial(\rho \boldsymbol{u})}{\partial t} + \mathbf{div}\left(\rho \boldsymbol{u} \otimes \boldsymbol{u} - \boldsymbol{\sigma}\right) = \rho \boldsymbol{f}, \quad \text{in } \Omega(t), \quad t > 0, \tag{26}$$

which componentwise reads

$$\frac{\partial(\rho u_i)}{\partial t} + \sum_{j=1}^{3} \frac{\partial}{\partial x_j}(\rho u_i u_j - \sigma_{ij}) = \rho f_i, \quad i = 1, 2, 3.$$

In contrast, Eq. (25) is generally said to be in the *gradient* or in *quasi-linear* form.

The equations have been here written in Eulerian formulation. In the case of a fixed computational domain Ω they can be used directly, just replacing $\Omega(t)$ with Ω (we recall that Ω is a subset of $\Omega(t)$ for all t). In the case of a moving domain it is preferable to use the ALE time-derivative instead of the Eulerian one. To this aim it is sufficient to employ Eq. (18). If instead one wants to use a full Lagrangian formulation it is necessary to transform also the space differential operators, in order to write the equations on $\widehat{\Omega}$ instead of $\Omega(t)$. The Piola formula (6) can then become handy.

To close the equations just derived, we need to make precise how the Cauchy stress tensor is linked to the kinematics. It is indeed at this point where the behavior of solids and fluids diverges.

As solids react to deformations, the Cauchy stress must depend on $\widehat{\boldsymbol{F}}$ (or on quantities which are directly related to $\widehat{\boldsymbol{F}}$). The reference configuration plays here an important role.

Fluids instead can adapt to a deformation: a fluid can fill freely a container of arbitrary shape. Yet it takes time to fill it. It means that fluids react mechanically not to the deformation itself but to its rate. More precisely, the relevant quantity is here the *strain rate tensor* $\boldsymbol{D}$; componentwise, the strain rate is defined as

$$D_{ij} = \frac{1}{2}\left(\frac{\partial u_i}{\partial x_j} + \frac{\partial u_j}{\partial x_i}\right), \quad i, j = 1, \ldots 3,$$

and its dimensions are $[\boldsymbol{D}] = \mathrm{s}^{-1}$. In a fluid then $\boldsymbol{\sigma}$ is a function of $\boldsymbol{D}$, while it is independent of $\widehat{\boldsymbol{F}}$. A consequence is that the reference configuration, even if it is a useful concept for the derivation of the equations, eventually does not play any particular role for a fluid.

The relation between the Cauchy stress tensor $\boldsymbol{\sigma}$ and the kinematic quantities is called *constitutive relation*, or constitutive law, and is a characteristic of the type of material under consideration.

In a *Newtonian incompressible fluid*, the Cauchy stress tensor depends linearly on the strain rate. More precisely, we have

$$\boldsymbol{\sigma} = \boldsymbol{\sigma}(\boldsymbol{u}, P) = -P\boldsymbol{I} + 2\mu\boldsymbol{D}(\boldsymbol{u}) = -P\boldsymbol{I} + \mu(\nabla\boldsymbol{u} + \nabla\boldsymbol{u}^T), \tag{27}$$

where P is the *pressure*, $\boldsymbol{I}$ is the identity matrix, μ is the *dynamic viscosity* of the fluid and is a positive quantity.

The term $2\mu\boldsymbol{D}(\boldsymbol{u})$ is often referred to as *viscous stress* component of the stress tensor. We have that $[P] = \text{Ne}/\text{m}^2$ and $[\mu] = \text{kg}/\text{ms}$.

The viscosity may vary, for example it may depend on the fluid temperature. The assumption of Newtonian fluid, however, implies that μ is *independent of kinematic quantities*. Simple models for non-Newtonian fluids, often used for blood flow simulations, express the viscosity as function of the strain rate, that is $\mu = \mu(\boldsymbol{D}(\boldsymbol{u}))$.

The momentum Eq. (26) may then be written as

$$\rho\frac{\partial \boldsymbol{u}}{\partial t} + \rho\,\mathbf{div}(\rho\boldsymbol{u}\otimes\boldsymbol{u}) + \nabla P - 2\,\mathbf{div}\,(\mu\boldsymbol{D}(\boldsymbol{u})) = \rho\boldsymbol{f}. \tag{28}$$

Since ρ is constant, it is sometimes convenient to introduce the kinematic viscosity $\nu = \mu/\rho$, with $[\nu] = \text{m}^2/\text{s}$, and write

$$\frac{\partial \boldsymbol{u}}{\partial t} + \mathbf{div}(\boldsymbol{u}\otimes\boldsymbol{u}) + \nabla p - 2\,\mathbf{div}\,(\nu\boldsymbol{D}(\boldsymbol{u})) = \boldsymbol{f}, \tag{29}$$

where $p = P/\rho$ is a scaled pressure (with $[p] = \text{m}^2/\text{s}^2$).

We have here considered the conservative form (26) of the momentum equation, clearly we can also write the equation in gradient form. Starting from Eq. (25) we have

$$\rho\frac{\partial \boldsymbol{u}}{\partial t} + \rho(\boldsymbol{u}\cdot\nabla)\boldsymbol{u} + \nabla P - 2\,\mathbf{div}\,(\mu\boldsymbol{D}(\boldsymbol{u})) = \rho\boldsymbol{f}. \tag{30}$$

5. Navier–Stokes Equations for Blood Flow in the ALE Frame

When dealing with a moving computational domain $\omega(t)$ we can use the Navier–Stokes equations in the ALE framework. By using Eq. (18) on the momentum Eq. (30) we derive

$$\rho \frac{\partial \boldsymbol{u}}{\partial t}_{|\widetilde{\mathcal{A}}} + \rho[(\boldsymbol{u} - \boldsymbol{w}) \cdot \nabla]\boldsymbol{u} + \nabla P - 2\,\mathbf{div}\,(\mu \boldsymbol{D}(\boldsymbol{u})) = \rho \boldsymbol{f}, \quad \text{in } \omega(t),$$
$$\mathbf{div}\,\boldsymbol{u} = 0, \quad \text{in } \omega(t). \tag{31}$$

The introduction of the ALE time-derivative has induced a correction in the transport term by subtracting to the "transport velocity" $\boldsymbol{u}$ the domain velocity $\boldsymbol{w}$ given by Eq. (12).

A conservation form may be devised as well. In analogy with what already done for the Lagrangian frame, we define the Jacobian of the ALE movement $\widetilde{J}_{\widetilde{\mathcal{A}}} = \det \partial\widetilde{\mathcal{A}}_t / \partial\widetilde{x}$ and with $J_{\widetilde{\mathcal{A}}}$ its composition with the ALE movement. Recasting the Euler expansion formula to the ALE mapping we obtain

$$\frac{\partial J_{\widetilde{\mathcal{A}}}}{\partial t}_{|\widetilde{\mathcal{A}}} = J_{\widetilde{\mathcal{A}}} \operatorname{div} \boldsymbol{w}.$$

We have then

$$J_{\widetilde{\mathcal{A}}} \frac{\partial \boldsymbol{u}}{\partial t}_{|\widetilde{\mathcal{A}}} = \frac{\partial (J_{\widetilde{\mathcal{A}}} \boldsymbol{u})}{\partial t}_{|\widetilde{\mathcal{A}}} - J_{\widetilde{\mathcal{A}}} \boldsymbol{u} \operatorname{div} \boldsymbol{w},$$

by which, with simple manipulations we get the following conservation form of Eq. (31)

$$J_{\widetilde{\mathcal{A}}}^{-1} \rho \frac{\partial (J_{\widetilde{\mathcal{A}}} \boldsymbol{u})}{\partial t}_{|\widetilde{\mathcal{A}}} + \mathbf{div}\left(\rho \boldsymbol{u} \otimes (\boldsymbol{u} - \boldsymbol{w})\right)$$
$$+ \nabla P - 2\,\mathbf{div}\,(\mu \boldsymbol{D}(\boldsymbol{u})) = \rho \boldsymbol{f}, \quad \text{in } \omega(t), \tag{32}$$
$$\operatorname{div} \boldsymbol{u} = 0, \quad \text{in } \omega(t).$$

6. Elastodynamics Equations for the Vessel Wall Deformation

We now use the Piola transform and Eq. (6) to get

$$\widehat{\rho}_0 \frac{\partial^2 \widehat{\boldsymbol{\eta}}}{\partial t^2} - \mathbf{div}_{\widehat{x}} \widehat{\boldsymbol{\Pi}}_\sigma = \widehat{\rho}_0 \widehat{\boldsymbol{f}}, \quad \text{in } \widehat{\Omega}, \quad t > 0. \tag{33}$$

The tensor $\widehat{\boldsymbol{\Pi}}_\sigma = \mathcal{P}_{\widehat{\varphi}}(\sigma) = \widehat{J}\widehat{\sigma}\widehat{\boldsymbol{F}}^{-T}$ is called the *first Piola–Kirchhoff tensor* and Eq. (33) is the *momentum equation* (or the equation of *elastodynamics*) written in the Lagrangian frame.

Unlike the Cauchy stress tensor $\boldsymbol{\sigma}$, the first Piola–Kirchhoff tensor $\widehat{\boldsymbol{\Pi}}_\sigma$ is non-symmetric. Since constitutive laws are often expressed in terms of symmetric stress tensor, it is natural to introduce the *second Piola–Kirchhoff tensor* $\widehat{\boldsymbol{\Sigma}}$

$$\widehat{\boldsymbol{\Sigma}} = \widehat{\boldsymbol{F}}^{-1}\widehat{\boldsymbol{\Pi}}_\sigma = \widehat{J}\widehat{\boldsymbol{F}}^{-1}\widehat{\boldsymbol{\sigma}}\widehat{\boldsymbol{F}}^{-T}, \tag{34}$$

which is symmetric.

For an *elastic material* the stress is a function of the deformation (and possibly of thermodynamic variables such as the temperature) but is independent of the deformation history and thus of time. The material characteristics may still vary in space. In a *homogeneous* material the mechanical properties do not vary with $\boldsymbol{x}$, whence the strain energy function depends only on the deformation. A material is mechanically *isotropic* if its response to deformation is the same in all directions.

The constitutive equation is then a function of $\widehat{\boldsymbol{F}}$. More precisely, it is usually written in terms of the *Green–Lagrange strain tensor*, defined by

$$\widehat{\boldsymbol{E}} = \frac{1}{2}(\widehat{\boldsymbol{F}}^T\widehat{\boldsymbol{F}} - \boldsymbol{I}), \tag{35}$$

being $\boldsymbol{I}$ the identity tensor. Applying Eqs. (1) and (2) we obtain

$$\widehat{\boldsymbol{E}} = \frac{1}{2}\left(\nabla_{\hat{x}}\hat{\eta} + \nabla_{\hat{x}}^T\hat{\eta}\right) + \frac{1}{2}\nabla_{\hat{x}}^T\hat{\eta}\nabla_{\hat{x}}\hat{\eta}, \tag{36}$$

which componentwise reads

$$\widehat{E}_{ij} = \frac{1}{2}\left(\frac{\partial\widehat{\eta}_i}{\partial\widehat{x}_j} + \frac{\partial\widehat{\eta}_j}{\partial\widehat{x}_i}\right) + \sum_{l=1}^{3}\frac{\partial\widehat{\eta}_l}{\partial\widehat{x}_i}\frac{\partial\widehat{\eta}_l}{\partial\widehat{x}_j}, \quad i, j = 1, 2, 3.$$

$\widehat{\boldsymbol{E}}$ is not affected by a superimposed rigid body motion, and in particular by rigid rotations. Indeed, from a geometric point of view $\widehat{\boldsymbol{E}}$ is directly related to the difference of the squared length of an elemental vector $d\widehat{\boldsymbol{x}}$ and its image. Indeed, by recalling

$$||d\boldsymbol{x}|| = \sqrt{d\widehat{\boldsymbol{x}}^T\widehat{\boldsymbol{F}}^T\widehat{\boldsymbol{F}}d\widehat{\boldsymbol{x}}}, \tag{37}$$

we have that

$$\frac{1}{2}(||d\boldsymbol{x}||^2 - ||d\widehat{\boldsymbol{x}}||^2) = d\widehat{\boldsymbol{x}}^T \widehat{\boldsymbol{E}} d\widehat{\boldsymbol{x}}.$$

Many constitutive laws can be devised for a solid. For a *hyperelastic material* we first define a *density of elastic energy* $W : \mathbb{R}^{3\times 3} \longrightarrow \mathbb{R}^+$, and then set

$$\widehat{\boldsymbol{\Sigma}}(\widehat{\boldsymbol{E}}) = \frac{\partial W}{\partial \widehat{\boldsymbol{E}}}(\widehat{\boldsymbol{E}}), \text{ or, componentwise, } \widehat{\Sigma}_{ij} = \frac{\partial W}{\partial \widehat{E}_{ij}}, \quad i, j = 1, 2, 3. \tag{38}$$

A simple example of energy density for a *homogeneous isotropic material* whose reference configuration is the natural state (i.e., a configuration where the Cauchy stress tensor is zero everywhere) is the St-Venant–Kirchhoff model, where

$$W(\widehat{\boldsymbol{E}}) = \frac{\lambda}{2}(\text{tr}\,\widehat{\boldsymbol{E}})^2 + \mu\, \text{tr}\,\widehat{\boldsymbol{E}}^2, \tag{39}$$

which componentwise reads (by exploiting the symmetry of $\widehat{\boldsymbol{E}}$)

$$W = \frac{\lambda}{2}\left(\sum_{i=1}^{3} \widehat{E}_{ii}\right)^2 + \mu \sum_{i=1}^{3}\sum_{j=1}^{3} \widehat{E}_{ij}^2.$$

Here, λ and μ denote the first and second *Lamé coefficients*.

Correspondingly, we have

$$\widehat{\boldsymbol{\Sigma}}(\widehat{\boldsymbol{E}}) = \lambda(\text{tr}\,\widehat{\boldsymbol{E}})\boldsymbol{I} + 2\mu\widehat{\boldsymbol{E}}. \tag{40}$$

More complex constitutive relations for hyperelastic materials may be found in Ref. 13, and in particular models specially tailored for biological tissues and blood vessels are reported in Refs. 11 and 14.

Often it is more convenient to characterize an elastic material by its *Young modulus* E and *Poisson coefficient* ξ. Indeed, these quantities are usually inferred from experiments more directly than the Lamé coefficients. We have the following relations

$$E = \mu\frac{3\lambda + 2\mu}{\lambda + \mu}, \quad \xi = \frac{1}{2}\frac{\lambda}{\lambda + \mu} \tag{41}$$

and

$$\lambda = \frac{E\xi}{(1-2\xi)(1+\xi)}, \quad \mu = \frac{E}{2(1+\xi)}. \tag{42}$$

The equations written so far are rather general. Yet, even if we employ a linear relation between $\widehat{\boldsymbol{\Sigma}}$ and $\widehat{\boldsymbol{E}}$, like for instance Eq. (40), they give rise to a non-linear problem in the displacement $\hat{\boldsymbol{\eta}}$, because of the presence of the deformation gradient in the relation between $\widehat{\boldsymbol{\Sigma}}$ and $\widehat{\boldsymbol{\Pi}}$ and the quadratic term in Eq. (36).

However, when both the strain and the displacements are small we may derive a simpler, linear form of the equation. In haemodynamics, the hypothesis of small displacements can be accepted only in smaller arteries. Yet, it is sometimes used also in large vessels when deriving reduced models of structure dynamics, since it is assumed that this approximation is of the same importance as the others introduced by the model reduction process.

However, the configuration $\widehat{\Omega}$ is usually not a natural one. In fact, a vessel when extracted from its natural site tends to shrink, and it opens up when cut longitudinally.[11] This is a clear sign that even when at rest the stresses in an artery are not zero. Therefore, the linearization procedure (and therefore the assumption of small displacements) has to take place with respect to a pre-stressed reference state, different than $\widehat{\Omega}$.

Since hyperelastic constitutive equations are written assuming always a natural (i.e., zero stress) reference state, it is clear that the problem is not straightforward. We proceed by assuming the existence of a natural configuration $\widehat{\Omega}_0$ from which the actual reference configuration $\widehat{\Omega}$ is recovered by the map $\hat{\boldsymbol{\eta}}_0 = \hat{\boldsymbol{\eta}}_0(\widehat{\boldsymbol{x}}_0)$, being $\widehat{\boldsymbol{x}}_0 \in \widehat{\Omega}_0$. The current configuration $\Omega(t)$ is then obtained as usual from $\widehat{\Omega}$ by applying the displacement $\hat{\boldsymbol{\eta}}$, which is assumed small.

Therefore, the total displacement from the natural configuration is given by $\hat{\boldsymbol{\eta}}_t = \hat{\boldsymbol{\eta}}_0 + \hat{\boldsymbol{\eta}}$, and $\hat{\boldsymbol{\eta}}_t$ is in general not small, and the motion of $\Omega(t)$ is the superposition of a time-independent deformation from $\widehat{\Omega}_0$ to $\widehat{\Omega}$ and the motion from $\widehat{\Omega}$ to $\Omega(t)$. That is, $\boldsymbol{x} = \widehat{\boldsymbol{x}} + \hat{\boldsymbol{\eta}}$ in the current configuration is associated to a point in the natural configuration by $\boldsymbol{x} = \widehat{\boldsymbol{x}} + \hat{\boldsymbol{\eta}}_0 + \hat{\boldsymbol{\eta}} = \widehat{\boldsymbol{x}} + \hat{\boldsymbol{\eta}}_t$.

We will then write the elastodynamics equations with respect to the domain $\widehat{\Omega}_0$ and then apply a linearization procedure around the reference configuration $\widehat{\Omega}$. We do not report the steps of this derivation here. Rather, we refer the interested reader to Ref. 10, Chap. 3.

The equations of linear elasticity in a pre-stressed state read

$$\widehat{\rho}_0 \frac{\partial^2 \hat{\eta}}{\partial t^2} - \mathbf{div}_{\hat{x}} \left[\widehat{\boldsymbol{\sigma}}^0 \boldsymbol{\varepsilon}(\hat{\eta}) + \boldsymbol{H}^{\mathrm{p}} : \boldsymbol{\varepsilon}(\hat{\eta}) \right] = \widehat{\rho}_0 \widehat{\boldsymbol{f}}, \quad \text{in } \widehat{\Omega}, \quad t > 0, \qquad (43)$$

where

$$\boldsymbol{H}^{\mathrm{p}} = \widehat{J}_0^{-1} \left(\widehat{\boldsymbol{F}}^0 \widehat{\boldsymbol{F}}^0 \right) : \boldsymbol{H} : \left(\widehat{\boldsymbol{F}}^{0T} \widehat{\boldsymbol{F}}^{0T} \right) \qquad (44)$$

is the linear elasticity fourth-order tensor in the reference pre-stressed configuration $\widehat{\Omega}$, $\widehat{\boldsymbol{F}}^0$ and $\widehat{\boldsymbol{\sigma}}^0 = \widehat{J}_0^{-1} \widehat{\boldsymbol{F}}^0 \widehat{\boldsymbol{\Sigma}}^0 \widehat{\boldsymbol{F}}^{0T}$ are, respectively, the deformation gradient and the Cauchy stress tensor from the reference configuration $\widehat{\Omega}_0$ in correspondence to the deformation $\hat{\eta}_0$, and

$$\boldsymbol{H} = \frac{\partial^2 W}{\partial \widehat{\boldsymbol{E}}^2} (\widehat{\boldsymbol{E}}_0(\hat{\boldsymbol{\eta}}_0)). \qquad (45)$$

In the case where $\widehat{\Omega}$ is in a natural state $\widehat{\boldsymbol{\sigma}}^0 = \mathbf{0}$, $\boldsymbol{H}^p$ reduces to the standard linear elasticity tensor $\boldsymbol{H}$. System (43) becomes then the standard system of equations of linear elastodynamics:

$$\hat{\rho}_{s,0} \frac{\partial^2 \hat{\eta}}{\partial t^2} - \mathbf{div}_{\hat{x}} (\widehat{\boldsymbol{\sigma}}(\hat{\eta})) = \widehat{\boldsymbol{f}}, \quad \text{in } \widehat{\Omega}, \qquad (46)$$

with

$$\widehat{\boldsymbol{\sigma}}(\hat{\eta}) = \lambda (\operatorname{tr} \boldsymbol{\varepsilon}(\hat{\eta})) \boldsymbol{I} + 2\mu \boldsymbol{\varepsilon}(\hat{\eta})). \qquad (47)$$

Even if the material is homogeneous and isotropic with respect to the natural configuration (for instance it obeys the St-Venant–Kirchhoff model (40)), the same material in the pre-stressed configuration $\widehat{\Omega}$ is, in general, neither isotropic nor homogeneous. Indeed, these two properties depend not only on the material but also on the chosen reference state. Homogeneity is retained whenever $\widehat{\boldsymbol{F}}^0$ (and thus $\widehat{\boldsymbol{\sigma}}^0$) is constant, while isotropy requires that $\widehat{\boldsymbol{F}}^0 = a\boldsymbol{I}$, for a non negative a, and (consequently) that $\widehat{\boldsymbol{\sigma}}^0$ be proportional to the identity tensor $\boldsymbol{I}$. For more discussion we refer to Ref. 10, Chap. 3.

7. Reduced Structural Models

Sometimes we can use reduced models, much simpler than those derived in Sec. 6, to describe the vessel wall deformation. This choice may reduce computational costs when we are interested in the effects of the structure mechanics on the fluid, rather than in an accurate description of the stresses inside the vessel tissue.

Of special interest are models based on a single spatial coordinate, the one along the longitudinal axis, which usually describes the radial deformation of the vessel wall. These models are based on the following further simplifying assumptions.

Small thickness and plain stresses. The vessel wall thickness h is sufficiently small to allow a shell-type representation of the vessel geometry. In addition, we will also suppose that it is constant in the reference configuration. The vessel structure is subjected to plain stresses.

Cylindrical reference geometry and radial displacements. The reference vessel configuration is described by a circular cylindrical surface with straight axes. Indeed, this assumption may be partially dispensed with, by assuming that the reference configuration is "close" to that of a circular cylinder. The model here derived may be supposed valid also in this situation. The displacements are only in the radial direction.

Small deformation gradients. We assume that the deformation gradients are small, so that the structure basically behaves like a linear elastic solid and $\partial R/\partial\theta$ and $\partial R/\partial z$ remain uniformly bounded during the motion.

Incompressibility. The vessel wall tissue is incompressible, i.e., it maintains its volume during the motion. This is a reasonable assumption since biological tissues are indeed nearly incompressible.

Under the above assumptions we can derive the following one dimensional model that describes the deformation $\boldsymbol{\eta} = \eta\mathbf{e}_r$ of the arterial wall (see Ref. 17):

$$\rho^s\frac{\partial^2\eta}{\partial t^2} - a\frac{\partial^2\eta}{\partial z^2} + b\eta - c\frac{\partial^3\eta}{\partial t\partial z^2} = g, \quad 0 < z < L,\ t > 0, \tag{48}$$

where z denotes the longitudinal space coordinate (aligned along the vessel axis), L the length of the vessel at rest, while a, b and c are suitable coefficients which depend on material properties. Precisely:

$$a = \frac{\sigma_z}{h}, \qquad b = \frac{E}{(1-\varsigma^2)R_0^2},$$

while c is a positive coefficient that accounts for viscoelastic effects, R_0 is the radius of the cylindrical vessel at rest and h is the thickness of the vessel wall at rest, ς is the Poisson ratio, E is the Young modulus, while σ_z is the magnitude of the longitudinal stress.

The first term in Eq. (48) models the inertia, the second one the shear, the third one the elasticity, the fourth one the viscoelastic damping. Finally, g accounts for the forcing terms.

For a thorough mathematical derivation of these (and further) models, the interested reader is referred to Ref. 10, Chaps. 3 and 10.

8. The Coupled Fluid-Structure Problem

In this section we describe the general non-linear fluid-structure system in large displacements arising in blood flows in large arteries. We consider as computational domain a model of a portion of an artery, see Fig. 5. It consists of a deformable structure, the vessel wall, which occupies a region that we denote by $\Omega_s(t)$ surrounding a moving domain, that we denote by $\Omega_f(t)$, filled by a fluid (the blood) under motion. The fluid structure interface, i.e., the common boundary between $\Omega_s(t)$ and $\Omega_f(t)$, is denoted by $\Gamma(t) = \partial\Omega_f(t) \cap \partial\Omega_s(t)$. In the sequel, variables with a sub-script s or f shall refer to quantities within the fluid or the solid domains, respectively.

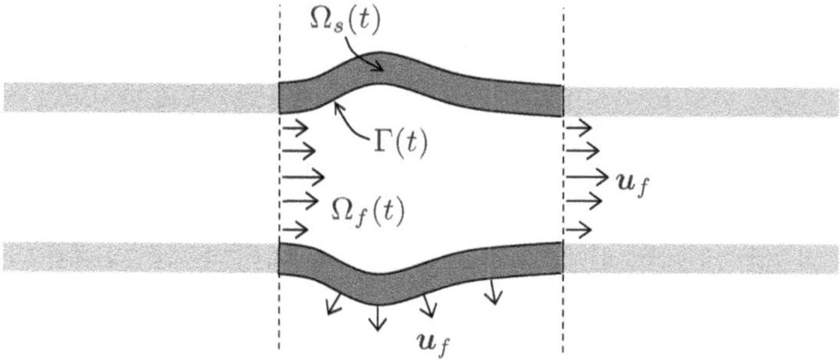

Fig. 5. Geometric configuration (2D section). (Picture taken from Ref. 10, Chap. 3.)

We will ignore body forces, i.e., we take $\boldsymbol{f} = \boldsymbol{0}$ both for the fluid and the structure. For haemodynamic applications this corresponds in practice to ignore the effects of gravity.

We assume the motion of $\Omega_f(t)$ to be parametrized by an ALE map $\widetilde{\mathcal{A}} : \hat{\Omega}_f \times \mathbb{R}^+ \longrightarrow \mathbb{R}^3$ (see Sec. 3), i.e., $\Omega_f(t) = \widetilde{\mathcal{A}}(\hat{\Omega}_f, t)$. The reference domain $\hat{\Omega}_f$ represents the position of the control volume at the initial time. We assume that the inlet $\Gamma_{f,D}$ and outlet $\Gamma_{f,N}$ boundaries are at a fixed axial position along the artery model of Fig. 6.

When dealing with moving domains it is natural to work with ALE time-derivatives. More precisely, we will use formulation (32).

The differential equations have to be completed with proper boundary conditions on $\partial\Omega_f(t)$. For instance, for $\boldsymbol{u}_D$ and $\boldsymbol{u}_\Gamma$ being given velocities and $\boldsymbol{g}_{f,N}$ a given density of surface load, the fluid dynamics system in the ALE frame (32) reads:

$$\begin{aligned}
\frac{\rho_f}{\widetilde{J}_{\widetilde{\mathcal{A}}}} \frac{\partial \widetilde{J}_{\widetilde{\mathcal{A}}} \boldsymbol{u}_f}{\partial t}\Big|_{\widetilde{\mathcal{A}}} + \mathbf{div}\left(\rho_f \boldsymbol{u}_f \otimes (\boldsymbol{u}_f - \boldsymbol{w}) - \boldsymbol{\sigma}_f(\boldsymbol{u}_f, P)\right) = \boldsymbol{0}, &\quad \text{in } \Omega_f(t),\\
\operatorname{div} \boldsymbol{u}_f = 0, &\quad \text{in } \Omega_f(t).\\
\boldsymbol{u}_f = \boldsymbol{u}_{f,D}, &\quad \text{on } \Gamma_{f,D},\\
\boldsymbol{\sigma}_f(\boldsymbol{u}_f, P)\boldsymbol{n}_f = \boldsymbol{g}_{f,N}, &\quad \text{on } \Gamma_{f,N},\\
\boldsymbol{u}_f = \boldsymbol{u}_\Gamma, &\quad \text{on } \Gamma(t).
\end{aligned} \tag{49}$$

Now we consider a Lagrangian description of the motion of the vessel structure in terms of its displacement field $\hat{\boldsymbol{\eta}}_s : \hat{\Omega}_s \times \mathbb{R}^+ \longrightarrow \mathbb{R}^3$ (see Fig. 7). By assuming, for the sake of simplicity, the structure to be clamped

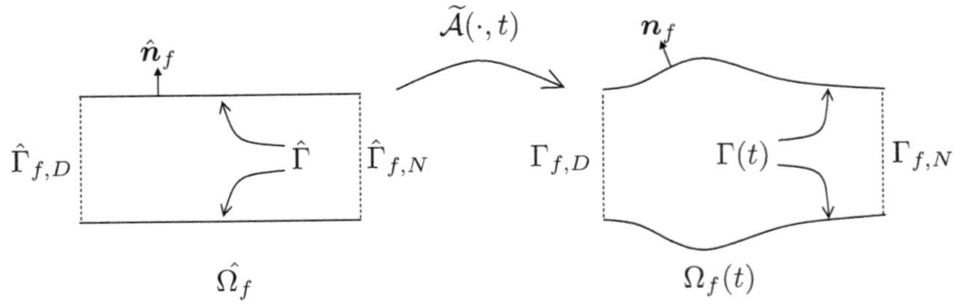

Fig. 6. Description of the motion of the computational domain for the fluid via the ALE map $\widetilde{\mathcal{A}}$. (Courtesy of Ref. 10, Chap. 3.)

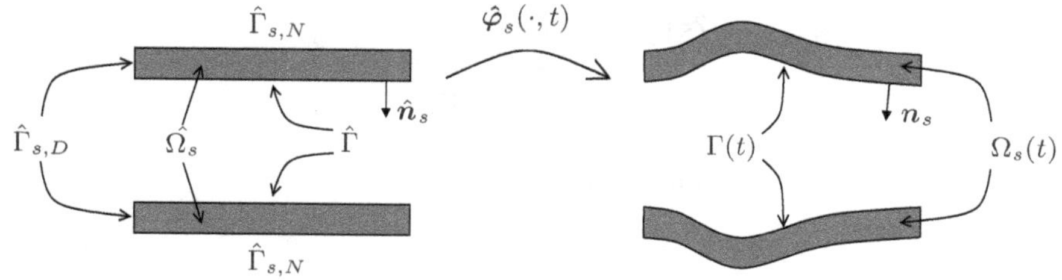

Fig. 7. Description of the motion of the solid (2D section). (Picture taken from Ref. 10, Chap. 3.)

on the boundaries $\widehat{\Gamma}_{s,D}$, the differential problem for the structure part then reads (see Eqs. (33) and (34))

$$
\begin{aligned}
\hat{\rho}_{s,0}\frac{\partial^2 \hat{\boldsymbol{\eta}}_s}{\partial t^2} - \mathbf{div}_{\hat{x}}(\widehat{\boldsymbol{F}}_s \widehat{\boldsymbol{\Sigma}}) = \mathbf{0}, &\quad \text{in } \hat{\Omega}_s,\\
\hat{\boldsymbol{\eta}}_s = \mathbf{0}, &\quad \text{on } \Gamma_{s,D},\\
\widehat{\boldsymbol{F}}_s \widehat{\boldsymbol{\Sigma}} \widehat{\boldsymbol{n}}_s = \widehat{J}_s |\widehat{\boldsymbol{F}}^{-T} \widehat{\boldsymbol{n}}_s| \widehat{\boldsymbol{g}}_{s,N}, &\quad \text{on } \widehat{\Gamma}_{s,N},\\
\widehat{\boldsymbol{F}}_s \widehat{\boldsymbol{\Sigma}} \boldsymbol{n}_s = \widehat{J}_s |\widehat{\boldsymbol{F}}^{-T} \widehat{\boldsymbol{n}}_s| \widehat{\boldsymbol{g}}_{\Gamma}, &\quad \text{on } \widehat{\Gamma},
\end{aligned} \tag{50}
$$

with $\widehat{\boldsymbol{\Sigma}}$ related to $\hat{\boldsymbol{\eta}}_s$ through a constitutive law of the form (38).

Should Eq. (33) be replaced by the pre-stressed model (46) or by anyone of the reduced models described in Sec. 7, this set of equations would modify in the obvious way.

The fluid and solid problems (49) and (50) must be coupled by imposing three interface coupling conditions: on geometry, velocity and stress.

We first enforce that the moving fluid domain follows the interface motion, i.e.,

$$\widetilde{\mathcal{A}} = \hat{\boldsymbol{\varphi}}_s, \quad \text{on } \hat{\Gamma}. \tag{51}$$

This is a geometry-coupling condition. Since we describe the motion of the solid in terms of its displacement $\hat{\boldsymbol{\eta}}_s$, it is also useful to describe the ALE map in terms of the displacement of the control volume, $\hat{\boldsymbol{\eta}}_f : \hat{\Omega}_f \times \mathbb{R}^+ \longrightarrow \mathbb{R}^3$, defined by

$$\hat{\boldsymbol{\eta}}_f(\hat{\boldsymbol{x}}, t) = \widetilde{\mathcal{A}}(\hat{\boldsymbol{x}}, t) - \hat{\boldsymbol{x}},$$

for all $\hat{\boldsymbol{x}} \in \hat{\Omega}_f$. Thus, Eq. (51) reduces to

$$\hat{\boldsymbol{\eta}}_f = \hat{\boldsymbol{\eta}}_s, \quad \text{on } \hat{\Gamma}. \tag{52}$$

By differentiating this equality with respect to t, it follows that

$$\hat{\boldsymbol{w}} = \hat{\boldsymbol{u}}_f, \quad \text{on } \hat{\Gamma}. \tag{53}$$

On the other hand, since the inlet and outlet boundaries remain fixed along the motion, we also have

$$\hat{\boldsymbol{\eta}}_f = \boldsymbol{0}, \quad \text{on } \hat{\Gamma}_{f,D} \cup \hat{\Gamma}_{f,N}. \tag{54}$$

Notice that Eqs. (52) and (54) provide the value of $\hat{\boldsymbol{\eta}}_f$ on the boundary of $\hat{\Omega}_f$. However, inside $\hat{\Omega}_f$, $\hat{\boldsymbol{\eta}}_f$ (and hence $\widetilde{\mathcal{A}}$) is arbitrary (this explains the name arbitrary-Lagrangian–Eulerian map): it can be any reasonable extension of $\hat{\boldsymbol{\eta}}_{s|\hat{\Gamma}}$ over $\hat{\Omega}_f$ (subjected to Eq. (54)). In the sequel we will denote this operation by

$$\hat{\boldsymbol{\eta}}_f = \mathrm{Ext}(\hat{\boldsymbol{\eta}}_{s|\hat{\Gamma}}). \tag{55}$$

Since the fluid is viscous, it perfectly sticks to the interface (or solid) boundary. This means that the whole velocity field must be continuous at the interface. Thus, in Eq. (49) we set $\boldsymbol{u}_\Gamma = \boldsymbol{u}_f$, which from Eq. (52) yields the coupling condition

$$\boldsymbol{u}_f = \boldsymbol{w}, \quad \text{on } \Gamma(t). \tag{56}$$

Finally, in order to ensure the balance of stresses on the interface, we set $\boldsymbol{g}_\Gamma = -\boldsymbol{\sigma}_f(\boldsymbol{u}_f, P)\boldsymbol{n}_f$ in Eq. (50). Using the properties of the Piola transform (Eq. (6)) we get the coupling condition

$$\widehat{\boldsymbol{F}}_s\widehat{\boldsymbol{\Sigma}}\hat{\boldsymbol{n}}_s = \hat{J}_s\hat{\boldsymbol{\sigma}}_f(\boldsymbol{u}_f, P)\widehat{\boldsymbol{F}}_s^{-T}\hat{\boldsymbol{n}}_s, \quad \text{on } \hat{\Gamma}. \tag{57}$$

Gathering now the coupling conditions (55) to (57) the coupled fluid-structure interaction problem reads: find $\hat{\boldsymbol{\eta}}_f : \hat{\Omega}_f \times \mathbb{R}^+ \longrightarrow \mathbb{R}^3$, $\boldsymbol{u}_f : \Omega_f(t) \longrightarrow \mathbb{R}^3$, $p : \Omega_f(t) \longrightarrow \mathbb{R}$ and $\hat{\boldsymbol{\eta}}_s : \hat{\Omega}_s \times \mathbb{R}^+ \longrightarrow \mathbb{R}^3$, such that

$$\hat{\boldsymbol{\eta}}_f = \mathrm{Ext}(\hat{\boldsymbol{\eta}}_{s|\hat{\Gamma}}), \quad \hat{\boldsymbol{w}} = \frac{\partial \hat{\boldsymbol{\eta}}_f}{\partial t}, \quad \Omega_f(t) = \underbrace{(I + \hat{\boldsymbol{\eta}}_f)}_{\widetilde{\mathcal{A}}}(\hat{\Omega}_f), \tag{58}$$

$$
\begin{aligned}
&\frac{\rho_f}{\widetilde{J}_{\widetilde{A}}}\frac{\partial \widetilde{J}_{\widetilde{A}} \boldsymbol{u}_f}{\partial t}_{|\widetilde{A}} + \mathbf{div}\left(\rho_f \boldsymbol{u}_f \otimes (\boldsymbol{u}_f - \boldsymbol{w}) - \boldsymbol{\sigma}_f(\boldsymbol{u}_f, P)\right) = \mathbf{0}, \quad \text{in } \Omega_f(t),\\
&\mathrm{div}\, \boldsymbol{u}_f = 0, \quad \text{in } \Omega_f(t),\\
&\boldsymbol{u}_f = \boldsymbol{u}_D, \quad \text{on } \Gamma_{f,D},\\
&\boldsymbol{\sigma}_f(\boldsymbol{u}_f, P)\boldsymbol{n}_f = \boldsymbol{g}_{f,N}, \quad \text{on } \Gamma_{f,N},\\
&\boldsymbol{u}_f = \boldsymbol{w}, \quad \text{on } \quad \Gamma(t),
\end{aligned}
\tag{59}
$$

$$
\begin{aligned}
&\hat{\rho}_{s,0}\frac{\partial^2 \hat{\boldsymbol{\eta}}_s}{\partial t^2} - \mathbf{div}_{\hat{x}}\left(\widehat{\boldsymbol{F}}_s \widehat{\boldsymbol{\Sigma}}\right) = \mathbf{0}, \quad \text{in } \hat{\Omega}_s,\\
&\hat{\boldsymbol{\eta}}_s = \mathbf{0}, \quad \text{on } \Gamma_{s,D},\\
&\widehat{\boldsymbol{F}}_s \widehat{\boldsymbol{\Sigma}} \hat{\boldsymbol{n}}_s = \hat{J}_s \left|\widehat{\boldsymbol{F}}_s^{-T} \hat{\boldsymbol{n}}_s\right| \hat{\boldsymbol{g}}_{s,N}, \quad \text{on } \hat{\Gamma}_{s,N},\\
&\widehat{\boldsymbol{F}}_s \widehat{\boldsymbol{\Sigma}} \hat{\boldsymbol{n}}_s = \hat{J}_s \hat{\boldsymbol{\sigma}}_f(\boldsymbol{u}_f, P) \widehat{\boldsymbol{F}}_s^{-T} \hat{\boldsymbol{n}}_s, \quad \text{on } \hat{\Gamma}.
\end{aligned}
\tag{60}
$$

We will not dwell here with the analysis of this coupled problem. A complete theory is not available yet. Several authors have obtained interesting existence results (of either strong or weak solutions) for simplified models. A (by far not exhaustive) list includes H. Beirão Da Veiga, M. J. Esteban, A. Chambolle, B. Desjardins, C. Grandmont and Y. Maday, M. Padula and V. A. Solonnikov, S. Čanić, D. Couthand and S. Shkoller, G. P. Galdi, E. H. Kim and G. Guidoboni. A review of partial results is presented by Y. Maday in Ref. 10, Chap. 8.

9. Algorithms of FSI

A great variety of strategies have been proposed to solve fluid-structure interaction (FSI) problems like, say, Eqs. (58)–(60). A first issue to be faced is how to deal with the non-linearity of the problem. In fact, not only the fluid (and in some cases the structure) equations are non-linear, but also the structure displacement modifies the fluid domain generating geometrical non-linearities. The fixed point technique (e.g., Ref. 5) is the simplest to linearize the FSI problem, however Newton (e.g., Ref. 15) and quasi-Newton (e.g., Ref. 12) methods have also been considered. The latter can be derived upon replacing the exact Jacobian by suitable block-triangular approximations, or by linear operators inspired by simpler fluid-structure models.

A classical restriction for fluid-structure algorithms is *modularity*. Most of the times the codes for the pure fluid problem and for the pure structure problem already exist and they are optimized for the specific mathematical features of the two different problems. Then the best way to solve the FSI problem would be to design algorithms involving only a separate use of the two codes, which communicate only through the coupling conditions on $\Gamma(t)$.

Explicit coupling algorithms well serve this purpose. By these algorithms the fluid and the structure subproblems are solved only once per time-step. Typically, when going from t^n to t^{n+1}, one solves for the fluid (59) on the previous domain configuration, say Ω^n, with known ALE velocity $\boldsymbol{w}^n$, then for the structure (60) (with normal stresses computed from the fluid iterate), to yield $\hat{\boldsymbol{\eta}}_s^{n+1}$, and finally use the latter to set the new ALE velocity $\boldsymbol{w}^{n+1}$ from Eq. (58).

Such an algorithm requires the solution of an Oseen equation in Ω_f^n, then an elastodynamic equation in $\hat{\Omega}_s$, finally an update Ω_f^{n+1} of the fluid domain by applying an extension operator (this requires the solution of a Laplace equation).

Due to the imperfect energy balance at the interface, these algorithms may become unstable, especially for FSI problems featuring a large *added-mass effect*. In fact, the load exerted by the fluid on the structure can be interpreted as an *added-mass* (see Ref. 4). When the structure density is much bigger than the fluid density, as it happens in aeroelasticity, the added-mass effect is negligible and the numerical approximation of the FSI problem through iterative procedures is less challenging. However, when the two densities feature the same order of magnitude, as in hemodynamics, the added mass effect becomes important and iterative procedures fail or are too slow. When the added-mass effect is critical, explicit algorithms can be unconditionally unstable, as shown in Ref. 4 on a simple model problem.

An alternative approach makes use of implicit coupling algorithms. In this case subiterations are carried out between Eqs. (59), (60) and then Eq. (58). This generates a subsequence $\{(\boldsymbol{u}_f{}^{n,k}, P^{n,k}), \hat{\boldsymbol{\eta}}_s^{n,k}, \hat{\boldsymbol{w}}^{n,k}\}$ that hopefully converges to the new solution at time-step $n+1$ as far as $k \to \infty$. Besides being more costly, this algorithm may require severe under-relaxation (with parameter heavily dependent on both the grid spacing and the time-step) for problems with large added-mass effect. We still refer to Ref. 4 for the analysis of a model coupled problem.

For critical values of the added mass, more efficient implicit coupling algorithms are *substructuring techniques* stemming from a domain decomposition viewpoint (see Ref. 8). Among these procedures, the classical Dirichlet–Neumann technique is one of the most widely used. Usually iterations are carried out using Richardson or conjugate gradient methods for the interface equation. These iterative procedures are minimization techniques that always converge but their convergence rate will depend on the relevance of the added-mass effect.

A radical alternative would be to solve the *monolithic fluid-structure system* (after linearization and discretization): since no coupling iterations are performed, the added-mass effect would not play any role. Furthermore, we need to develop a global FSI solver, which is not modular, and the computational cost for solving the monolithic system may become prohibitive for real applications.

A further alternative is offered by *semi-implicit schemes*, introduced in Ref. 9. The idea is to decouple the fluid velocity computation from the strongly coupled fluid-structure system, which only involves pressure and structure unknowns, with the double advantage of reducing computational costs and ensuring stability. In fact, since the pressure is still coupled to the structure, the stability of the schemes is independent of the added-mass effect. In Ref. 9 the FSI system is solved through the Chorin–Temam projection scheme (see Refs. 6 and 18).

The same idea can be pursued if we derive semi-implicit schemes from *algebraic* splitting methods rather than from differential ones. Algebraic splitting methods are based on the inexact factorization of the matrix arising from the full discretization (both in time and space) of the given initial-boundary value problem. In this way, the boundary conditions are already incorporated in the discretized operator and no further boundary conditions have to be selected. The aim is to take advantage of the good accuracy properties shown by many of these techniques (which do not have a differential counterpart, see Refs. 1 and 2) when solving the incompressible Navier–Stokes equations. This approach has been adopted for the first time in Ref. 16. An incremental version is investigated in Ref. 3: in the same paper, the algebraic pressure-correction method is adopted to the coupled fluid-structure problem, and inexact factorizations of the FSI system matrix are used as a preconditioner, leading to predictor-corrector methods that converge to the solution of the monolithic FSI system.

In general, semi-implicit schemes enjoy better convergence properties than explicit or fully-implicit schemes. The price to pay is relatively low, since one has to resort to the solution of a coupled problem involving the fluid pressure P^{n+1} and the wall deformation $\hat{\eta}_s^{n+1}$. The overall efficiency depends on how well this coupled problem is solved. A discussion based on the use of the Schur-complement matrices at the interface $\Gamma(t)$ is presented in Ref. 3. For a further discussion and for more details, see the contribution by M. Fernández and J. F. Gerbeau in Ref. 10, Chap. 9.

References

1. F. Saleri, A. Quarteroni and A. Veneziani, Analysis of the Yosida method for the incompressible Navier–Stokes equations, *J. Math. Pures Appl.* **78** (1999):473–503.
2. F. Saleri, A. Quarteroni and A. Veneziani, Factorization methods for the numerical approximation of Navier–Stokes equations, *Comput. Meth. Appl. Mech. Eng.* **188** (2000):505–526.
3. S. Badia, A. Quaini and A. Quarteroni, Splitting methods based on algebraic factorization for fluid-structure interaction, *SIAM, J. Sci. Comput.* **30** (2008):1778–1805.
4. P. Causin, J.-F. Gerbeau and F. Nobile, Added-mass effect in the design of partitioned algorithm for fluid-structre interaction, *Comp. Meth. Appl. Mech. Eng.* **194** (2005):4506–4527.
5. M. Cervera, R. Codina and M. Galindo, On the computational efficiency and implementation of block-iterative algorithms for nonlinear coupled problems, *Eng. Comput.* **13** (1996):4–30.
6. A. J. Chorin, Numerical solutions of the Navier–Stokes equations, *Math. Comput.* **22** (1968):745–762.
7. P. G. Ciarlet, *Mathematical Elasticity*, Vol. I, in *Studies in Mathematics and its Applications*, Vol. 20 (North-Holland Publishing Co., Amsterdam, 1988).
8. S. Deparis, M. Discacciati, G. Fourestey and A. Quarteroni, Fluid-structure algorithms based on Steklov-Poincaré operators, *Comput. Meth. Appl. Mech. Eng.* **195** (2006):5797–5812.
9. M. A. Fernández, J.-F. Gerbeau and C. Grandmont, A projection algorithm for fluid-structure interaction problems with strong added-mass effect, *C. R. Acad. Sci. Paris.* **342** (2006):279–284.
10. L. Formaggia, A. Quarteroni and A. Veneziani (eds.), *Cardiovascular Mathematics* (Springer, Heidelberg, 2009).

11. Y. C. Fung, *Biomechanics: Mechanical Properties of Living Tissues* (Springer-Verlag, New York, 1993).
12. J.-F. Gerbeau and M. Vidrascu, A quasi-Newton algorithm based on a reduced model for fluid-structure interaction problems in blood flows, *Math. Model. Num. Anal.* **37** (2003):631–648.
13. G. A. Holzapfel, *Non Linear Solid Mechanics. A Continuum Approach for Engineering* (John Wiley & Sons, Chichester, UK, 2000).
14. G. A. Holzapfel, Determination of material models for arterial walls from uniaxial extension tests and histological structure, *J. Theor. Biol.* **238** (2006):290–302.
15. M. A. Fernández and M. Moubachir, A Newton method using exact Jacobians for solving fluid-structure coupling, *Comp. Struct.* **83** (2005).
16. A. Quaini and A. Quarteroni, A semi-implicit approach for fluid-structure interaction based on an algebraic fractional step method, *Math. Models Meth. Appl. Sci* **17** (2007):957–983.
17. A. Quarteroni and L. Formaggia, Mathematical modelling and numerical simulation of the cardiovascular system, in *Computational Models for the Human Body* ed. N. Ayache (Elsevier, Amsterdam, 2004), pp. 3–129.
18. R. Temam, Sur l'approximation de la solution des équations de Navier–Stokes par la méthod de pas fractionaires (II), *Arch. Rat. Mech. Anal.* **33** (1969):377–385.

INDEX

www.ingramcontent.com/pod-product-compliance
Lightning Source LLC
Chambersburg PA
CBHW050543190326
41458CB00007B/1898